## MEDICAL INTELLIGENCE UNIT

# Mechanisms of Insulin Action

Alan R. Saltiel
Departments of Internal Medicine
and Physiology
Life Sciences Institute
The University of Michigan Medical Center
Ann Arbor, Michigan, U.S.A.

Jeffrey E. Pessin
Department of Pharmacological Sciences
State University of New York at Stony Brook
Stony Brook, New York, U.S.A.

LANDES BIOSCIENCE
AUSTIN, TEXAS
U.S.A.

SPRINGER SCIENCE+BUSINESS MEDIA
NEW YORK, NEW YORK
U.S.A.

# MECHANISMS OF INSULIN ACTION

## Medical Intelligence Unit

Landes Bioscience
Springer Science+Business Media, LLC

ISBN: 978-0-387-72203-0    Printed on acid-free paper.

Copyright ©2007 Landes Bioscience and Springer Science+Business Media, LLC

All rights reserved. This work may not be translated or copied in whole or in part without the written permission of the publisher, except for brief excerpts in connection with reviews or scholarly analysis. Use in connection with any form of information storage and retrieval, electronic adaptation, computer software, or by similar or dissimilar methodology now known or hereafter developed is forbidden.

The use in the publication of trade names, trademarks, service marks and similar terms even if they are not identified as such, is not to be taken as an expression of opinion as to whether or not they are subject to proprietary rights.

While the authors, editors and publisher believe that drug selection and dosage and the specifications and usage of equipment and devices, as set forth in this book, are in accord with current recommendations and practice at the time of publication, they make no warranty, expressed or implied, with respect to material described in this book. In view of the ongoing research, equipment development, changes in governmental regulations and the rapid accumulation of information relating to the biomedical sciences, the reader is urged to carefully review and evaluate the information provided herein.

Springer Science+Business Media, LLC, 233 Spring Street, New York, New York 10013, U.S.A.
http://www.springer.com

Please address all inquiries to the Publishers:
Landes Bioscience, 1002 West Avenue, 2nd Floor, Austin, Texas 78701, U.S.A.
Phone: 512/ 637 6050; FAX: 512/ 637 6079
http://www.landesbioscience.com

Printed in the United States of America.

9 8 7 6 5 4 3 2 1

## Library of Congress Cataloging-in-Publication Data

Mechanisms of insulin action / [edited by] Alan R. Saltiel, Jeffrey E. Pessin.
   p. ; cm. -- (Medical intelligence unit)
 Includes bibliographical references and index.
 ISBN 978-0-387-72203-0 (alk. paper)
  1. Insulin--Physiological effect. I. Saltiel, Alan R. II. Pessin, Jeffrey E. III. Series: Medical intelligence unit (Unnumbered : 2003)
  [DNLM: 1. Insulin--pharmacology. WK 820 M486 2007]
  QP572.I5M43 2007
  612.3'4--dc22
                                                                                              2007013421

# CONTENTS

**Preface** .................................................................................. xi

1. **Insulin and IGF-I Receptor Structure and Binding Mechanism** ............. 1
   *Pierre De Meyts, Waseem Sajid, Jane Palsgaard, Anne-Mette Jensen,
   Hassan Aladdin and Jonathan Whittaker*
   Evolutionary Biology of the Insulin Peptide Family
     and Their Receptors ...................................................................... 1
   Structure of the Insulin and IGF-I Receptor Genes
     and Predicted Protein Tertiary Structure ............................................. 4
   Modular Receptor Structures Elucidated by X-Ray Crystallography ......... 8
   Ligand Binding Properties ................................................................ 9
   Receptor Crosslinking with Bifunctional and Photoreactive Ligands ....... 10
   Definition of Ligand Binding Specificity Using Chimeric
     Insulin/IGF-I Receptors ................................................................. 11
   Natural Receptor Mutations that Affect Insulin Binding
     in Syndromes of Extreme Insulin Resistance ..................................... 11
   Mapping of Ligand Binding Sites on the Insulin and IGF-I
     Receptors by Site-Directed and Alanine-Scanning Mutagenesis ......... 12
   Reconstitution of Modular Minimized Receptor Constructs
     with Low and High Affinity ............................................................ 13
   Attempts at Insulin Receptor Structure Definition by Electron
     Microscopy ................................................................................. 16
   Mapping of Receptor-Binding Sites on the Insulin
     and IGF-I Molecules .................................................................... 17
   Mechanism of Ligand Binding and Receptor Activation ..................... 19
   Conclusions and a Word of Caution ................................................ 22

2. **Subcellular Compartmentalization of Insulin Signaling Processes
   and GLUT4 Trafficking Events** ............................................................ 33
   *Robert T. Watson, Alan R. Saltiel, Jeffrey E. Pessin
   and Makoto Kanzaki*
   The Insulin Receptor and Its Immediate Downstream
     Substrate Proteins ....................................................................... 34
   The PI3-Kinase Is Necessary for Insulin-Stimulated GLUT4
     Translocation ............................................................................. 34
   Is There a Second Signaling Pathway Required
     for Insulin-Stimulated Glucose Uptake? .......................................... 37
   The APS-CAP-Cbl Pathway Is Compartmentalized Within Plasma
     Membrane Microdomains ............................................................ 38
   TC10 Generates Spatially Compartmentalized Signals
     that Contribute to the Specificity of Insulin Action .......................... 39
   Downstream Targets of TC10 ........................................................ 40
   Sorting GLUT4 Into and Out of the Insulin-Responsive
     Storage Compartment ................................................................. 43
   Does Insulin Regulate the Intrinsic Transport Activity of GLUT4? ..... 45
   Conclusions and Future Directions ................................................ 46

3. **Regulation of Insulin Action and Insulin Secretion by SNARE-Mediated Vesicle Exocytosis** ..................................................... 52
   *Debbie C. Thurmond*
   Vesicle Exocytosis .................................................................................. 52
   Insulin Action: GLUT4 Vesicle Translocation ................................... 55
   Insulin Exocytosis in Pancreatic Beta Cells ....................................... 58
   Perspectives ............................................................................................ 62

4. **Control of Protein Synthesis by Insulin** ............................................... 71
   *Joseph F. Christian and John C. Lawrence, Jr.*
   mRNA ..................................................................................................... 71
   Ribosomes .............................................................................................. 74
   Initiation ................................................................................................. 75
   Elongation .............................................................................................. 80
   Termination ........................................................................................... 81
   The mTOR Signaling Pathway ............................................................ 81

5. **Hepatic Regulation of Fuel Metabolism** .............................................. 90
   *Catherine Clark and Christopher B. Newgard*
   Glucose Transport ................................................................................. 90
   Glycolysis ................................................................................................ 91
   Gluconeogenesis .................................................................................... 97
   Glycogen Metabolism ......................................................................... 101
   New Developments ............................................................................. 104

6. **Insulin Action Gene Regulation** ........................................................... 110
   *Calum D. Sutherland, Richard M. O'Brien and Daryl K. Granner*
   Insulin Signal Transduction and Gene Expression ........................ 113
   Key Insulin-Regulated Gene Promoters .......................................... 120
   Coordinated Regulation of PEPCK, G6Pase, IGFBP-1
     and TAT Gene Expression? ............................................................ 123

7. **Insulin Action in the Islet β-Cell** ......................................................... 133
   *Rohit N. Kulkarni*
   Embryonic and Early Post-Natal Development
     of the Endocrine Pancreas ............................................................. 135
   Global and Conditional Knockouts of Insulin, IGF-I, IGF-II,
     and Proteins in Their Signaling Pathways .................................. 137
   Maintenance of Adult β-Cell Mass ................................................... 140
   Growth and Development of Islet α-Cells ...................................... 143
   The Liver-Pancreas Connection ........................................................ 144
   Future Insights .................................................................................... 145

8. **Central Regulation of Insulin Sensitivity** .............................................. 152
   *Silvana Obici and Luciano Rossetti*
   Insulin Action in the Hypothalamus ................................................. 154
   Central Effects of Leptin on Insulin Sensitivity ................................. 158
   Hypothalamic Lipid Sensing ............................................................ 161

9. **Transgenic Models of Impaired Insulin Signaling** ............................ 168
   *Francesco Oriente and Domenico Accili*
   Insulin Receptor Knockout ............................................................... 168
   Conditional *Insulin Receptor* Knockouts ........................................ 169
   Mutations Affecting Insulin Receptor Signaling ............................... 173
   Gene Knockouts Associated with Increased Insulin Sensitivity ........ 177

10. **Insulin Resistance** ................................................................................ 185
    *C. Hamish Courtney and Jerrold M. Olefsky*
    Methods of Assessing Insulin Sensitivity .......................................... 185
    Insulin Resistance in Type 2 Diabetes Mellitus and Obesity ............. 187

**Index** ............................................................................................................ 211

# EDITORS

**Alan R. Saltiel**
Departments of Internal Medicine
and Physiology
Life Sciences Institute
The University of Michigan Medical Center
Ann Arbor, Michigan, U.S.A.
*Chapter 2*

**Jeffrey E. Pessin**
Department of Pharmacological Sciences
State University of New York at Stony Brook
Stony Brook, New York, U.S.A.
*Chapter 2*

# CONTRIBUTORS

Domenico Accili
Berrie Research Pavilion
New York, New York, U.S.A.
*Chapter 9*

Hassan Aladdin
Receptor Biology Laboratory
Hagedorn Research Institute
Gentofte, Denmark
*Chapter 1*

Joseph F. Christian
Department of Pharmacology
University of Virginia Health System
Charlottesville, Virginia, U.S.A.
*Chapter 4*

Catherine Clark
Departments of Pharmacology,
  Cancer Biology, Medicine
  and Biochemistry
Sarah W. Stedman Nutrition
  and Metabolism Center
Duke University Medical Center
Durham, North Carolina, U.S.A.
*Chapter 5*

C. Hamish Courtney
University of California San Diego
San Diego, California, U.S.A.
*Chapter 10*

Pierre De Meyts
Receptor Biology Laboratory
Hagedorn Research Institute
Gentofte, Denmark
*Chapter 1*

Daryl K. Granner
Department of Molecular Physiology
  and Biophysics
Vanderbilt University Medical School
Nashville, Tennessee, U.S.A.
*Chapter 6*

Anne-Mette Jensen
Insulin Chemical Control
Fuglebakken, Denmark
*Chapter 1*

Makoto Kanzaki
Tohoku University Biomedical
  Engineering Research Organization
Sendai City, Miyagi, Japan
*Chapter 2*

Rohit N. Kulkarni
Division of Cell and Molecular
  Physiology
Joslin Diabetes Center
and Department of Medicine
Harvard Medical School
Boston, Massachusetts, U.S.A.
*Chapter 7*

John C. Lawrence, Jr.
Department of Pharmacology
University of Virginia Health System
Charlottesville, Virginia, U.S.A.
*Chapter 4*

Christopher B. Newgard
Departments of Pharmacology,
  Cancer Biology, Medicine
  and Biochemistry
Sarah W. Stedman Nutrition
  and Metabolism Center
Duke University Medical Center
Durham, North Carolina, U.S.A.
*Chapter 5*

Silvana Obici
Departments of Medicine and Molecular
  Pharmacology
Diabetes Research Center
Albert Einstein College of Medicine
Bronx, New York, U.S.A.
*Chapter 8*

Richard M. O'Brien
Department of Molecular Physiology
  and Biophysics
Vanderbilt University Medical School
Nashville, Tennessee, U.S.A.
*Chapter 6*

Jerrold M. Olefsky
University of California San Diego
San Diego, California, U.S.A.
Email: olefsky@popmail.ucsd.edu
*Chapter 10*

Francesco Oriente
Department of Medicine
College of Physicians and Surgeons
  of Columbia University
New York, New York, U.S.A.
*Chapter 9*

Jane Palsgaard
Receptor Biology Laboratory
Hagedorn Research Institute
Gentofte, Denmark
*Chapter 1*

Luciano Rossetti
Departments of Medicine and Molecular
  Pharmacology
Diabetes Research Center
Albert Einstein College of Medicine
Bronx, New York, U.S.A.
*Chapter 8*

Waseem Sajid
Receptor Biology Laboratory
Hagedorn Research Institute
Gentofte, Denmark
*Chapter 1*

Calum D. Sutherland
Department of Pathology
  and Neurosciences
University of Dundee
Ninewells Medical School
Dundee, Scotland, U.K.
*Chapter 6*

Debbie C. Thurmond
Department of Biochemistry
  and Molecular Biology and Center
  for Diabetes Research
Indiana University School of Medicine
Indianapolis, Indiana, U.S.A.
*Chapter 3*

Robert T. Watson
Department of Pharmacological Sciences
State University of New York
  at Stony Brook
Stony Brook, New York, U.S.A.
*Chapter 2*

Jonathan Whittaker
Departments of Nutrition
  and Biochemistry
Case Western Reserve University
Cleveland, Ohio, U.S.A.
*Chapter 1*

# PREFACE

We now know that the pathophysiology of Type 2 diabetes involves defects in three organ systems that conspire together to produce abnormal glucose and lipid metabolism. While there is some uncertainty regarding the primary lesion, or relative importance of different tissues, metabolic defects in liver, peripheral target tissues such as fat and muscle and pancreatic b cells all contribute to the syndrome. Insulin resistance, which is defined as a state of reduced responsiveness to normal circulating concentrations of insulin, is now recognized as a characteristic trait of Type 2 diabetes, and contributes to abnormalities in all of these tissues. Even in the absence of diabetes, insulin resistance is a key feature of other human disease states. These findings suggest that studies on the molecular mechanisms underlying insulin action are crucial to further the understanding of this devastating disease. Thus, we have gathered together several renowned experts in this field to produce this monograph *Mechanisms of Insulin Action*.

These articles provide novel insight into the key issues underlying the molecular biology of insulin action and insulin resistance. De Meyts et al cover the structure and function of insulin and insulin-like growth factor receptors. Watson and colleagues outline the important events at the intersection of signal transduction and vesicle trafficking that are crucial to the stimulation of glucose uptake by insulin. Thurmond reviews the important events that occur when glut4 vesicles dock and fuse at the plasma membrane. Christian and Lawrence comment on the mechanisms responsible for the regulation of protein synthesis by insulin, while Clark and Newgard review those mechanisms responsible for changes in hepatic fuel metabolism. Because gene expression is so important in metabolism, Sutherland et al outline the steps that are involved in the regulation of transcription by insulin. As the primary event in metabolic control, Kulkarni describes how the beta cell is regulated by insulin, and in related work, Obici and Rossetti cover the central control of peripheral insulin sensitivity. Much progress on studies in insulin action have been made in animal models, and Oriente and Accili review transgenic and knock out models of insulin action and resistance. Finally, Courtney and Olefsky review the occurrence and treatment of insulin resistance. Together, these authors have provided a comprehensice summary of our understanding of insulin action from cellular physiology to the integration of tissue specific signaling events that are responsible for whole body glucose homeostasis.

*Alan Saltiel*
*Jeffrey Pessin*

# CHAPTER 1

# Insulin and IGF-I Receptor Structure and Binding Mechanism

Pierre De Meyts,* Waseem Sajid, Jane Palsgaard, Anne-Mette Theede, Lisbeth Gauguin, Hassan Aladdin and Jonathan Whittaker

## Abstract

The insulin and IGF-I receptors are members of the superfamily of receptor tyrosine kinases (RTKs). Unlike most RTKs that are single-chain monomeric transmembrane polypeptides, the insulin and IGF-I receptors are covalent dimers composed of two extracellular α subunits and two transmembrane β subunits containing the tyrosine kinase domains. The α subunits contain the ligand binding sites, of which at least three subdomains have been defined by photoaffinity crosslinking, alanine-scanning mutagenesis or minimized receptor constructs. All RTKs are dimeric or oligomeric in the ligand-activated form. The residues of insulin involved in receptor binding have been mapped by alanine-scanning mutagenesis. They form at least two major epitopes that partially overlap with the dimer- and hexamer-forming surfaces of the insulin molecule, and we propose that insulin is using those surfaces to asymmetrically cross-link the two receptor α subunits. This mechanism provides a structural basis for high affinity binding and negative cooperativity, and probably also operates in the IGF-receptor interaction. It also provides a structural basis for the approximation and transphosphorylation of the kinase domains and triggering of the signalling cascade.

## Evolutionary Biology of the Insulin Peptide Family and Their Receptors

Insulin and the insulin-like growth factors (IGF)-I and -II belong to a phylogenetically ancient family of peptide hormones and growth factors[1-3] that play a fundamental role in the control of essential cellular and physiological processes such as the cell cycle, survival or apoptosis, cell migration, proliferation and differentiation, and body growth, metabolism, reproduction, and longevity.

Proteins of the insulin superfamily are synthesized as prepro-proteins consisting of 4 domains (pre, B, C, A). These are then processed by proteolytic removal of the pre domain and in some cases the C- domain and fold to form mature proteins, in which the A and B domains are covalently linked by two disulfide bonds. The IGF precursors have additional C-terminal D and E peptides; the latter is removed proteolytically post-translationally. The basic fold is shared for all molecules in the superfamily whose structure is known although the lengths of the secondary structural elements vary and the connecting loops can exhibit several conformations; the B domain contains a single α helix which lies across the 2 helices of the A domain (Fig. 1).

*Corresponding Author: Pierre De Meyts—Receptor Systems Biology Laboratory, Hagedorn Research Institute, Niels Steensens Vej 6, DK-2820 Gentofte, Denmark. Email: pdm@novonordisk.com

*Mechanisms of Insulin Action*, edited by Alan R. Saltiel and Jeffrey E. Pessin. ©2007 Landes Bioscience and Springer Science+Business Media.

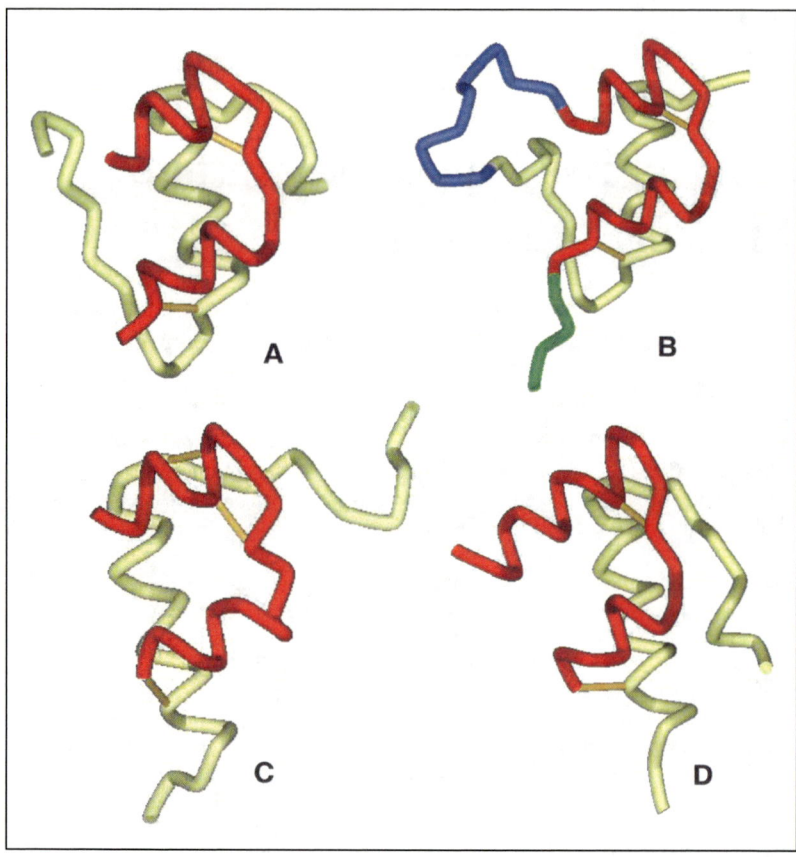

Figure 1. The insulin peptide family protein folds. A) Insulin. The A-chain is red, the B-chain yellow. Adapted from Baker E et al. Phil Trans R Soc Lond 1988; B19:369-456. B) IGF-I. The A domain is red, the B domain yellow, the C domain blue and the D domain green. Adapted from Brzozowski et al. Biochemistry 2002; 41:9389-9397. C) Bombyxin-II. The A domain is red, the B domain yellow. Adapted from Nagata K et al. J Mol Biol 1995; 253:749-758. D) Human relaxin 2. The A domain is red, the B domain yellow. Adapted from Eigenbrot C et al. J Mol Biol 1991; 221:15-21.

These properties have lead to the development of criteria for the identification of superfamily members from data obtained from genomic and cDNA cloning studies.

Candidate open reading frames should have at least a putative signal peptide, a B domain and an A domain defined as follows:

1. A domain predicted to contain two helices joined by a loop.
2. B domain with an extended N-terminal coil followed by a tight turn and a central helix.
3. A hydrophobic core forming the interface between the A and B domains.
4. At least three disulfide bonds formed by conserved cysteine residues.

In addition in most family members the B and A domains are encoded by individual exons separated by an intron at the 3' end of the coding region for the B domain. These criteria encompass the plethora of vertebrate and non vertebrate insulin family members identified during the last decade.

In humans, the insulin-like peptide family comprises ten members, the closely related insulin and insulin-like growth factors (IGF)-I and II, and the seven peptides related to relaxin (INSL/RLFs).[4,5] While insulin and the IGFs bind to receptors belonging to the superfamily of receptor tyrosine kinases (RTKs), the relaxin-like peptides were recently shown unexpectedly to bind to leucine-rich repeat-containing G protein-coupled receptors (LGRs),[4] involved in the development and physiology of the reproductive tract.[6,7]

The sequence of the zebrafish and fugu genomes has revealed the existence of two insulin genes; evidence suggests that a genome duplication event occurred on the lineage leading to teleost fish.[8]

Several insulin-like family members have been identified in invertebrates as well, including bombyxins in the silkworm (38 putative genes), *Bombyx mori*,[9] and related peptides in the sweet potato hornworm *Agrius convolvuli*,[10] a locust insulin-related peptide (LIRP) in the grasshopper *Locusta migratoria*,[11,12] seven different molluscan insulin-like peptides (MIP) in the pond snail, *Lymnea stagnalis*, an insulin-like gene in the sea slug *Aplysia californica*,[13] and seven insulin-like genes in the fruit fly *Drosophila melanogaster*.[14] In addition, three new gene families of insulin-like peptides with atypical disulphide bond pattern (38 putative peptides in all) have been recently identified in the nematode *C. elegans* that may be ligands (both agonists and antagonists) for the *daf-2* insulin-like receptor, involved in reproduction, growth and adult life span.[15] Surprisingly, one of the *C. elegans* peptides (INS-6) was reported to bind and activate the human insulin receptor, despite absolute lack of conservation of any of the aminoacids involved in mammalian receptor binding (see below),[16] a finding which warrants further investigation. In contrast to the abovementioned insulin-related peptides, the reported sequence of an insulin-like peptide in the sponge *Geodia cydonium*[17] is suspiciously too close to human insulin (60-80%) to represent a true phylogenetically ancient ancestor[1] and probably represents contamination by rodent material, as later acknowledged by the authors (Muller WEG, personal communication quoted in ref. 18).

Evidence for the existence of receptors related to the insulin receptor has also been reported in various invertebrate species such as *D. melanogaster*,[19] *B. mori*,[20] the mosquitos *Aedes aegypti*[21] and *Anopheles gambiae*,[22] the molluscs, *L. stagnalis*,[23] *Biomphalaria glabrata*,[24] *Aplysia californica*[25] and the Pacific oyster *Crassostrea gigas*,[26] *C. elegans*,[27] the human blood flukes *Schistosoma mansoni*[28] and *Schistosoma japonicum*,[29] the tapeworm *Echinococcus multilocaris*,[30] and the cnidarian *Hydra vulgaris*[31] as well as sponges, although in this last case identification was based only on homologies to the insulin receptor kinase domain.[32,33]

Interestingly, the protochordate amphioxus (*Branchiostoma californiensis*) contains a single gene for an insulin-like peptide and a related receptor, both of which have hybrid characterictics of the insulin and IGF systems and may therefore represent the ancestral genes,[34,35] both have undergone duplications in the transition from protochordates to vertebrates. This has been corroborated by the sequencing of the genome of the sea squirt *Ciona intestinalis*, another protochordate.[36]

The concept that the Amphioxus insulin-like peptide represents the common ancestral gene of insulin and IGFs was challenged by the discovery of both an insulin and an IGF molecule in the urochordate tunicate *Chelyosoma productum*;[37] however, the predicted aminoacid sequences for the tunicate insulin and IGF peptides share a remarkably high sequence identity including in the evolutionarily highly variable C domain, and Chan and Steiner therefore suggested that the tunicate sequences are the product of a recent gene duplication event.[1]

While a detailed discussion of the evolutionary biology of the insulin signaling system is beyond the scope of the review and the evolutionary record is of necessity very incomplete, several points deserve further comment. Firstly prior to the emergence of the vertebrates, the full spectrum of the biological effects of the signaling system (regulation of metabolism, growth, cell survival, feeding, longevity and reproductive function and developmental timing) appear to be mediated by a single receptor. Secondly there has been a high degree of conservation of function from the insects, as manifested in *Drosophila*, to mammals. However the nematode *C. elegans* seems to be an exception to this general rule. Firstly the existence of the very large number of genes encoding insulin-like peptides (38) in this species is atypical, although a similar number exists in *Bombyx*. The genes for many of these are organized into clusters of 3-7. This clustering, the tandem arrangement and sequence similarity between genes in a given cluster suggest that these clusters may have arisen relatively recently by gene duplication. However sequence diversity within the clusters suggests that they are continuing to rapidly evolve. The completion of the sequencing of the *C. briggsae* genome[38] should provide further insights into this issue. Recent database mining suggests the presence of multiple insulin receptor-like proteins as well, with a very short C-rich region between the two L domains.[39] Secondly *C. elegans* seems to be the only species where both insulin-like peptides function as both agonists and antagonists. Finally signaling by the insulin receptor homolog in this species appears to inhibit growth and anabolism in constrast to the stimulation universally seen in other species.[40,41] However, inhibiting the insulin/IGF system appears to prolong lifespan in *C.elegans* as well as in *Drosophila*,[42] rodents[43] and possibly humans.[44]

Within vertebrate species, multiple types of receptor are observed. In fish there are genes encoding multiple receptor types with high homology to either mammalian insulin or IGF-I receptors but not mammalian IRR (insulin receptor-related receptor).[1] The *Fugu rubripes* genome[45] has four genes encoding receptors from this family, two with high homology to the mammalian insulin receptors and two with high homology to mammalian IGF-I receptors. The absence of IRR type receptors in fish is corroborated by the failure to identify DNAs encoding such proteins by molecular cloning studies in multiple species of fish, where insulin- and IGFI-like receptor have been identified by such techniques. In reptiles and amphibian only insulin and IGF-I receptors have been identified.[46-51] However this cannot be considered as definitive until genomes of examples of each have been sequenced. In mammals there are insulin, IGF-I and IRR receptors and all the proteins appear to be highly conserved. The same three types of receptor have also been identified in birds.

Throughout evolution, the prototypical domain structure of these proteins (see below) appears to have been well conserved. However certain homologs have either additional N terminal or C terminal domains or both. These can vary considerably in size from 30 to 40 amino acids (N and C- terminal extensions of *Bombyx* IR) to 200 to 400 amino acids (N and C terminal extensions of *Drosophila* IR and C terminal extension of Daf-2). The primary structures of these domains vary considerably. Of particular interest are the C-terminal domains of the *Drosophila* and *C. elegans* homologs. These contain several YXXM motifs, potential docking sites for PI3 kinase when phosphorylated, suggesting that they may play a role in signal transduction by these receptors.

The signal transduction pathways used by these receptors also appear to be well conserved throughout evolution.[14,40,44,52-54] While detailed discussion of this topic is beyond the scope of this review, it should be mentioned that there is extensive genetic evidence that the insulin receptor homologs of *C. elegans* and *Drosophila* both utilize signal transduction pathways closely related to the mammalian signal transduction pathway mediated by PI3 kinase activation in response to receptor autophosphorylation. Genetic evidence also suggests that the biological effects mediated by these receptors in mammals, nematodes and fruit flies are highly conserved. Insulin receptor and IGF receptor knockout studies in mice indicate that in addition to regulation of intermediary metabolism and growth, signaling via these receptors is involved in the regulation of energy homeostasis, regulation of reproductive function and of longevity.[55] Similar biological roles for signaling by the appropriate receptor homologs have been demonstrated in mutational studies in *Drosophila*[40] and *C. elegans*.[56]

It has been hypothesized that the ancestral insulin-like gene functioned primarily as a "mitogenic" growth factor, and that its duplication allowed insulin to develop as a metabolic regulator while the mitogenic activity was retained by the IGF-I gene.[1,57]

## Structure of the Insulin and IGF-I Receptor Genes and Predicted Protein Tertiary Structure

As mentioned above, the insulin, IGF-I and IRR receptors are members of the RTK family. In humans, the RTK superfamily comprises ~60 members distributed into ~20 subfamilies depending on the modular architecture of their extracellular domains and the degree of identity in their intracellular tyrosine kinase domains (Fig. 2). The importance of these receptors is demonstrated by the discovery of a growing number of congenital genetic syndromes linked to

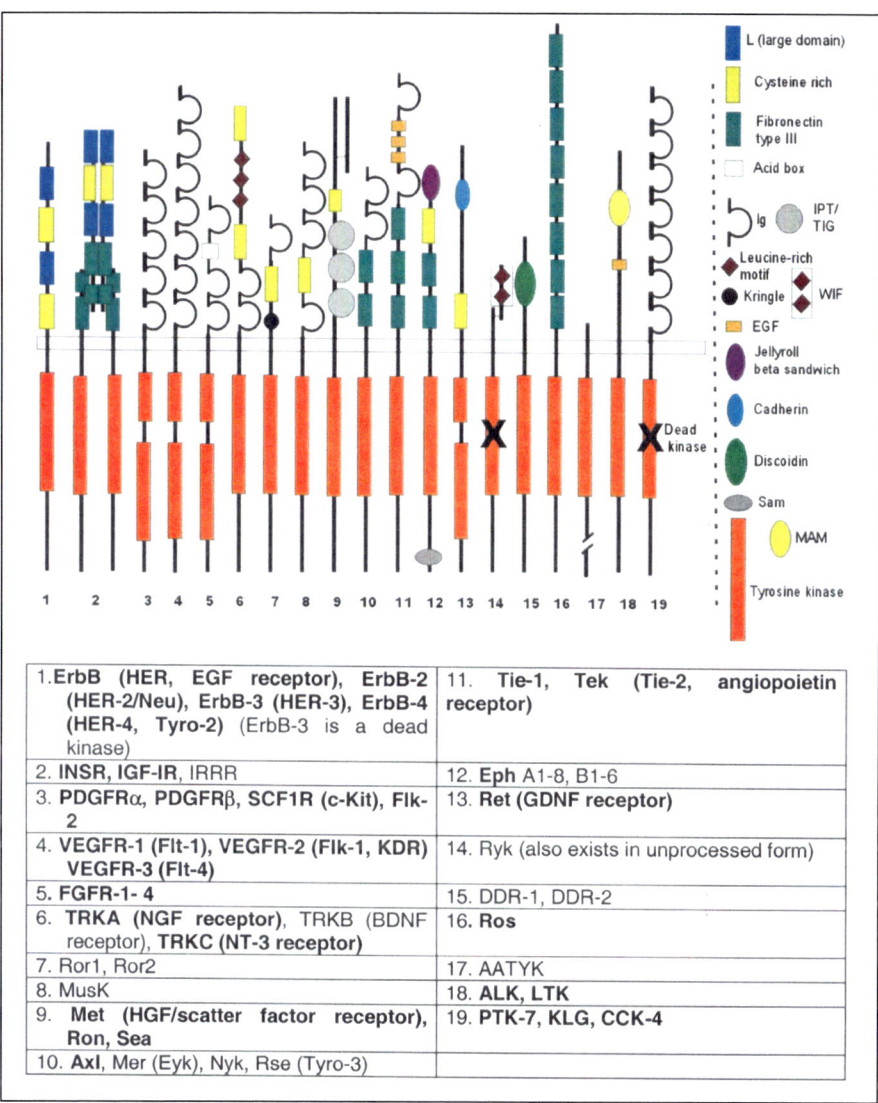

Figure 2. Modular structure of the receptor tyrosine kinases. This figure is adapted from reference 113, updated with information compiled from references 58, 59, 223-225 and various databases such as Pfam. We found that all published figures on the domain organisation of RTKs contain some inaccuracies and discrepancies, which we have tried to correct and reconciliate. We would be grateful if any remaining error or omission was reported to us. A 20th RTK, SuRTK106 has been reported from the Sugen database but a number of critical substitutions make it doubtful that it would have kinase activity (T. Hunter, personal communication 2003). The nomenclature of the RTKs is rather complex with alternative names existing besides those shown above (see e.g., ref. 223). The RTKs in bold type have been implicated in human malignancies. Reprinted with permission from Gray SG, Stenfeldt-Mathiasen I, De Meyts P. Horm Metab Res 2003; 35:857-871.

gain-of-function (constitutive activation) or loss-of-function (inactive or dominant-negative receptor) mutations.[58] Some of the RTKs are oncogenes[59] and at least a dozen of the RTK families, including the IGF-I and insulin receptors[59,60] have been implicated in human cancers due to amplification, overexpression, loss of parental imprinting or somatic gene mutations. Consequently, RTKs are major targets for anticancer therapy.[61-64] The principles for designing structure-based agonists and antagonists of RTKs have been recently reviewed.[65]

The receptor for insulin was first characterised biochemically by direct binding studies using radiolabelled insulin in the early seventies.[66-69] Evidence for its subunit structure was provided in the early eighties,[70,71] as was the demonstration that it is a receptor tyrosine kinase,[72-74] which catalyzes the transfer of the γ phosphate of ATP to tyrosine residues on protein substrates, the first being the receptor itself. Cloning of insulin receptor complementary DNA (cDNA) was achieved in the mid-eighties,[75-77] soon followed by that of the IGF-I receptor[78] and by the sequencing of both genes,[79,80] giving valuable insight into the receptor structure and organisation.

The gene for a mammalian related receptor (insulin receptor-related receptor or IRR) was identified in humans and guinea pigs in 1989,[81] with expression in heart, skeletal muscle, kidney, liver, pancreas and sympathetic ganglia.[82] No ligand for this receptor has been so far identified. The mouse knockout has no phenotype, so the physiological relevance of this receptor has been questioned. However, it has recently been shown that it is required for testis determination in mice since XY mice that are mutant for all three insulin related receptors (*Ir, Igf1r and Irr*) develop ovaries and show a completely female phenotype.[83] Also, it has been reported that coexpression of IRR and IGF-IR correlates with enhanced apoptosis and differentiation in human neuroblastomas.[84] Thus, it is quite possible that the IRR plays a role in modulating the activity of the insulin and IGF-I receptors, possibly through the formation of hybrid receptors, similar to the role of the ligand-less ErbB2 receptor in the EGF receptor family.

The insulin and IGF-I receptors differ from the majority of RTKs in that they are covalent dimeric structures, although all the RTKs dimerize or oligomerize upon ligand binding,[85] resulting in activation of the kinase by transphosphorylation. A claim that monomeric insulin receptor mutants could autophosphorylate[86] was recently found to be artifactual.[87]

The insulin receptor has a modular structure encoded by a gene with 22 exons and 21 introns.[88] The receptor sequence modules encoded by the various exons is shown in Figure 3. Exon 11 is alternatively spliced, resulting in two isoforms (A and B) of the insulin receptor differing by the absence or presence of a 12 residue-sequence (717-729). The physiological significance of this alternative splicing is still unclear; it is absent in the IGF-I receptor which has no equivalent to exon 11. The two isoforms differ slightly in affinity for insulin,[89-91] but the A isoform has significantly higher affinity for IGF-I (40 nM vs 350 nM)[90] and IGF-II (close to that of insulin).[92] The IGF-I receptor binds IGF-II with a lower affinity than IGF-I and insulin with a 500-fold lower affinity.[93]

The deduced amino acid sequences of the receptors indicate that they are synthesized as single chain prereceptors with a 30-residue signal peptide, which is cleaved cotranslationally; the single chain precursor is glycosylated cotranslationally, and folded and dimerized under the guidance of the chaperones calnexin and calreticulin[94] prior to transport to the Golgi apparatus where it is processed at a tetrabasic RKRR furin protease cleavage site (732-735) to yield the mature $\alpha_2\beta_2$ receptor. Interestingly, the B-isoform of the proreceptor binds insulin with high affinity but not the A-isoform.[95]

In cells expressing both insulin and IGF-I receptors, heterodimeric hybrid receptors are formed consisting of one half of each. They have been reported to bind IGF-I with high affinity and insulin with low affinity;[96] the relative affinities are dependent on the insulin isoform involved.[97] Their physiological role is unknown.

Comparative sequence analysis of the insulin/IGF-I receptors and the related EGF receptor families had generated a hypothetical tertiary structure of the extracellular part of the receptor, suggesting that the N-terminal half consists of two homologous globular domains (L1, residues 1-120 and L2, residues 211-428), separated by a cysteine-rich region.[98] The Cys-rich region was predicted to consist of a series of disulfide-linked modules similar to those found in the tumor-necrosis factor (TNF) receptor and later in laminin.[99] All these

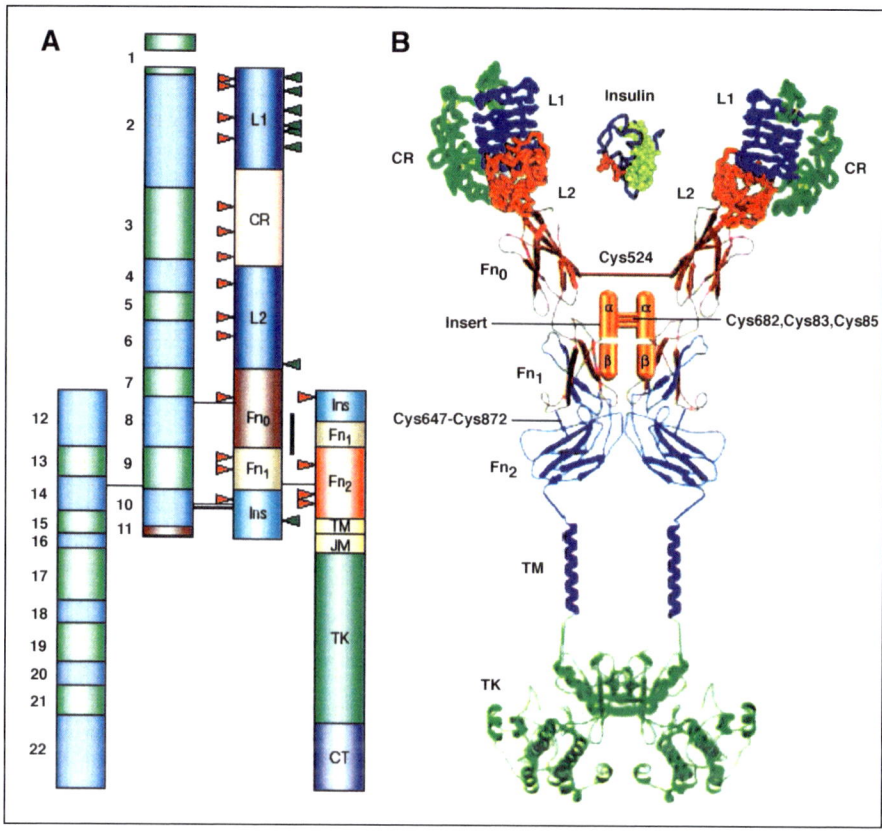

Figure 3. Modular structure of the insulin receptor. Left: cartoon of the α2β2 structure of the insulin receptor, drawn to scale. On the left half of the receptor, spans of the 22 exon-encoded sequences. On the right half, spans of predicted protein modules. Orange arrowheads: N-glycosylation sites. Green arrowheads: ligand binding "hotspots" identified by single amino acids site-directed mutagenesis (see text and for details). Right: Supra-domain organisation of the insulin receptor. Stretched-out model of predicted or actual modular structures (L1CL2 and TK domains based on X-ray analysis data[110,114]). The 3D structure of the insulin molecule[187] is also shown. The classical binding surface is shown in yellow as van der Waals spheres; A13 and B17 Leu in the new binding surface are shown in red. Modelling using DS ViewerLite 5.0 (Accelrys). Reprinted with permission from De Meyts P and Whittaker J. Nat Rev Drug Disc 2002; 1:769-783.

predictions were confirmed by the determination of the crystal structure of this domain (see below). The C-terminal half of the receptors was predicted to consist of three fibronectin type III (FnIII) domains, each with a seven-stranded β-sandwich structure.[100-102] The second FnIII domain comprises the C-terminal part of the α-subunit and the N-terminal part of the β-subunit and contains a large insert domain of ~120-130 residues of unknown structure containing the site of cleavage between α- and β-subunits. The intracellular portion of the β-subunit contains the kinase catalytic domain (980-1255) flanked by two regulatory regions, a juxtamembrane region involved in docking insulin receptor substrates (IRS) 1-4 and Shc as well as in receptor internalization, and a C-terminal tail containing two phosphotyrosine binding sites. The detailed organization of the modular domains of the insulin receptor is shown in Figure 3. The IGF-I receptor has a very similar organization with sequence homology varying from 41% to 84% depending on the domain, being maximal in the kinase domain (see ref. 103 for details).

## Localization of Disulfide Bridges

The α-subunit of the insulin receptor has 37 cysteine residues, 25 of which are in the Cys-rich region. The β-subunit has 4 extracellular Cys and 6 intracellular ones including two free thiols. The two α-subunits are linked by a disulfide bond between the two Cys 524 in the first FnIII domain.[104] One to three of the triplet Cys at 682, 683 and 685 in the insert within the second FnIII domain are also involved in α-α disulfide bridges.[105] There is a single disulfide bridge between α and β subunits between Cys 647 in the second FnIII domain and Cys 872 (nomenclature of B isoform[75]) in the third FnIII domain.[86,105,106]

## Localization of Glycosylation Sites

The insulin receptor is heavily glycosylated.[107] The α-chain contains 14 potential sites for N-linked glycosylation and the β-chain four (Fig. 3). O-linked glycosylation has been demonstrated only in the β-subunit. Although almost all of these sites can be mutated individually without altering cell-surface expression, receptor processing and ligand binding, the major domains of the receptor require at least one intact glycosylation site to ensure correct folding and processing.[108] However, mutation of N-linked glycosylation sites in the β subunit has been shown to impair receptor signalling without detectable perturbation of receptor processing or ligand binding.[109]

## Modular Receptor Structures Elucidated by X-Ray Crystallography

The 3D structure of the L1/Cys-rich/L2 (L1CL2) domain fragment (amino acids 1-460) of the IGF-I receptor has been solved at 2.6 Å resolution[103,110] (Fig. 4). An extended bilobed structure (40 x 48 x 105 Å) comprises the two globular L domains with a new type of right-handed β-helix fold flanking the Cys-rich domain. They appear to be part of the leucine-rich repeat superfamily.[111] While L1 (1-150) contacts the Cys-rich domain along its length, there is minimal contact with L2 (300-460). The Cys-rich domain comprises an array of disulfide-linked modules resembling as predicted those in TNF and laminin. The different orientations of L1 and L2 relative to the Cys-rich domain are probably an artifact of crystal packing and the position of L2 is probably more parallel to L1 in the native structure, as shown in the recently determined structures of the structurally related EGF receptor complex with EGF or TGFα (for review see ref. 112). The flexibility between the Cys-rich domain and L2 may be important for ligand binding.[113] A cavity of ~30 Å diameter occupies the centre of the molecule and represents a potential binding pocket, although this construct does not bind IGF-I (see below for further discussion). Figure 4 shows also the equivalent domains (3-468) of the insulin receptor modelled on the IGF-I structure.

The 3D structures of the tyrosine kinase domain of the insulin receptor, both in the inactive state (unphosphorylated)[114] and in the active state (tris-phosphorylated in combination with an ATP analogue and a peptide substrate)[115] have been determined. The structure of the IGF-I receptor kinase in complex with an ATP analogue and a specific peptide substrate has also been solved.[116] The structures have revealed the determinants of substrate preference for tyrosine rather than serine or threonine and a novel autoinhibition mechanism whereby Tyr 1162, one of the three tyrosines that are autophosphorylated in the activation loop in response to insulin (1158, 1162, 1163) is bound in the active site, hydrogen bonded to a conserved Asp 1132 in the catalytic loop. Tyr 1162 in effect competes with protein substrates before autophosphorylation. In the activated state, Tyr 1163 becomes hydrogen-bonded to a conserved Arg 1155 in the beginning of the activation loop, which stabilizes the repositioned tris-phosphorylated loop. Mutation of the Asp 1161 (which participates in several hydrogen bonds that stabilize the closed conformation of the loop) to Ala substantially increases the ability of the unphosphorylated kinase to bind ATP. The structure of this mutant kinase has also been solved.[117] Covalent dimerization of the soluble insulin and IGF-I receptor kinase domains fused to the homodimeric glutathion-S-transferase (GST) resulted in 10-100-fold increases in the phosphotransferase activity in both the auto-and

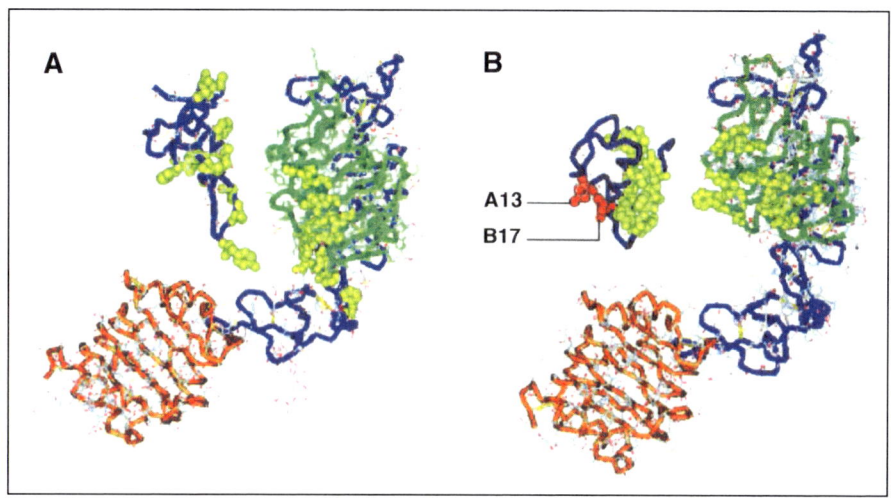

Figure 4. Structures of the insulin and IGF-I receptors and their ligands, and mapping of binding domains. Left: 3-D structure of the L1CL2 domain of the IGF-I receptor determined by X-ray crystallography. The amino acids determined by alanine scanning mutagenesis to be important for ligand binding are shown in yellow as van der Waals spheres (see text and ref. 65 for details). The 3D structure of the IGF-I molecule[201] is shown to scale. The amino acids determined by site-directed mutagenesis to be important for receptor binding are shown in yellow as van der Waals spheres. Since Arg 36 and Arg 37 are lacking in the structure, residues 35 and 38 have been highlighted instead to show approximate location in the middle of the C-peptide domain. Right: The insulin receptor L1-CR-L2 domain has been modelled on the corresponding IGF-I domain coordinates[110] using the program SwissModel. The amino acids determined by site-directed mutagenesis to be important for receptor binding are shown in yellow as van der Waals spheres (see text and ref. 65 for details). The 3D structure of the insulin molecule[187] is shown to scale. The classical binding surface is shown in yellow as van der Waals spheres; A13 and B17 Leu in the new binding surface are shown in red. Modelling using DS ViewerLite 5.0 (Accelrys). Reprinted with permission from De Meyts P and Whittaker J. Nat Rev Drug Disc 2002; 1:769-783.

substrate phosphorylation reactions and rendered the autophosphorylation reaction concentration-independent.[118]

## Ligand Binding Properties

The binding of insulin to its receptor, whether studied on whole cells, purified membranes or purified receptor protein in solution, is complex (Fig. 5), as has been appreciated since the very first binding studies.[57,119-121] Scatchard plots are curvilinear, indicating the coexistence of high and low affinity binding sites. Dissociation studies have shown that the dissociation rate of prebound ligand is accelerated by the presence of unlabelled ligand, consistent with the existence of negative cooperativity between binding sites. Dose-response curves for the accelerated dissociation are bell-shaped (self-antagonism) with a loss of the acceleration of dissociation at concentrations of insulin over 100 nM. See reference 57 for a detailed discussion. IGF-I binding to its receptor shows a similar phenomenology, except that the dose-response curves for accelerated dissociation are sigmoid with no loss of response at high concentration.[122]

Alterations in the various components of this complex kinetic behaviour have been observed with certain amino acid substitutions or deletions in the insulin or receptor molecule, or with various minimized receptor constructs, and have provided essential clues as to the nature of the ligand binding mechanism and established the reality and specificity of the negatively cooperative behaviour (see below).

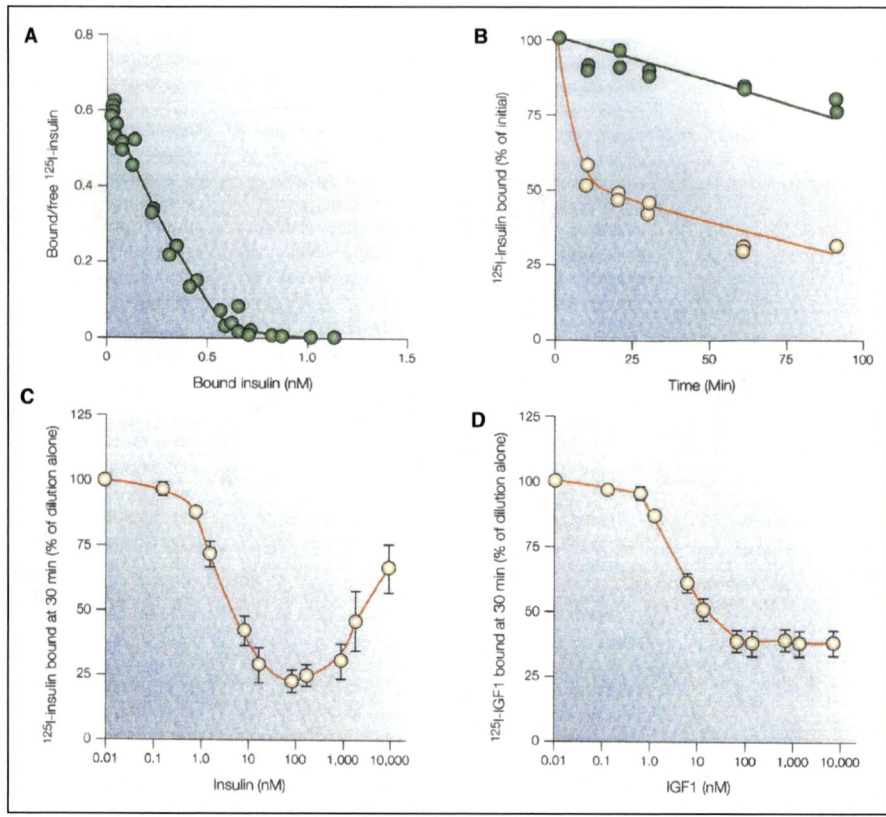

Figure 5. Ligand binding properties of the insulin and IGF-I receptors. A) Curvilinear Scatchard plot of insulin binding (IGF-I's is similar). B) Accelerated dissociation of labelled insulin, the hallmark of negative cooperativity (IGF-I gives similar data). C) Dose-response curve for negative cooperativity (insulin). D) Dose-response-curve for negative cooperativity (IGF-I). Reprinted with permission from De Meyts P and Whittaker J. Nat Rev Drug Disc 2002; 1:769-783.

## Receptor Crosslinking with Bifunctional and Photoreactive Ligands

Early crosslinking experiments using radiolabelled insulin and bifunctional N-hydroxysuccinimide esters demonstrated that insulin's B-chain was crosslinked to a 55 kDa chymotryptic N-terminal fragment of the receptor α-subunit.[123]

Photoaffinity labelling of affinity-purified insulin receptor with a radioactive insulin photoprobe derivatized at Lys B29 labelled a 23-kDa fragment suggested to comprize residues 205-316, i.e., most of the Cys-rich region, on the basis of its immunoreactivity.[124] In contrast, insulin with a different photoaffinity reagent at B29 labelled a 14-kDa tryptic fragment shown by N-terminal sequencing to start at Leu 20 and assumed to extend to approximately Asn 120,[125] but due to glycosylation more probably to Arg 86 or Lys 102.[103] Subsequent studies with chimeric or alanine-mutated insulin receptors have not supported a role for the Cys-rich domain in insulin binding (unlike IGF-I binding) but confirmed the importance of the N-terminal sequence (see below). Insulin derivatized at Phe B1 labelled a tryptic fragment of the soluble receptor ectodomain starting at Gly 390 and estimated to extend to Arg 488, covering the domains encoded by exon 6 and 7 (second half of L2 and first FnIII domain).[126] The importance of this region in insulin binding is supported by other experimental evidence (see below). Two groups showed that insulin derivatized at Phe B25 crosslinks covalently to the insulin receptor,[127,128] the contact site was mapped to the 15-residue sequence Thr 704 to Lys 718 (nomenclature for B isoform) near the end of the A-chain by sequencing of a photo-labeled peptide generated by Lys-C-endoproteinase digestion, in the insert within the second FnIII domain.[128] The fact that photoaffinity labels only 4 residues apart at the C-terminus of the insulin B-chain label segments of the receptor located respectively at the very N-terminus and at the very C-terminus of the a subunit more than 550 residues apart, strongly suggests that these receptor regions must be located close together in space despite their wide separation in the primary sequence.

All the available evidence suggests that all the insulin contact sites are located in the α subunit.

## Definition of Ligand Binding Specificity Using Chimeric Insulin/ IGF-I Receptors

Extensive studies using a variety of chimeric receptor constructs with swapped domains of the insulin and IGF-I receptors have unequivocally demonstrated that the determinants of ligand binding specificity reside in different segments of the two receptors.[129-131] Both whole receptor constructs and soluble ectodomains have been used with relatively consistent results. These data have been recently exhaustively reviewed.[103,132] In brief, the Cys-rich region plus flanking regions from L1 and L2 are prime requirements for IGF-I binding, especially the residues 253-266 in the variable loop of module 6 in the Cys-rich region. Alanine scanning studies suggest that this loop does not directly participate in ligand binding implying that its role in modulating affinity must be indirect. The loop in the insulin receptor exhibits significant charge differences from that of the IGFR; two lysine and two arginine residues in the insulin receptor loop compared to two glutamates and one aspartate in the IGFR loop, suggesting it might produce an unfavourable charge environment in the putative binding cavity. In addition the IR loop is significantly longer than that of the IGFR, possibly limiting steric access of the bulkier IGF molecule to the binding site. This mechanism is supported by the finding that a mini-IGF-I with a shortened C-domain and a deleted D domain exhibits a 5 fold increase in its affinity for the insulin receptor.[133] Further support is provided by the finding that shortening the insulin receptor loop increases its affinity for IGF-I (Whittaker and Groth, unpublished observations). In contrast the N-terminal residues 1-137 in the L1 domain of the insulin receptor (in agreement with the photoaffinity labelling data of Wedekind et al[125]) and residues 325-524, comprising most of the L2 domain and part of the first FnIII domain, are important determinants of insulin binding. Studies with chimeric receptors have localized the N-terminal element contributing to insulin specificity to residues 1-68 and Kjeldsen et al have further suggested that F39 in the insulin receptor may play a significant role in this.[134]

## Natural Receptor Mutations That Affect Insulin Binding in Syndromes of Extreme Insulin Resistance

Naturally occuring mutations in the insulin receptor have been characterized in several dozen patients with genetic syndromes associated with extreme insulin resistance: acanthosis nigricans type A, leprechaunism (Donohue syndrome), Rabson-Mendenhall syndrome[135] and congenital fiber-type dysproportion myopathy.[136-138] Most of the patients carry compound heterozygous mutations, exceptionally homozygous. The functional consequences of the mutations fall into five classes: impaired biosynthesis of full-length receptor due to premature termination mutations, impaired transport to the cell surface due to perturbed protein folding, impaired tyrosine kinase activity, alterations in binding affinity and/or negative cooperativity, and accelerated receptor degradation.[135] One mutation can affect several of these parameters. A large number of these mutations alter the tyrosine kinase domain (reviewed in ref. 114).

The mutations that affect binding affinity or other ligand binding properties provide clues to the localization of the binding domains that complement the in vitro studies. The majority of mutations that impair ligand binding have been found in the exon 2-encoded L1 domain. A natural mutation at Asn 15 (to Lys) decreased insulin binding 5-fold,[139] and the importance of this residue was confirmed later by alanine scanning mutagenesis (see below). A natural mutation of Asp 59 to Gly decreased binding affinity 4-fold.[140] A natural mutation of Arg 86 to Pro results in lack of transport to the plasma membrane, loss of binding affinity and a constitutively activated kinase.[141,142] A natural mutation of Leu 87 to Pro reduced the affinity 6-fold,[143] also confirmed by alanine mutagenesis. A natural deletion of Lys 121 created a temperature-sensitive alteration in insulin binding and negative cooperativity.[144] A mutation of Val 140 to Leu has been reported, but no assessment of the binding affinity was done.[145]

A mutant receptor with Arg 252 mutated to Cys in the Cys-rich domain also showed reduced affinity (as well as endocytosis) but showed normal signalling; the abnormal Cys 252 is close enough to the first Cys at position 8 in the L1 domain to create an abnormal disulfide that may lock the L1 domain in a conformation unfavorable for high affinity binding.[146] Two patients with a mutation to His at the same position have been described;[147,148] both cell surface expression and binding affinity were decreased.

Studies of a receptor with Ser 323 mutated to Leu in the L2 domain (found in two patients with Rabson-Mendenhall syndrome[149,150] and one leprechaun patient[147]) have generated somewhat conflicting results, Roach et al finding that it decreases binding affinity of the full length receptor without perturbing intracellular transport and membrane insertion while Whittaker and Mynarcik found that both binding affinity and secretion of the recombinant extracellular domain were impaired when that mutation was introduced, suggesting a disturbed structure. In the latter study these effects were found to be more profound with mutation of the A isoform and furthermore mutation to Ala had no significant effect on binding.[151]

Another leprechaun patient had a homozygous Asn 431 to Asp mutation, in the L2 domain; this decreased processing and receptor signalling but retained substantial binding affinity, indicating that no major disruption of the binding site was caused by this mutation.[147]

Finally, an extensively studied mutation of Lys 460 to Glu which retarded ligand dissociation at acid pH and enhanced negative cooperativity, and accelerated receptor degradation, pointed to a possible key role of the C-terminal region of the L2 domain in negative cooperativity, leading to further exploration of this domain (see below).[152]

## Mapping of Ligand Binding Sites on the Insulin and IGF-I Receptors by Site-Directed and Alanine-Scanning Mutagenesis

Alanine scanning studies of the secreted recombinant insulin receptor ectodomain have provided the most definitive demonstration that the domains implicated in ligand binding by affinity labeling studies and chimeric receptor experiments are in fact components of a ligand binding site and have also provided a more detailed functional characterization of this binding site. Alanine scanning studies of the L1 domain of the insulin receptor[153] (Fig. 4) indicate that side chains of residues in the first five turns of the β helix form two functional epitopes with discrete footprints on the base of the domain. The first of these is composed of Asp 12, Arg 14 and Asn 15; Gln 34, Leu 36 and Leu 37; Phe 64; Val 94, Glu 97; Glu 120 and Lys 121 which are located in the second β strands of turns 1-5 of the β helix, respectively. Those amino acids, providing the majority of the free energy of binding (Arg 14, Asn 15 and Phe 64, whose mutation to alanine inactivates insulin binding), are located centrally within the footprint and those making more minor contributions are more peripheral, as has been observed for interactions of growth hormone with its receptor.[154] The second functional epitope is located in the bulge between the first and second β sheets of the β helix and consists of Leu 87, Phe 89, Asn 90 and Tyr 91. This epitope had been earlier identified by site-directed mutagenesis of Phe 89 to Leu.[155]

Alanine scanning mutagenesis of the secreted receptor also confirms that the C-terminus of the α subunit, amino acids 704-715, is also an essential component of this ligand binding site.[156,157] In this region, alanine mutations of Phe 705, Glu 706, Tyr 708, His 710 and Asn 711 inactivate insulin binding and mutations of Leu 709 and Phe 714 produce a 30-100 fold reduction in affinity, while mutations of Asp 707, Val 713 and Val 715 were without effect. A conflicting result has been reported regarding mutation of Asp 707 to Ala, found in a leprechaun patient, with abolition of binding.[158] Analogue binding studies reveal that the free energy contribution of this region to insulin binding is considerably greater than that of the L1 domain, consistent with the observation that both the L1 and L1/CRD/L2 domains are incapable of binding insulin when expressed on their own. In the insulin receptor, the cysteine rich domain does not appear to be involved in insulin binding. Alanine scanning of ligand accessible residues just adjacent to the L1 domain fail to demonstrate a functional epitope for insulin binding[159] and more distal residues in the Cys-rich domain are too far from the L1 hotspot residues for simultaneous contacts with the putative insulin binding site to occur, even with bridging by water molecules.

Alanine scanning of the A and B isoforms extracellular domains[159] showed that a number of mutations produced differential effects on the two isoforms. Mutation of Asn 15 in the L1 domain and Phe 714 at the C terminus of the alpha subunit inactivated the A isoform but only reduced the affinity of the B isoform 40- to 60-fold. At the C terminus of the alpha subunit, mutations of Asp 707, Val 713 and Val 715 produced 7- to 16-fold reductions in affinity of the A isoform but were without effect on the B isoform. In contrast, alanine mutations of Tyr 708 and Asn 711 inactivated the B isoform but only reduced the affinities of the A isoform 11- and 6-fold, respectively. This indicates that the energetic contributions of certain side chains differ in each isoform, suggesting that different molecular mechanisms are used to obtain the same affinity.

Alanine scanning studies of the secreted recombinant IGF-I receptor[160] (Fig. 4) indicates that the L1 and Cys-rich domains and the C-terminal peptide of the α subunit 692-702) corresponding to 705-715 of the insulin receptor are components of the IGF-I binding site. In the L1 the side chains of residues in the first 4 turns of the β helix (Asp 8 and Asn 11; Tyr 28, His 30 and Leu 33; Leu 56 and Arg 59; Phe 90 respectively) appear to form a discrete functional epitope on the base of the domain. Trp 79 which is situated in a bulge in the fourth turn of the L1 β helix and Arg 240, Phe 241, Glu 242 and Phe 251 in the cysteine rich domain form a second functional epitope. None of these mutations in L1 or the cysteine rich domain caused a greater than 25 fold decrease in affinity for IGF-I. Alanine mutations of all residues in the C-terminal peptide of the α subunit with the exception of Phe 695, Ser 699 and Val 702 compromised IGF-I binding. However in contrast to the corresponding insulin receptor mutants, only mutation of Phe 701 (corresponding to Phe 714 of the insulin receptor) to alanine appeared to inactivate ligand binding.

## Reconstitution of Modular Minimized Receptor Constructs with Low and High Affinity

It should first be mentioned that discussion of the affinity of receptor constructs should carefully consider the conditions of the experiment since the affinity of the wild type receptor varies widely depending on the assay conditions. In physiological buffer on whole cells, the high affinity component of insulin binding is usually in the sub-nM range (e.g., 0.3 nM on IM-9 cells) while low affinity binding is typically ~10 nM. However, when detergent solubilized holoreceptor or some other dimeric constructs are studied in solution, the high affinity can be optimized to the low pM range.[161] In that context, nM is "low affinity".

Studies with purified receptors have established that the holoreceptor binds only one insulin molecule with high affinity, and at least one additional molecule with low affinity.[161] It has the same negatively cooperative binding as the cell-bound receptor. In contrast, a purified secreted dimeric ectodomain comprising the two α-subunits and the extracellular portions of the β-subunits binds two molecules of insulin, but with low affinity; it shows a linear Scatchard plot and a fast dissociation rate that is not accelerated by unlabeled insulin.[162] The bivalency of the insulin receptor was also demonstrated by photoaffinity labelling.[163] Monomeric αβ receptor halves prepared by disulfide bond reduction also bind insulin but only with low

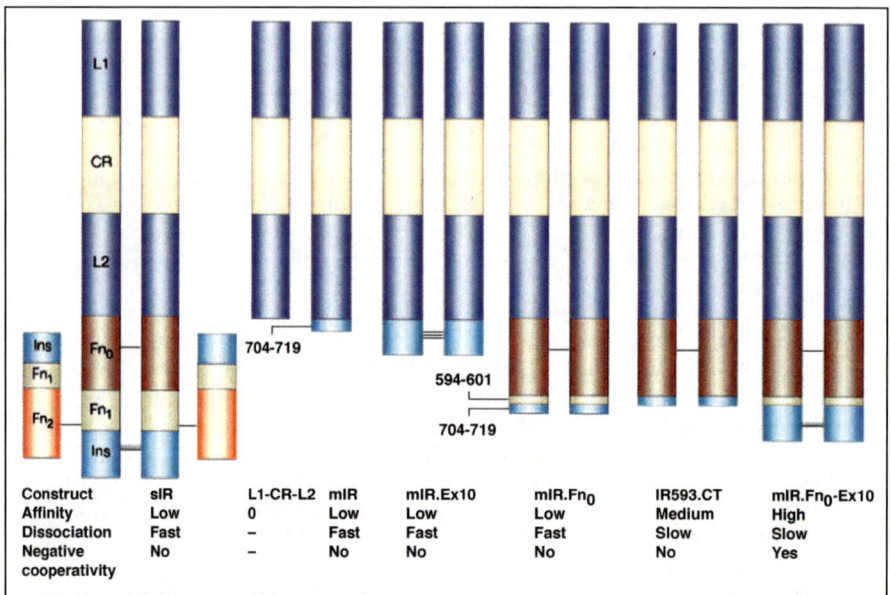

Figure 6. Progressive reconstitution of a minimal dimeric receptor structure with high affinity and negative cooperativity: ligand binding properties of minimized receptor constructs. See text for explanation. Reprinted with permission from De Meyts P and Whittaker J. Nat Rev Drug Disc 2002; 1:769-783.

affinity.[164,165] When the soluble ectodomain is reduced to αβ form, no significant change in affinity is observed.[161] These experiments have established that each αβ receptor half contains partial binding sites for insulin but that cooperation between the two receptor halves is required to create one site with high affinity. High affinity and negative cooperativity could be restored in the soluble ectodomain by fusing it to dimeric constant domains from immunoglobulin Fc and λ subunits[166] or leucine zipper,[167] or by extending the construct to include the transmembrane domains.[168,169] These data show that it is the improper orientation of the β-subunits that disturbs the structure of the soluble ectodomain, although the beta subunits do not contain binding determinants, and that this is corrected by constraints imposed by membrane insertion or TM domain interactions.

A variety of attempts at deleting various modules from the insulin receptor extracellular domain have been made, with the aims of mapping domains essential for ligand binding as well as making minimized receptor structures more amenable to crystallization (some of which are illustrated in Figure 6.

Kadowaki et al (1990) deleted the sequence 486-569 (i.e., 67% of what was later identified as the first FnIII domain including Cys 524 involved in one of the α-α disulfide bonds) and found that the construct had decreased binding affinity, three times faster dissociation rate (measured only at 5 minutes) and altered negative cooperativity as shown by near linearization of the Scatchard plot.[152] This was not the main emphasis of the paper and the properties of this construct were not extensively quantitated. In contrast, Sung et al made a somewhat more extensive deletion of this FnIII domain (485-599) and reported that it did not affect insulin binding, but abolished the interaction of various anti-receptor antibodies,[170] in agreement with the finding that region 450-601 is a major immunogenic determinant. The results of the two studies regarding the role of $FnIII_0$ in insulin binding affinity are thus conflicting, but recent studies on purified receptor constructs have established that the first FnIII domain is unquestionably a major determinant of high affinity binding and negative cooperativity.

Kristensen et al deleted the region 487-599 from a high affinity soluble ectodomain fused to the Fc region of IgG.[171] This led to complete loss of high affinity binding (from pM to 3.9 nM, i.e., more than a 1,000-fold loss of affinity). However, in Table 1 of that paper, only the low affinity of the wild-type receptor is tabulated for comparison with the various low affinity deletion mutants. The statement in the discussion that deletion of 234 aminoacids in the central part of the IR α-subunit does not compromise ligand binding implicitly refers to the most extensive deletion yielding a monomeric fragment (due to loss of all the α-α disulfides)—referred to as "minimized insulin receptor" or mIR, see below—as compared to the *low affinity ectodomain*. The statement does not apply to the dimeric FnIII$_0$ deletion mutant compared to the Fc-fused high affinity wild type, which is the appropriate control. This has been apparently misunderstood by some authors who have reviewed the data by Kristensen et al as being in agreement with Sung et al.[172] Adding to the confusion, another paper with some of the same authors quote Sung et al and Kadowaki et al as both showing *decreased* affinity and linear Scatchard plots in the FnIII$_0$ deletion mutants.[173] The issue has been clarified in two later studies[174,175] which have shown unequivocally that the presence of both exon 10 (which contains the cysteines 682, 683 and 685 involved in α-α disulfide bonds, as well as the CT sequence 704-717 essential for insulin binding) and the FnIII$_0$ domain (which contains the 524 cysteine) are required for high affinity binding. Introducing only FnIII$_0$ into the minireceptor does not improve the nM affinity, and the ligand dissociation rate is fast without acceleration by unlabeled insulin, despite the fact that the construct forms dimers. When both domains are introduced in the mini-receptor, the affinity of the dimeric construct increases 1,000 fold to the pM range, dissociation of tracer is slow, and the negative cooperativity is reintroduced as shown by accelerated dissociation.

Compared to the full ectodomain, this high affinity construct lacks only the 47 amino acid sequence 602-649 encoded by exon 9 (first half of the second FnIII domain just before the insert) and the alternatively spliced exon 11 plus 3 C-terminal residues from the α subunit, and of course the β-subunit extracellular domain; this again suggests that the low affinity of the soluble ectodomain is due to structure disruption by misoriented β-subunits. It also shows that the exon 9-encoded domain has no other structural function than to provide the bond with the β-subunit, and that ~90% of the α-subunit native structure is required to maintain the integral binding properties of the native receptor, without contributions from the β-subunit.

An interesting finding of Surinyia et al investigating similar constructs,[175] is that while the FnIII$_0$ module introduced by Brandt et al[174] is extended by 9 amino acids from the N-terminal sequence of the second FnIII domain, the introduction of a FnIII$_0$ module terminating at the exact boundary of FnIII$_0$ (amino acid 593) also restores the high affinity even in the absence of exon 10, and also slow dissociation, but not the accelerated dissociation (meaning that the receptor is locked in the high affinity state). This 9-amino acid sequence contains three prolines, thus possibly with some degree of secondary structure; this linker may thus be critical in providing an orientation of the CT domain that is essential for negative cooperativity. If the amino acids encoded by exon 10 are present, the presence or absence of the 9-amino acid sequence does not make a difference and negative cooperativity is observed in both cases (Siddle K, Brandt J, personal communication).

The binding properties of the minireceptor construct of Kristensen et al[171] were quite unexpected. As mentioned above, the L1CL2 fragment (1-468) equivalent to the IGF-IR fragment crystallized by Garret et al[110] does not bind the ligand. Fusion of a 16-amino acid peptide (amino acids 704-719 from the C-terminus of the α-subunit to its C-terminus resulted in a protein with low (nM) affinity for insulin; these amino acids have been shown to be an insulin contact site by photoaffinity labelling and alanine scanning mutagenesis. Similar findings were obtained by Molina et al.[173] Molina et al mentioned that this construct binds with only 10-fold lower affinity than the native receptor, but their assay conditions were not optimized to detect very high affinity binding (low tracer, long incubation time). The reconstitution of nM affinity with the corresponding CT peptide was true also for the IGF-I receptor. Further a construct with only the L1C sequence (1-308) and the CT peptide binds insulin. In fact, the peptide does not need to be covalently attached to the receptor domain, coincubation of L1CL2 or L1C with a synthetic CT peptide reconstitutes the nM affinity, even if the N-terminal domain is further shortened to 1-255.[176] This, together with the alanine scanning and photoaffinity data, suggest that one binding site on the receptor comprises residues of the L1 domain and the CT domain, which are probably close to each other in the native structure as discussed above.

The minireceptor paradigm was also used to investigate the role of the L1CL2 and CT domains in the specificity of ligand binding in the insulin/IGF-I receptor family.[177] It was shown that the CT domain of the insulin-receptor related receptor (IRR), an orphan receptor related to the above that binds neither insulin nor IGF-I, abolishes binding to both insulin and IGF-I minireceptors, surprisingly since there are only 4 aminoacid differences. It was shown that mutating Phe 714 to Ala in the insulin receptor CT peptide abolishes the ability of the CT peptide to confer binding to L1CL2, confirming the alanine scanning of the full receptor. The constructs with L1CL2 of the insulin receptor with either IR or IGF-IR CT sequences show low selectivity for ligands, binding both insulin and IGF-I with affinities in the nM range. In contrast, the IGF-IR L1CL2 with either the IGF-IR or IR CT domains discriminates more than 1000 fold against insulin.

## Attempts at Insulin Receptor Structure Definition by Electron Microscopy

While the crystal structure of the N-terminal fragment of the IGF-I receptor and the homology modelling studies of the FnIII repeats provide reasonably accurate predictions of the structures of the modules forming the extracellular domain of the receptors, they fail to provide any indication of their overall organization in the receptor dimer. Some clues to the answer to this question have been provided by single molecule electron microscopic imaging of either purified recombinant receptor extracellular domain[178,179] or of purified full length receptor using a variety of different techniques.[180-183] The results of such studies have been somewhat variable, but suggested T or Y shapes for the receptor (see ref. 65 for review).

Tulloch et al have reported the structure of EM images of the insulin receptor ectodomain decorated with Fab' fragments of conformation-specific monoclonal anti-receptor antibodies directed towards defined epitopes of the receptor, using negative staining techniques.[179] This approach has the advantages of providing both controls for the structural integrity of the molecules imaged and also, within the limits of the resolution of the technique, clues to the localization of subdomains. In contrast to the Y shaped structure deduced from the studies discussed above, the consensus image of the molecular envelope of the receptor ectodomain, in this study, was that of a U shaped prism of overall dimensions 90 x 80 x 120 Å; despite the differences in the overall image, the dimensions are similar to those found in the other studies. The prism has a central cleft 30-40 Å wide, adequate to accommodate a single molecule of insulin. Fab fragments directed towards epitopes in the first FnIII domain or in the insert region of the second FinIII domain appear to bind to the base of the U. A third Fab fragment directed towards an epitope in the distal part of the cysteine rich domain binds half way up the uprights of the U on different sides at each end. This suggests an arrangement of the receptor dimer in which the 2 L1CL2 fragments are arranged in an antiparallel configuration and form the uprights and the base of the cavity of the U and the FnIII domains form the membrane proximal base, implying that the L2 domain occupies a similar position to the L1 domain relative to the cysteine rich domain, in contrast to what is observed in the crystal structure. The dimensions of the crystallized fragment can be accommodated within such a structural model.

More recently Luo and coworkers have reported the quarternary structure of the insulin-insulin receptor complex based on 3D reconstructions from low-dose, low temperature, dark-field scanning electron microscopic studies of the complex.[183] The complexes were found to be approximately spherical with a diameter of approximately 150 Å. Covalent labeling of the receptor molecules with nanogold labelled insulin indicated that the majority of the receptors only bound a single molecule of insulin although binding of two insulin molecules was occasionally observed. This procedure also identified the location of the L1CL2 fragment within the reconstruction. Using the published coordinates of the IGF-I receptor N-terminal fragment, the insulin receptor tyrosine kinase catalytic domain, and of Type III repeats from fibronectin, they produced a model of the domain structure within the protein envelope of the receptor dimer obtained from the EM images. In this model the L1CL2 structure occupies the central core of the extracellular domain dimer with the FnIII domains surrounding them just proximal to the membrane.

A number of assumptions made in this reconstruction and the details of the proposed insulin-receptor interface inferred in this and subsequent publications[184,185] have been questioned.[65] Nevertheless, these studies make it likely that the insulin-receptor complex is a globular molecule with a diameter of 150Å. In the extracellular component the L1CL2 domains form the membrane distal core and the FnIII domains form the membrane proximal part.

## Mapping of Receptor-Binding Sites on the Insulin and IGF-I Molecules

Insulin aggregates as a function of concentration, first into dimers, and in the presence of zinc, three dimers form a hexamer (the storage form in the granules of the β-cell). The insulin molecule has one obvious flat surface studded with aromatic and aliphatic residues, which aggregates primarily through nonpolar forces. This dimer-forming surface is made of mostly B-chain residues: Gly B8, Ser B9, Val B12, Tyr B16, Gly B23, Phe B24, Phe B25, Tyr B26, Thr B27, Pro B28, Asn A21. The surfaces of the dimer buried in the hexamer comprize both polar and nonpolar residues from the A and B chains: Leu A13, Tyr A14, Glu A17, Phe B1, Val B2, Gln B4, Gln B13, Ala B14, Leu B17, Val B18, Cys B19, Gly B20.[186,187]

There has been a consensus for many years that a number of surface residues that have been widely conserved during the evolution of vertebrates are probably involved in receptor binding: Gly A1, Gln A5, Tyr A19, Asn 21, Val B12, Tyr B16, Gly B23, Phe B24, Phe B25, Tyr B26 ("classical binding surface").[186-188] De Meyts et al showed that a subset of residues from this surface (A21, B23-26) was essential for the negative cooperativity in binding.[121] In addition, Ile A2 and Val A3 (a residue mutated in a diabetic patient: insulin Wakayama[189]), which are not on the surface of the molecule, probably become exposed and interact with the receptor upon displacement of the B-chain C-terminus during the receptor binding process,[190] although the extent of this rearrangement has been debated.[191]

The concept that the classical binding surface was the only one involved in receptor binding was challenged by studies of the properties of phylogenetically ancient insulins from the hagfish (reviewed in ref. 57 and more recently from the lamprey[192] (Holst PA, Conlon JM, Whittaker J, De Meyts P, unpublished data)). Despite absolute conservation of the classical binding surface, these insulins display different binding behaviour from most mammalian insulins: low affinity for the holoreceptor, but better relative affinity for the soluble low affinity ectodomain, low metabolic potency, slow association kinetics, decreased negative cooperativity with a sigmoid (not bell-shaped) dose-response curve. This strongly suggested that certain residues outside the classical binding surface must contact the receptor and their mutation in hagfish and lamprey insulins explain the abnormal behaviour. The evolutionarily divergent hystricomorph insulins exhibit similar binding behaviour but the interpretation of its underlying mechanism was less clearcut since they also have a few substitutions in the classical binding surface.[57] This concept was validated when we found that insulin analogues with mutations at two residues in the hexamer-forming surface, Leu A13 to Ser and Leu B17 to Gln, bound to the holoreceptor with similar characteristics to hagfish insulin.[57]

These findings prompted a reexamination of the structure-function relationship of insulin by alanine scanning mutagenesis. Kristensen et al produced 21 new insulin analogue constructs with single alanine substitutions and tested their binding to the soluble ectodomain fused to the IgG fusion protein.[193] Since it is difficult to accurately determine the high affinity (pM) binding component of such constructs in competition assays due to ligand depletion, requirement for minimal tracer amounts and long incubation time, the authors focused on the low affinity component of binding. They showed that mutation of residues A2, A19, B23 and B24 from the classical binding surface and of B13 were most disruptive of binding. Mutations of B6 and B8 were also disruptive, probably due to conformational alterations.

Since high affinity binding is more amenable to precise determination in the context of the cell-bound receptor, we tested the binding affinity of the alanine-scanned analogues using IM-9 human lymphocytes, a "gold standard" for insulin receptor studies for nearly three decades. We reconfirmed the importance of the residues from the classical binding surface A2, A19, B16, B26. Others have shown that mutations of A3[194] (Aladdin H and De Meyts P, unpublished data) and B12[195] to Ala are also very disruptive. In addition, we found that a number of residues outside the classical surface disrupted binding, validating the concept of a second receptor binding surface: Ser A12, Leu A13, Glu A17, His B10, Glu B13, Leu B17[196] (Theede A-M, Aladdin H, De Meyts P, manuscript in preparation). Interestingly, all of these are involved in the formation of the hexamer. Mutation of Thr A8 to Ala (as in ox insulin) impacted binding moderately; previous studies of analogues mutated in this position had already suggested that it may extend the classical binding surface[197] and its mutation to His explains the enhanced affinity of chicken and turkey insulins (Piron MA and De Meyts P, unpublished data).

Moreover, we recently showed by introducing substitutions from hystricomorph[198] or hagfish (Sajid W, Andersen AS and De Meyts P, unpublished data) insulins into human insulin, alone or in combination, that the aberrant behaviour of these species is largely explained by a small number of deleterious mutations in the novel binding surface.

To date, alanine scanning studies of insulin and its receptor have only determined the effect of mutations on equilibrium binding affinity. This may underestimate the impact of the mutations. We have performed direct binding kinetic studies of several of the analogues and found that some mutations affect association and dissociation rates in compensatory ways; thus, A13 Ala and B17 Ala had a nearly 20-fold decrease in association rates, but a slower dissociation rate as well (similar to hagfish and hystricomorph insulins). This may apply to insulin receptor mutations as well but has not been tested. This contrasts with the growth hormone-receptor system where the majority of alanine mutations primarily affected dissociation rates with proportional effects on affinity.[199] However in this system, mutation of several charged residues and a mutation resulting in a localized structural perturbation decreased association rates but not to the extent observed with A13 and A17 Ala mutations of insulin.

From the above, it therefore appears, perhaps unsurprisingly for a small compact molecule, that insulin uses more or less the same functional surfaces for receptor binding as for self-aggregation.

The structure-function relationships of the IGF-I and II molecules have not been as extensively mapped as insulin's (see refs. 200, 103 for detailed review). Unlike insulin, the IGFs conserve a permanent C-peptide and feature a D-peptide extension at the C-terminus (Fig. 1). 3D structures have been determined by solution NMR or crystallography in the presence of detergents.[201-203] Mutagenesis data of IGF-I have shown Ala 8, Asp 12, Phe 23 and Tyr 24 in the B domain; Tyr 31, Arg 36, Arg 37 in the C-peptide and Met 59, Tyr 60 and Ala 62 in the A domain to be important for high affinity binding to the IGF-I receptor (see refs. 103, 200 for review). Chimeric receptors show that the determinants that favor binding of Arg 36 and Arg 37 lie between residues 217 and 286 in the Cys-rich region.[204] Zhang et al showed that Ala replacement of both Lys 65 and Lys 68 in the D domain caused a 10-fold drop in affinity for the IGF-I receptor;[204] however, Bayne et al showed that removal of the entire D domain had an insignificant effect on IGF-I receptor binding.[133] Mutation of the highly conserved Val 43 in IGF-II to Leu, equivalent to the key conserved insulin-binding residue Val A3, resulted in a 16-fold reduction in IGF-I receptor binding. Moreover, a patient with a dwarfism phenotype due to a homozygous missense mutation at the equivalent Val 44 (to Met) in IGF-I has recently been found (Karperien M and Walenkamp MJE, personal communication). This position is homologous to insulin's Val A3 mutated in the diabetic patient with insulin Wakayama,[189] suggesting that this aminoacid is critical for receptor binding of both ligands (the affinity of both mutants for their respective receptor is reduced to less than 1%). The phenotype of the IGF-I mutant patient is comparable to that of a patient with a partial IGF-I gene deletion.[205]

The C domain of IGF-I plays a major role in IGF-I receptor binding since its deletion or replacement by short linkers alter the binding affinity dramatically.[133,206,207] The most recent X-ray structure determination of IGF-I has provided more details of the structure of the C-peptide and shows that when compared with the structure of insulin it produces a rearrangement of the B22-29 segment corresponding to the C-terminal insulin sequence important for receptor binding.[201] Interestingly, a chimeric molecule with the A and B chains of insulin fused to the IGF-I C-peptide (ICP) binds with high affinity to both insulin and IGF-I receptors, while proinsulin's C-peptide causes a 20-fold drop in affinity for the insulin receptor.[208]

Thus, while it is likely that the low affinity of insulin for the IGF-I receptor largely results from the lack of the C-peptide, more studies are needed to fully understand the low affinity of IGF-I for the insulin receptor (see Note Added in Proof).

## Mechanism of Ligand Binding and Receptor Activation

We have now laid all the pieces of the puzzle on the table, how do we assemble them? In other words, how do the bits and pieces of structural information discussed above help us match the putative binding sites on insulin with those on the receptor? An important clue to the possible topography of the receptor complex was the discovery that human growth hormone (hGH) forms a 1:2 complex with the extracellular domain of its receptor, i.e., that a single GH molecule is a bivalent ligand with two opposite surfaces binding to basically the same site on the receptor.[209] Such a mechanism could also explain how the insulin receptor generates one high affinity binding site from low affinity parts.

Twenty five years ago, Martin Raff was the first to speculate that the complex insulin binding kinetics may be the result of insulin having more than one binding site for the receptor and its consequent ability to crosslink the receptor (see discussion p.116 in ref. 210). Later, Lee et al[211] and Yip[212] also proposed that insulin may contact both halves of the receptor. Schäffer was the first to put forward a model that attempted to integrate all the available kinetic information, proposing that it may be two different receptor sites (labelled site 1 and site 2) that contact an insulin molecule, in contrast to the GH-GH receptor interaction.[161] Schäffer found that insulin analogues with mutations in the hexamer-forming surface e.g., at A13 and B17 had a biological activity lower than was expected from affinity, slow binding kinetics and a higher relative affinity for the soluble receptor ectodomain than for the holoreceptor, and concluded that these residues are probably involved in receptor binding. De Meyts had independently found that the same analogues had very slow association and dissociation rates in IM-9 cells, and reduced negative cooperativity with a sigmoid rather than bell-shaped dose-response curve (properties reminiscent of hagfish and hystricomorph insulins), and came to the same conclusion.[57] Schäffer proposed that the insulin molecule would have two binding surfaces labelled site 1 (the classical binding surface overlapping the dimer-forming surface) and site 2 (the new binding site overlapping the hexamer-forming surface), and that a single insulin molecule would crosslink the insulin receptor's site 1 and site 2. This would explain the 1:2 stoichiometry and the coexistence of high and low affinity sites (curved Scatchard plot). While the affinity of the crosslinked insulin molecule is in the pM range, Schäffer determined the affinity of site 1, using the high affinity analogue X92, to be 1 nM.

However, a model with a single crosslink does not predict the rapid acceleration of tracer dissociation induced by cold ligand[213] (Shymko and De Meyts, unpublished modelling). This is also demonstrated experimentally by the absence of negative cooperativity in the hGH receptor system. To explain accelerated dissociation, Schäffer proposed a *cis*-induced conformational change disturbing the crosslink upon binding of a second molecule to the unoccupied

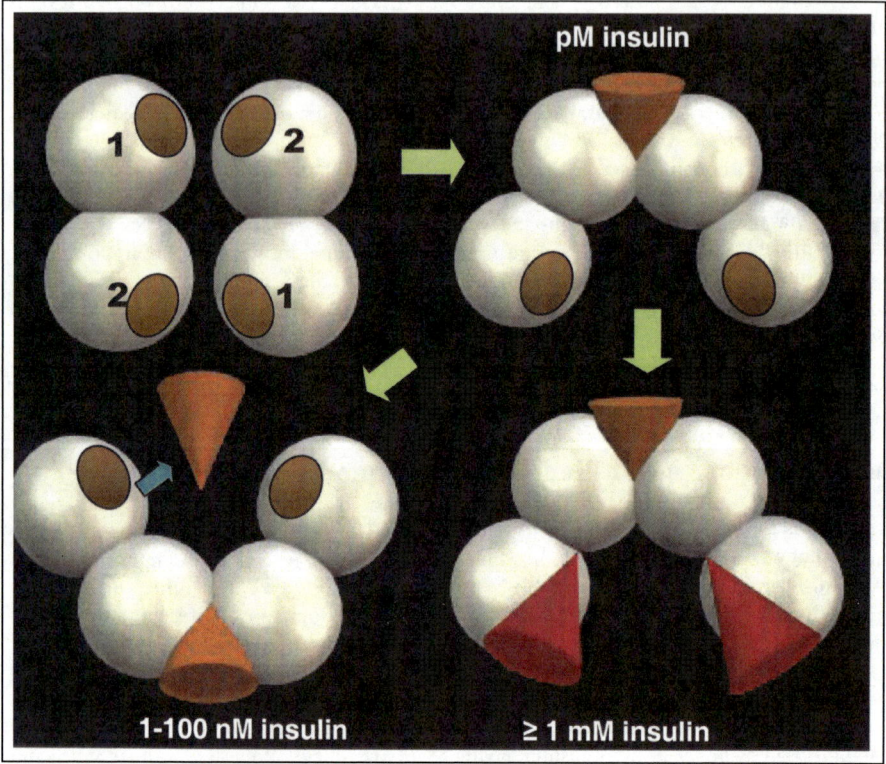

Figure 7. Symmetrical model of bivalent crosslinking insulin binding mechanism. Both α-subunit N-terminal pairs of binding sites (1 and 2) are represented in a symmetrical antiparallel arrangement[57] as also suggested by Tulloch et al.[179] The first insulin molecule (shown as a red cone) binds with high affinity by crosslinking sites 1 and 2. Upon partial dissociation of the first bound molecule, a second insulin molecule is able to crosslink the remaining sites 1 and 2, causing complete dissociation of the first bound insulin (accelerated dissociation, hallmark of the negative cooperativity). At high insulin concentrations, monovalent binding of two additional insulin molecules saturate the leftover sites 1 and 2 and stabilize the binding of the prebound insulin in the first crosslink, explaining the bell-shaped curve for negative cooperativity (see ref. 57 for detailed explanation).

site 1. Also, the single crosslink per se does not explain the bell-shaped curve for negative cooperativity. Schäffer postulates an additional competing mechanism such as the binding of a molecule to the unoccupied binding site 2, "putting a lid" on the complex, sterically impairing dissociation.

While not questioning the plausibility of these postulates, De Meyts proposed that all three parameters of the negative cooperativity (high and low affinity sites, accelerated dissociation and bell-shaped dose-response curve) could be explained by the crosslinking mechanism per se without need for additional mechanisms, if the receptor binding sites 1 and 2 were disposed in an antiparallel symmetry[57] (Fig. 7). A second insulin molecule bridging both leftover sites 1 and 2 upon partial dissociation of the first crosslink would prevent its rebinding and accelerate dissociation of the first bound insulin (Fig. 7). A bell-shaped curve is the hallmark of any effect that requires crosslinking (as demonstrated by the dose-response curve for biological effects of GH.[214] The concept of antiparallel symmetry was supported by the EM studies of Tulloch et al.[179]

Whichever modality of the crosslinking model proves to be correct, the general concept proposed by Schäffer and De Meyts is widely supported by the body of evidence discussed above and it also provides a molecular basis for activation of the receptor tyrosine kinase and signalling pathways. The motion involved in simultaneous interaction of the bivalent ligand with the two half-receptors may approximate the two kinase domains and permit transphosphorylation. This is analogous to the mechanism demonstrated for the dimeric bacterial aspartate receptor, which also exhibits negative cooperativity.[215] In addition, in the cytokine family, previously thought to be activated by ligand-induced dimerization, structural and functional evidence has been presented that demonstrates that the erythropoietin receptor exists as a preformed dimer in the absence of ligands, in which the stems of the extracellular domain and thus presumably the two janus kinase domains are too far apart (73 Å) for transactivation.[216,217] The binding of an EPO-like peptide ligand changes the structure and brings the stems closer together (39 Å).

It remains to identify sites 1 and 2 on the insulin receptor. We should clarify here that the nomenclature we use for site 1 and 2 is the opposite of that used by Schäffer,[161] who labelled site 1 (classical binding surface) and site 2 (new surface) according to their order of discovery. If one postulates an ordered, sequential mechanism as is the case for hGH, then it is mutations in the site that bind last that create antagonism.[218] Since mutations at the C-terminus of the B-chain create antagonists for negative cooperativity,[57] we have postulated that this site binds last (and is therefore site 2), and that site 1 is the site containing Leu A13 and Leu B17, in agreement with the fact that mutations of these two residues slow down the initial association 20-fold, suggesting that they normally make the initial contact.

Photoaffinity labelling data suggest that insulin's site 2 binds to a receptor site composed of the L1 domain and the CT domain,[125,128] whose functional epitopes have been identified by alanine scanning.[153,156] This is consistent with the L1C-CT construct binding insulin with nM affinity.[171]

The insulin receptor binding site 1 has not been mapped by alanine scanning yet, but several lines of evidence suggest that is is located within the structures encoded by exons 6-8, i.e., the C-terminal part of L2 and the first FnIII domain ($FnIII_0$):

1. The B1-photoaffinity analogue covalently labels a peptide mapping to 390-488 generated by tryptic digestion of the receptor.[126]
2. All autoimmune anti-receptor antibodies from patients with extreme insulin resistance recognize an epitope contained within residues 450-601.[219] Such antibodies lock the patient's insulin receptors into a low affinity state with fast dissociation rate and lack of negative cooperativity; plasmapheresis restores high affinity, slow dissociation and negative cooperativity.[220] Monoclonal anti-receptor antibodies that inhibit insulin binding bind to the same epitope.[103,219] Chimeric insulin receptors with residues 450-601 replaced by the corresponding residues from the IGF-I receptor had decreased ability to bind insulin.[219] Monoclonal antibodies that activate the receptor bind to an epitope within residues 469-592.[103]
3. A mutation of Lys 460 to Glu found in a leprechaun patient enhances negative cooperativity and alters pH dependence of binding; mutation of Lys 460 to Arg suppresses negative cooperativity and linearizes the Scatchard plot[152] (Wallach and De Meyts, unpublished). Interestingly, the equivalent residue in the IGF-I receptor is Arg and its mutation does not affect binding kinetics (Wallach and De Meyts, unpublished).
4. Substitution of the insulin receptor domain encoded by exons 6 and 7 into the IGF-I receptor induces a bell-shaped instead of sigmoid dose-response curve for negative cooperativity.

In summary, the studies described above support the concept that insulin binds asymmetrically to two discrete sites in the receptor dimer, crosslinking the constituent monomers. One of these sites, the binding site of the secreted recombinant receptor, has been characterized in detail and is partly located in the L1 domain and also contains a peptide from the C-terminus of the α-subunit. The second site, in contrast, is considerably less well studied but is located in the L2-FnIII$_0$ region. These findings are consistent with the topology of the extracellular portion of the receptor that has been proposed on the basis of crystallographic and EM studies.

## Conclusions and a Word of Caution

In the absence of a definitive three-dimensional structure of a high affinity ligand-receptor complex, circumstantial evidence generated by a multitude of biochemical and molecular biological approaches have generated a plausible working model of the insulin and IGF-I receptor interaction that provides a rational explanation for a variety of complex equilibrium and kinetic properties, including a solution to the three decades-old riddle of the aberrant behaviour of hagfish and hystricomorph insulins. This model has also provided a rational basis for the design of agonist and antagonist mimetic peptides,[221,222] (reviewed in ref. 65).

However, the concept described above of a single bivalent ligand sitting in the middle of a head-to-head dimeric receptor complex (which appears to be the rule in the RTK or cytokine ligand-receptor structures that have been solved so far), is called into question by the recently determined structure of the EGF receptor (which has significant structural homology to the insulin and IGF-I receptors) complexed to EGF or TGFα (reviewed in ref. 112). In those structures, two ligand molecules bind to the external surfaces of a back-to-back receptor dimer with a 2:2 stoichiometry (Fig. 8) rather than the 1:2 we have advocated for the insulin and IGF-I receptor. This means that we have to remain open-minded to the fact that insulin and IGF-I may link the two α-subunit binding epitopes in a cis-fashion within the same subunit

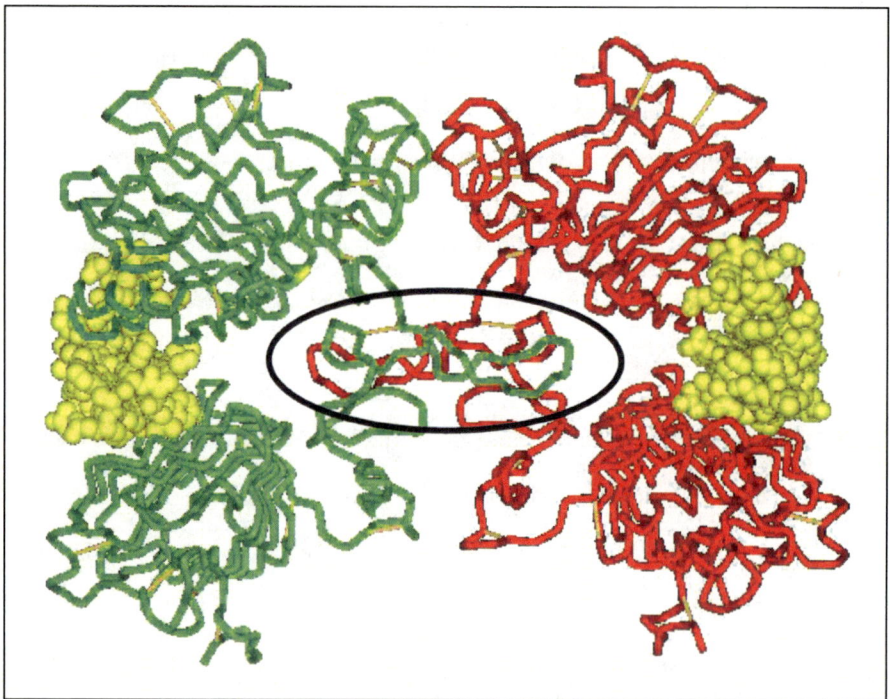

Figure 8. Crystal structure of the complex of human epidermal growth factor and receptor extracellular domains.[226] The two dimerized L1-CR-L2 domains are shown as tube models respectively in red and green. The two bound EGF molecules are shown as CPK models in yellow. The area of contact between the two receptor Cys-rich domain β-hairpins is highlighted in the ellipse.

rather than a trans-fashion between two subunits as we have proposed. However, the majority of the free energy for the formation of the EGF receptor back-to-back dimer is contributed primarily by contacts between two protruding β-hairpin arms in the CR domains that are entirely absent from the insulin and IGF-I receptors. Therefore it is still plausible that the insulin/IGF-I receptor dimer structure will rather resemble the head-to-head structures also present in the EGF receptor crystal structures.

It is clear therefore that the hypotheses and models generated over the last three decades of insulin and IGF-I ligand-receptor binding studies still need to be confirmed or invalidated by three-dimensional structural data of the high affinity ligand-receptor complexes.

## Note Added in Proof

Since this chapter was submitted, two new reviews have discussed the insulin receptor binding mechanism.[227,228]

Sørensen et al[229] have examined the molecular mechanisms underlying the lower affinity of IGF-II as compared to IGF-I for the IGF-I receptor by evaluating the effect of alanine mutations of residues forming the ligand binding site[160] on the affinity of the secreted recombinant receptor for IGF-II. In contrast to the functional epitope for IGF-I, which is composed of residues in the L1, cysteine rich and the Fn1 insert domain, there was no evidence for involvement of cysteine rich domain residues in binding IGF-II. Furthermore while identical residues in the L1 and insert domains were used to bind both ligands, there were small but significant differences in the free energy contributions of many of these residues to the binding of the two ligands.

Whittaker et al have now characterized the functional binding epitope of the native insulin receptor[230] (previous studies have focused on a secreted ectodomain). Alanine mutations Arg14, Phe 64 in the L1 domain and Phe 705, Glu 706, Tyr 708, His 710, and Asn 711 in the insert domain inactivate the insulin binding function of both forms of the receptor. Surprisingly, alanine mutation of Val 715, which has no impact on the affinity of the secreted receptor, inactivates the full-length receptor. Significant differences in effect on affinity of the holo-receptor for insulin were also observed for alanine mutations of Asp 12, Leu 37, Phe 89, Glu 97, and Glu 120 in the L1 domain and of Leu 709 and Phe 714 in the insert domain. It is possible that these differences are a consequence of the proposed sequential binding mechanism.

Chakravarty et al[231] have characterized the binding properties of a hybrid receptor dimer composed of a wild type IGF-I receptor monomer and an inactivating mutant monomer. The mutation of Phe 701 to alanine abolished the ability of both the secreted and full length receptors to bind IGF-I while the hybrid receptor bound IGF-I with an affinity indistinguishable from that of the wild type receptor. However, in dissociation experiments, no ligand-induced acceleration of dissociation was observed. These results are consistent with the proposed bivalent cross-linking model of ligand binding.[57]

Additional photo-affinity cross-linking experiments[232-235] have shown that Val A3 and Thr A8 in insulin's A-chain N-terminal α-helix crosslink to the CT domain in the insulin receptor's insert (like insulin's Phe B25) while Tyr B16 in the central α-helix of insulin's B-chain and Phe B24 in its B chain C-terminus crosslink to L1 domain of the receptor. These results confirm the role of the receptor's L1CT domain (our site 2) in binding the "classical" insulin binding surface.

Two studies have addressed the mechanisms for the higher affinity of IGF-II than IGF-I for the A isoform of the insulin receptor. Denley et al[236] have used chimeras where the C and D domains were exchanged between IGF-I and IGF-II either singly or together, and concluded that the C domain and to a lesser extent the D domain represent the principal determinants of the binding difference between IGF-I and IGF-II to IR-A. This is somewhat surprising since a single chain fusion protein containing insulin fused to C-peptide of IGF-I has high affinity for both the A-isoform of the insulin receptor's and the IGF-I receptor.[208] In contrast, Gauguin and De Meyts (manuscript in preparation[237]) found that the substitutions at the IGF-I and IGF-II positions equivalent to insulin's A8, A10, B5 and B16 explain the low affinities of the IGFs for the A-isoform of the insulin receptor, while two substitutions in IGF-II at positions equivalent to insulin's A18 and B14 have a positive effect on affinity that explains why IGF-II binds better than IGF-I.

Finally, the articles describing the Val 44 IGF-I mutation in a dwarf patient have been published.[238,239]

## Acknowledgements

The Receptor Systems Biology Laboratory at the Hagedorn Research Institute is an independent basic research component of Novo Nordisk A/S. The laboratory is also supported by grants from the Danish Medical Research Council through the Center for Growth and Regeneration, Medicon Valley Academy and Øresund IT Academy, and grants from the National Institutes of Health (5R01 DK065890) and the Juvenile Diabetes Research Foundation (1-2000-198) to J.W. Extensive discussions on the ligand binding mechanism with Lauge Schäffer and Jacob Brandt, and on insulin structure with Michael A. Weiss, are gratefully acknowledged, as well as the collaboration with Claus Kristensen on alanine scanning insulin mutants and Asser S. Andersen in making other recombinant mutated insulins.

## References

1. Chan SJ, Steiner DF. Insulin through the ages: Phylogeny of a growth promoting and metabolic regulatory hormone. Am Zool 2000; 40:213-222.
2. Conlon JM. Evolution of the insulin molecule: Insights into structure-activity and phylogenetic relationships. Peptides 2000; 22:1183-1193.
3. Steiner DF, Chan SJ, Welsh JM et al. Structure and evolution of the insulin gene. Annu Rev Genet 1985; 19:463-484.
4. Hsu SY. New insights into the evolution of the relaxin-LGR signaling system. Trends Endocrinol Metab 2003; 14:303-309.
5. Hsu SY, Nakabayashi K, Nishi S et al. Relaxin signaling in reproductive tissues. Mol Cell Endocrinol 2003; 202:165-170.
6. Kumagai J, Hsu SY, Matsumi H et al. INSL3/Leydig insulin-like peptide activates the LGR8 receptor important in testis descent. J Biol Chem 2002; 277:31283-31286.
7. Ferlin A, Simonato M, Bartoloni L et al. The INSL3-LGR8/GREAT ligand-receptor pair in human cryptorchidism. J Clin Endo Metab 2003; 88:4273-4279.
8. Irwin DM. A second insulin gene in fish genomes. Gen Comp Endo 2003; 135:150-158.
9. Kondo H, Ino M, Suzuki A et al. Multiple gene copies for Bombyxin, an insulin-related peptide of the sikmoth Bombyx mori: Structural signs for gene rearrangement and duplication responsible for generation of multiple molecular forms of bombyxin. J Mol Biol 1996; 259:926-937.
10. Iwami M, Furuya I, Kataoka H. Bombyxin-related peptides: cDNA structure and expression in the brain of the hornworm Agrius convolvuli. Insect Biochem Mol Biol 1996; 26:25-32.
11. Lagueux M, Lwoff L, Meister M et al. cDNAs from neurosecretory cells of brains of Locusta migratoria encoding a novel member of the superfamily of insulins. Eur J Biochem 1990; 187:249-254.
12. Smit AB, Geraerts PM, Meester I et al. Characterization of a cDNA clone encoding molluscan insulin-related peptide II of Lymnaea stagnalis. Eur J Biochem 1991; 199:699-703.
13. Floyd PD, Li L, Rubakhin SS et al. Insulin prohormone processing, distribution, and relation to metabolism in Aplysia californica. J Neurosci 1999; 19:7732-7741.
14. Brogiolo W, Stocker H, Ikeya T et al. An evolutionarily conserved function of the Drosophila insulin receptor and insulin-like peptides in growth control. Curr Biol 2001; 11:213-221.
15. Pierce SB, Costa M, Wisotzkey R et al. Regulation of DAF-2 receptor signaling by human insulin and ins-1, a member of the unusually large and diverse C. elegans insulin gene family. Genes Dev 2001; 15:672-686.
16. Hua QX, Nakagawa SH, Wilken J et al. A divergent INS protein in Caenorhabditis elegans structurally resembles human insulin and activates the human insulin receptor. Genes Dev 2003; 17:826-831.
17. Robitzki A, Schroder HC, Ugarkovic D. Demonstration of an endocrine signaling circuit for insulin in the sponge Geodia cydonium. EMBO J 1989; 8:2905-2909.
18. Duret L, Guex N, Peitsch MC et al. New insulin-like proteins with atypical disulfide bond pattern characterized in Caenorhabditis elegans by comparative sequence analysis and homology modeling. Genome Res 1998; 8:348-353.
19. Fernandez R, Tabarini D, Azpiazu N et al. The Drosophila insulin receptor homolog: A gene essential for embryonic development encodes two receptor isoforms with different signaling potential. EMBO J 1995; 14:3373-3384.

20. Fullbright G, Lacy ER, Büllesbach EE. The prothoracicotropic hormone bombyxin has specific receptors on insect ovarian cells. Eur J Biochem 1997; 245:774-780.
21. Graf R, Neuenschwander S, Brown MR et al. Insulin-mediated secretion of ecdysteroids from mosquito ovaries and molecular cloning of the insulin receptor homologue from ovaries of bloodfed Aedes aegypti. Insect Mol Biol 1997; 6:151-163.
22. Holt RA, Subramanian GM, Halpern A et al. The genome sequence of the malaria mosquito Anopheles gambiae. Science 2002; 298:129-149.
23. Roovers E, Vincent ME, van Kesteren E et al. Characterization of a putative molluscan insulin-related peptide receptor. Gene 1995; 162:181-188.
24. Lardans V, Coppin JF, Vicogne J et al. Characterization of an insulin receptor-related receptor in Biomphalaria glabrata embryonic cells. Bioch Biophys Acta 2001; 1510:321-329.
25. Jonas EA, Knox RJ, Kaczmarek LK et al. Insulin receptor in Aplysia neurons: Characterization, molecular cloning, and modulation of ion currents. J Neurosci 1996; 16:1645-1658.
26. Gricourt L, Bonnec G, Boujard D et al. Insulin-like system and growth regulation in the Pacific oyster Crassostrea gigas: hrIGF-1 effect on protein synthesis of mantle edge cells and expression of an homologous insulin receptor-related receptor. Gen Comp Endocrinol 2003; 134:44-56.
27. Kimura KD, Tissenbaum HA, Liu Y et al. Daf-2, an insulin receptor-like gene that regulates longevity and diapause in Caenorhabditis elegans. Science 1997; 277:942-946.
28. Verjovski-Almeida S, DeMarco R et al. Transcriptome analysis of the acoelomate human parasite Schistosoma mansoni. Nat Genetics 2003; 35:148-157.
29. Hu W, Yan Q, Shen DK et al. Evolutionary and biomedical implications of a Schistosoma japonicum complementary DNA resource. Nat Genetics 2003; 35:139-147.
30. Konrad C, Kroner A, Spiliotis M et al. Identification and molecular characterisation of a gene encoding a member of the insulin receptor family in Echinococcus multilocularis. Int J Parasitol 2003; 33:301-312.
31. Steele RE, Lieu P, Mai NH. Response to insulin and the expression pattern of a gene encoding an insulin receptor homolog suggest a role for an insulin-like molecule in regulating growth and patterning in Hydra. Dev Genes Evol 1996; 206:247-259.
32. Schäcke H, Schröder HC, Gamulin V et al. Molecular cloning of a tyrosine kinase gene from the marine sponge Geodia cydonium: A new member belonging to the receptor tyrosine kinase class II family. Mol Membr Biol 1994; 11:101-107.
33. Skorokhod A, Gamulin V, Gundacker D et al. Origin of insulin receptor-like tyrosine kinases in marine sponges. Biol Bull 1999; 197:198-206.
34. Chan SJ, Cao QP, Steiner DF. Evolution of the insulin superfamily: Cloning of a hybrid insulin/insulin-like growth factor cDNA from amphioxus. Proc Natl Acad Sci USA 1990; 87:9319-9323.
35. Pashmforoush M, Chan SJ, Steiner DF. Structure and expression of the insulin-like peptide receptor from amphioxus. Mol Endo 1996; 10:857-866.
36. Dehal P, Satou Y, Campbell RK et al. The draft genome of Ciona intestinalis: Insights into chordate and vertebrate origins. Science 2002; 298:2157-2167.
37. McRory JE, Sherwood NM. Ancient divergence of insulin and insulin-like growth factor. DNA Cell Biol 1997; 16:939-949.
38. Stein LD, Bao Z, Blasiar D et al. The genome sequence of Caenorhabditis briggsae: A platform for comarative genomics. PLoS Biology 2003; 1:166-192.
39. Dlakic M. A new family of putative insulin receptor-like proteins in C. elegans. Curr Biol 2002; 12:R155-R157.
40. Garofalo RS. Genetic analysis of insulin signaling in Drosophila. Trends Endocrinol Metab 2002; 13:156-162.
41. Efstratiadis A. Genetics of mouse growth. Int J Dev Biol 1998; 42:955-976.
42. Tatar M, Kopelman A, Epstein D et al. A mutant Drosophila insulin receptor homolog that extends life-span and impairs neuroendocrine function. Science 2001; 292:107-110.
43. Holzenberger M, Dupont J, Ducos B et al. IGF-1 receptor regulates lifespan and resistance to oxidative stress in mice. Nature 2003; 421:182-187.
44. Barbieri M, Bonafe M, Franceschi C et al. Insulin/IGF-I-signaling pathway: An evolutionarily conserved mechanism of longevity from yeast to humans. Am J Physiol Endocrinol Metab 2003; 285:E1064-E1071.
45. Aparicio S, Chapman J, Stupka E et al. Whole-genome shotgun assembly and analysis of the genome of Fugu rubripes. Science 2002; 297:1301-1310.
46. Scavo L, Shuldiner AR, Serrano J et al. Genes encoding receptors for insulin and insulin-like growth factor I are expressed in Xenopus oocytes and embryos. Proc Natl Acad Sci USA 1991; 88:6214-6218.
47. Hainaut P, Kowalski A, Giorgetti S et al. Insulin and insulin-like-growth-factor-I (IGF-I) receptors in Xenopus laevis oocytes. Comparison with insulin receptors from liver and muscle. Biochem J 1991; 273:673-678.

48. Foty RA, Liversage RA. Detection of insulin receptors in newt liver and forelimb regenerates and the effects of local insulin deprivation on epimorphic regeneration. J Exp Zool 1993; 266:299-311.
49. Shemer J, Penhos JC, LeRoith D. Insulin receptors in lizard brain and liver: Structural and functional studies of alpha and beta subunits demonstrate evolutionary conservation. Diabetologia 1986; 29:321-329.
50. Shemer J, Raizada M, LeRoith D. Structural and functional studies on insulin receptors from alligator brain and liver. Comp Biochem Physiol 1987; 86:55-61.
51. Hart C, Shemer J, Penhos JC et al. Frog brain and liver show evolutionary conservation of tissue-specific differences among insulin receptors. Gen Comp Endocrinol 1987; 68:170-178.
52. Kozma SC, Thomas G. Regulation of cell size in growth, development and human disease: PI3K, PKB and S6K. Bio Essays 2002; 24:65-71.
53. Leevers SJ. Growth control: Invertebrate insulin surprises! Curr Biol 2001; 11:R209-R212.
54. Puig O, Marr MT, Ruhf ML et al. Control of cell number by Drosophila FOXO: Downstream and feedback regulation of the insulin receptor pathway. Genes Dev 2003; 17:2006-2020.
55. Kitamura T, Kahn CR, Accili D. Insulin receptor knockout mice. Annu Rev Physiol 2003; 65:313-332.
56. Kenyon C, Chang J, Gensch E et al. C. elegans mutant that lives twice as long as wild type. Nature 1993; 366:461-464.
57. De Meyts P. The structural basis of insulin and insulin-like growth factor-I (IGF-I) receptor binding and negative cooperativity, and its relevance to mitogenic versus metabolic signaling. Diabetologia 1994; 37(suppl 2):S135-S148.
58. Robertson SC, Tynan JA, Donoghue DJ. RTK mutations and human syndromes. When good receptors turn bad. Trends in Genetics 2000; 16:265-271.
59. Blume-Jensen P, Hunter T. Oncogenic kinase signalling. Nature 2001; 411:355-365.
60. Gray A, Stenfeldt Mathiasen I, De Meyts P. The insulin-like growth factors and insulin-signalling systems: An appealing target for breast cancer therapy? Hormone Metab Res 2003; 35:857-871.
61. Drevs J, Medinger M, Schmidt-Gersbach C et al. Receptor tyrosine kinases: The main targets for new anticancer therapy. Curr Drug Targets 2003; 4:113-121.
62. Zwick E, Bange J, Ullrich A. Receptor tyrosine kinases as targets for anticancer drugs. Trends Mol Med 2002; 8:17-23.
63. Levitzki A. Tyrosine kinases as targets for cancer therapy. Eur J Cancer 2002; 38(Suppl 5):S11-S18.
64. Brunelleschi S, Penengo L, Santoro MM et al. Receptor tyrosine kinases as target for anti-cancer therapy. Curr Pharm Des 2002; 8:1959-1972.
65. De Meyts P, Whittaker J. Structural biology of insulin and IGF1 receptors: Implications for drug design. Nat Rev Drug Discov 2002; 1:769-783.
66. House PD, Weidemann MJ. Characterization of an [125 I]-insulin binding plasma membrane fraction from rat liver. Biochem Biophys Res Comm 1970; 41:541-548.
67. Freychet P, Roth J, Neville DM. Insulin receptors in the liver: Specific binding of I-125 insulin to the plasma membrane and its relation to insulin bioactivity. Proc Natl Acad Sci USA 1971; 68:1833-1837.
68. Cuatrecasas P. Insulin-receptor interactions in adipose tissue cells: Direct measurement and properties. Proc Natl Acad Sci USA 1971; 68:1264-1268.
69. Gammeltoft S, Gliemann J. Binding and degradation of 125I-labelled insulin by isolated rat fat cells. Biochim Biophys Acta 1973; 320:16-32.
70. Yip CC, Yeung CW, Moule ML. Photoaffinity labeling of insulin receptor proteins of liver plasma membrane preparations. Biochemistry 1980; 19:70-76.
71. Massague J, Czech MP. The subunit structures of two distinct receptors for insulin-like growth factors I and II and their relationship to the insulin receptor. J Biol Chem 1982; 257:5038-5045.
72. Shia MA, Pilch PF. The beta subunit of the insulin receptor is an insulin-activated protein kinase. Biochemistry 1983; 22:717-721.
73. Roth RA, Cassell DJ. Insulin receptor: Evidence that it is a protein kinase. Science 1983; 219:299-301.
74. Kasuga M, Fujita-Yamaguchi Y, Blithe DL et al. Tyrosine-specific protein kinase activity is associated with the purified insulin receptor. Proc Natl Acad Sci USA 1983; 80:2137-2141.
75. Ebina Y, Ellis L, Jarnagin K et al. The human insulin receptor cDNA: The structural basis for hormone-activated transmembrane signalling. Cell 1985; 40:747-758.
76. Ullrich A, Bell JR, Chen EY et al. Human insulin receptor and its relationship to the tyrosine kinase family of oncogenes. Nature 1985; 313:756-761.
77. Whittaker J, Okamoto AK, Thys R et al. High-level expression of human insulin receptor cDNA in mouse NIH 3T3 cells. Proc Natl Acad Sci USA 1987; 84:5237-5241.
78. Ullrich A, Gray A, Tam AW et al. Insulin-like growth factor-I receptor primary structure: Comparison with insulin receptor suggests structural determinants that define functional specificity. EMBO J 1986; 5:2503-2512.

79. Seino S, Seino M, Nishi S et al. Structure of the human insulin receptor gene and characterization of its promoter. Proc Natl Acad Sci USA 1989; 86:114-118.
80. Abbott AM, Bueno R, Pedrini MT et al. Insulin-like growth factor I receptor gene structure. J Biol Chem 1992; 267:10759-10763.
81. Shier P, Watt V. Primary structure of a putative receptor for a ligand of the insulin family. J Biol Chem 1989; 264:14605-14608.
82. Zhang B, Roth RA. The insulin receptor-related receptor: Tissue expression, ligand binding specificity, and signaling capabilities. J Biol Chem 1992; 267:18320-18328.
83. Nef S, Verma-Kurvari S, Merenmles J et al. Testis determination requires insulin receptor family function in mice. Nature 2003; 426:291-295.
84. Weber A, Huesken C, Bergmann E et al. Coexpression of insulin receptor-related receptor and insulin-like growth factor 1 receptor correlates with enhanced apoptosis and dedifferentiation in human neuroblastomas. Clin Cancer Res 2003; 9:5683-5692.
85. Heldin CH, Ostman A. Ligand-induced dimerization of growth factor receptors: Variations on the theme. Cytokine Growth Factor Rev 1996; 7:3-10.
86. Lu K, Guidotti G. Identification of the cysteine residues involved in the class I disulfide bonds of the human insulin receptor: Properties of insulin receptor monomers. Mol Biol Cell 1996; 7:679-691.
87. Jianping Wu J, Guidotti G. Construction and characterization of a monomeric insulin receptor. J Biol Chem 2002; 31:27809-27817.
88. Seino S, Seino M, Bell GI. Human insulin receptor gene. Diabetes 1990; 39:129-133.
89. McClain DA. Different ligand affinities of the two human insulin receptor splice variants are reflected in parallel changes in sensitivity for insulin action. Mol Endo 1991; 5:734-739.
90. Yamaguchi Y, Flier JS, Yokota A et al. Functional properties of two naturally occuring isoforms of the human insulin receptor in chinese hamster ovary cells. Endocrinology 1991; 129:2058-2066.
91. Gu JL, De Meyts P. The structure and function of the insulin receptor. In: Zai-Ping L, Zi-Xian L, You-Shang-Z, eds. Retrospect and Prospect of Protein Research. Singapore: World Scientific Publishing Co, 1991:120-125.
92. Frasca F, Pandini G, Scalia P et al. Insulin receptor isoform A, a newly recognized, high-affinity insulin-like growth factor II receptor in fetal and cancer cells. Mol Cell Biol 1999; 19:3278-3288.
93. De Meyts P, Wallach B, Christoffersen CT et al. The insulin-like growth factor-I receptor. Structure, ligand binding mechanism and signal transduction. Horm Res 1994; 42:152-169.
94. Bass J, Chiu G, Argon Y et al. Folding of insulin receptor monomers is facilitated by the molecular chaperones calnexin and calreticulin and impaired by rapid dimerization. J Cell Biol 1998; 141:637-646.
95. Pashmforoush M, Yoshimasa Y, Steiner DF. Exon 11 enhances insulin binding affinity and tyrosine kinase activity of the human insulin proreceptor. J Biol Chem 1994; 269:32639-32648.
96. Soos M, Field CE, Siddle K. Purified hybrid insulin/insulin-like growth factor-i receptors bind insulin-like growth factor-I, but not insulin, with high affinity. Biochem J 1993; 290:419-426.
97. Pandini G, Frasca F, Mineo R et al. Insulin/IGF-I hybrid receptors have different biological characteristics depending on the insulin receptor isoform involved. J Biol Chem 2002; 277:39684-39695.
98. Bajaj M, Waterfield MD, Schlessinger J et al. On the tertiary structure of the extracellular domains of the epidermal growth factor and insulin receptors. Biochim Biophys Acta 1987; 916:220-226.
99. Ward CW, Hoyne PA, Flegg RH. Insulin and epidermal growth factor receptors contain the cysteine repeat motif found in the tumor necrosis factor receptor. Proteins 1995; 22:141-153.
100. Marino-Buslje C, Mizuguchi K, Siddle K et al. A third fibronectin type III domain in the extracellular region of the insulin receptor family. FEBS Lett 1998; 441:331-336.
101. Mulhern TD, Booker GW, Cosgrove L. A third fibronectin-type-III domain in the insulin-family receptors. TIBS 1998; 23:465-466.
102. Ward CW. Members of the insulin receptor family contain three fibronectin type III domains. Growth factors 1999; 16:315-322.
103. Adams TE, Epa VC, Garrett TP et al. Structure and function of the type 1 insulin-like growth factor receptor. Cell Mol Life Sci 2000; 57:1050-1093.
104. Schaffer L, Ljungqvist L. Identification of a disulfide bridge connecting the alpha-subunits of the extracellular domain of the insulin receptor. Biochem Biophys Res Commun 1992; 189:650-653.
105. Sparrow LG, McKern NM, Gorman JJ et al. The disulfide bonds in the C-terminal domains of the human insulin receptor ectodomain. J Biol Chem 1997; 272:29460-29467.
106. Cheatham B, Kahn CR. Cysteine 647 in the insulin receptor is required for normal covalent interaction between alpha- and beta-subunits and signal transduction. J Biol Chem 1992; 267:7108-7115.
107. Hedo JA, Kasuga M, Van Obberghen E. Direct demonstration of glycosylation of insulin receptor subunits by biosynthetic and external labeling: Evidence for heterogeneity. Proc Natl Acad Sci USA 1981; 78:4791-4795.

108. Elleman TC, Frenkel MJ, Hoyne PA et al. Mutational analysis of the N-linked glycosylation sites of the human insulin receptor. Biochem J 2000; 347:771-779.
109. Leconte I, Auzan C, Debant A et al. N-linked oligosaccharide chains of the insulin receptor beta subunit are essential for transmembrane signaling. J Biol Chem 1992; 267:17415-17423.
110. Garrett TP, McKern NM, Lou M et al. Crystal structure of the first three domains of the type-1 insulin-like growth factor receptor. Nature 1998; 394:395-399.
111. Ward CW, Garrett TP. The relationship between the L1 and L2 domains of the insulin and epidermal growth factor receptors and leucine-rich repeat modules. BMC Bioinformatics 2001; 2:4.
112. Jorissen RN, Walker F, Pouliot N et al. Epidermal growth factor receptor: Mechanisms of activation and signalling. Exp Cell Res 2003; 284:31-53.
113. Hubbard SR, Till JH. Protein tyrosine kinase structure and function. Annu Rev Biochem 2000; 69:373-398.
114. Hubbard SR, Wei L, Ellis L et al. Crystal structure of the tyrosine kinase domain of the human insulin receptor. Nature 1994; 372:746-754.
115. Hubbard SR. Crystal structure of the activated insulin receptor tyrosine kinase in complex with peptide substrate and ATP analog. EMBO J 1997; 16:5572-5581.
116. Favelyukis S, Till JH, Hubbard SR et al. Structure and autoregulation of the insulin-like growth factor 1 receptor kinase. Nat Struct Biol 2001; 8:1058-1063.
117. Till JH, Ablooglu AJ, Frankel M et al. Crystallographic and solution studies of an activation loop mutant of the insulin receptor tyrosine kinase: Insights into kinase mechanism. J Biol Chem 2001; 276:10049-10055.
118. Baer K, Al Hasani H, Parvaresch S et al. Dimerization-induced activation of soluble insulin/IGF-1 receptor kinases: An alternative mechanism of activation. Biochemistry 2001; 40:14268-14278.
119. De Meyts P, Roth J, Neville DM et al. Insulin interactions with its receptors: Experimental evidence for negative cooperativity. Biochem Biophys Res Comm 1973; 55:154-161.
120. Kahn CR, Freychet P, Roth J et al. Quantitative aspects of the insulin-receptor interaction in liver plasma membranes. J Biol Chem 1974; 249:2249-2257.
121. De Meyts P, Van Obberghen E, Roth J et al. Mapping of the residues of the receptor binding region of insulin responsible for the negative cooperativity. Nature 1978; 273:504-509.
122. Christoffersen CT, Bornfeldt KE, Rotella CM et al. Negative cooperativity in the insulin-like growth factor-I (IGF-I) receptor and a chimeric IGF-I/insulin receptor. Endocrinology 1994; 135:472-475.
123. Waugh SM, DiBella EE, Pilch PF. Isolation of a proteolytically derived domain of the insulin receptor containing the major site of crosslinking/binding. Biochemistry 1989; 28:3448-3455.
124. Yip CC, Hsu H, Patel RG et al. Localization of the insulin-binding site to the cysteine-rich region of the insulin receptor alpha-subunit. Biochem Biophys Res Comm 1988; 157:321-329.
125. Wedekind F, Baer Pontzen K, Bala Mohan S et al. Hormone binding site of the insulin receptor: An analysis using photoaffinity-mediated avidin complexing. Biol Chem Hoppe Seyler 1989; 370:251-258.
126. Fabry M, Schaefer E, Ellis L et al. Detection of a new hormone contact site within the insulin receptor ectodomain by the use of a novel photoreactive insulin. J Biol Chem 1992; 267:8950-8956.
127. Shoelson SE, Lee J, Lynch CS et al. BpaB25 insulins. Photoactivatable analogues that quantitatively cross-link, radiolabel, and activate the insulin receptor. J Biol Chem 1993; 268:4085-4091.
128. Kurose T, Pashmforoush M, Yshimasa Y et al. Cross-linking of a B25 azidophenylalanine insulin derivative to the carboxyl-terminal region of the α-subunit of the insulin receptor. J Biol Chem 1994; 269:29190-29197.
129. Kjeldsen T, Andersen AS, Wiberg FC et al. The ligand specificities of the insulin receptor and the insulin-like growth factor-I receptor reside in different regions of a common binding site. Proc Natl Acad Sci USA 1991; 88:4404-4408.
130. Andersen AS, Kjeldsen T, Wiberg FC et al. Identification of determinants that confer ligand specificity on the insulin receptor. J Biol Chem 1992; 267:13681-13686.
131. Schumacher R, Mosthaf L, Schlessinger J et al. Insulin and insulin-like growth factor-I binding specificity is determined by distinct regions of their cognate receptors. J Biol Chem 1991; 266:19288-19295.
132. Tavare JM, Siddle K. Mutational analysis of insulin receptor function: Consensus and controversy. Biochim Biophys Acta 1993; 1178:21-39.
133. Bayne ML, Applebaum J, Underwood D et al. The C region of human insulin-like growth factor (IGF) I is required for high affinity binding to the type I IGF receptor. J Biol Chem 1988; 264:11004-11008.
134. Kjeldsen T, Wiberg FC, Andersen AS. Chimeric receptors indicate that phenylalanine 39 is a major contributor in insulin specificity of the insulin receptor. J Biol Chem 1994; 269:32942-32946.
135. Taylor SI, Cama A, Accili D et al. Mutations in the insulin receptor gene. Endocrine Rev 1992; 13:566-595.

136. Vorwerk P, Christoffersen CT, Muller J et al. Alternative splicing of exon 17 and a missense mutation in exon 20 of the insulin receptor gene in two brothers with a novel syndrome of insulin resistance (congenital fiber-type disproportion myopathy). Horm Res 1999; 52:211-220.
137. Vestergaard H, Klein HH, Hansen T et al. Severe insulin-resistant diabetes mellitus in patients with congenital muscle fiber type disproportion myopathy. J Clin Invest 1995; 95:1925-1932.
138. Klein HH, Muller R, Vestergaard H et al. Implications of compound heterozygous insulin receptor mutations in congenital muscle fibre type disproportion myopathy for the receptor kinase activation. Diabetologia 1999; 42:245-249.
139. Kadowaki T, Kadowaki H, Accili D et al. Substitution of lysine for asparagine at position 15 in the alpha-subunit of the human insulin receptor. A mutation that impairs transport of receptors to the cell surface and decreases the affinity of insulin binding. J Biol Chem 1990; 265:19143-19150.
140. Rouard M, Bass J, Grigorescu F et al. Congenital insulin resistance associated with a conformational alteration in a conserved beta-sheet in the insulin receptor L1 domain. J Biol Chem 1999; 274:18487-18491.
141. Longo N, Langley SD, Griffin LD et al. Activation of glucose transport by a natural mutation in the human insulin receptor. Proc Natl Acad Sci USA 1993; 90:60-64.
142. Grønskov K, Vissing H, Shymko RM et al. Mutation of arginine 86 to proline in the insulin receptor alpha subunit causes lack of transport of the receptor to the plasma membrane, loss of binding affinity and a constitutively activated tyrosine kinase in transfected cells. Biochem Biophys Res Comm 1993; 192:905-911.
143. Nakae J, Morioka H, Ohtsuka E et al. Replacements of leucine 87 in human insulin receptor alter affinity for insulin. J Biol Chem 1995; 270:22017-22022.
144. Liu R, Zhu J, Jospe N et al. Deletion of lysine 121 creates a temperaturesensitive alteration in insulin binding by the insulin receptor. J Biol Chem 1995; 270:476-482.
145. Rique S, Nogues C, Ibanez L et al. Identification of three novel mutations in the insulin recepor gene in type A insulin resistant patients. Clinical Genetics 2000; 57:67-69.
146. Hamer I, Foti M, Emkey R et al. An arginine to cysteine 252 mutation in insulin receptors from a patient with severe insulin resistance inhibits receptor internalisation but preserves signalling events. Diabetologia 2002; 45:657-667.
147. Maassen JA, Tobias ES, Kayserilli H et al. Identification and functional assessment of novel and known insulin receptor mutations in five patients with syndromes of severe insulin resistance. J Clinical Endo Metab 2003; 88:4251-4257.
148. Nakashima N, Umeda F, Yanase T et al. Insulin resistance associated with substitution of histidine for arginine 252 in the alpha-subunit of the human insulin receptor: Trial of insulin-like growth factor I injection therapy to enhance insulin sensitivity. J Clin Endo Metab 1995; 80:3662-3667.
149. Roach P, Zick Y, Formisano P et al. A novel human insulin receptor gene mutation uniquely inhibits insulin binding without impairing posttranslational processing. Diabetes 1994; 43:1096-1102.
150. Krook A, Soos MA, Kumar S et al. Functional activation of mutant human insulin receptor by monoclonal antibody. Lancet 1996; 347:1586-1590.
151. Whittaker J, Mynarcik DC. Phenotype of the Ser-323 to Leu mutation of the insulin receptor is isoform dependent. Diabetes 1998; 47(Suppl 1):264, (abstract).
152. Kadowaki H, Kadowaki T, Cama A et al. Mutagenesis of lysine 460 in the human insulin receptor. Effects upon receptor recycling and cooperative interactions among binding sites. J Biol Chem 1990; 265:21285-21296.
153. Williams PF, Mynarcik DC, Qin YG et al. Mapping of an NH2-terminal ligand binding site of the insulin receptor by alanine scanning mutagenesis. J Biol Chem 1995; 270:1-5.
154. Clackson T, Wells JA. A hot spot of binding energy in a hormone-receptor interface. Science 1995; 267:383-386.
155. De Meyts P, Gu JL, Shymko RM et al. Identification of a ligand-binding region of the human insulin receptor encoded by the second exon of the gene. Mol Endo 1990; 4:409-416.
156. Mynarcik DC, Yu GQ, Whittaker J. Alanine-scanning mutagenesis of a C-terminal ligand binding domain of the insulin receptor α subunit. J Biol Chem 1996; 271:2439-2442.
157. Mynarcik DC, Williams PF, Schäffer L et al. Analog binding properties of insulin receptor mutants. Identification of amino acids interacting with the COOH terminus of the B chain of the insulin molecule. J Biol Chem 1997; 272:2077-2081.
158. Hart LM, Lindhout D, Van der Zon GC et al. An insulin receptor mutant (Asp707 --> Ala), involved in leprechaunism, is processed and transported to the cell surface but unable to bind insulin. J Biol Chem 1996; 271:18719-18724.
159. Whittaker J, Sørensen H, Gadsboll VL et al. Comparison of the functional insulin binding epitopes of the A and B isoforms of the insulin receptor. J Biol Chem 2002; 277:47380-47384.

160. Whittaker J, Groth AV, Mynarcik DC et al. Alanine scanning mutagenesis of a type 1 insulin-like growth factor receptor ligand binding site. J Biol Chem 2001; 276:43980-43986.
161. Schäffer L. A model for insulin binding to the insulin receptor. Eur J Biochem 1994; 221:1127-1132.
162. Markussen JM, Halstrom J, Wiberg F et al. Immobilized insulin for high capacity affinity chromatography of insulin receptors. J Biol Chem 1991; 266:18814-18818.
163. Yip CC, Jack E. Insulin receptors are bivalent as demonstrated by photoaffinity labeling. J Biol Chem 1992; 267:13131-13134.
164. Sweet LJ, Morrison BD, Pessin J. Isolation of functional α-β heterodimers from the purified human placental α2-β2 heterotetrameric insulin receptor concept. Structural basis for high affinity ligand binding. J Biol Chem 1987; 262:6939-6942.
165. Boni-Schnetzler M, Scott W, Waugh SM et al. The insulin receptor. Structural basis for high affinity ligand binding. J Biol Chem 1987; 262:8395-8401.
166. Bass J, Kurose T, Pashmforoush M et al. Fusion of insulin receptor ectodomains to immunoglobulin constant domains reproduces high-affinity insulin binding in vitro. J Biol Chem 1996; 271:19367-19375.
167. Hoyne PA, Cosgrove LJ, McKern NM et al. High affinity insulin binding by soluble insulin receptor extracellular domain fused to a leucine zipper. FEBS Lett 2000; 479:15-18.
168. Whittaker J, Garcia P, Yu GQ et al. Transmembrane domain interactions are necessary for negative cooperativity of the insulin receptor. Mol Endo 1994; 8:1521-1527.
169. Florke RR, Schnaith K, Passlack W et al. Hormone-triggered conformational changes within the insulin-receptor ectodomain: Requirement for transmembrane anchors. Biochem J 2001; 360:189-198.
170. Sung CK, Wong KY, Yip CC et al. Deletion of residues 485-599 from the human insulin receptor abolishes antireceptor antibody binding and influences tyrosine kinase activation. Mol Endo 1994; 8:315-324.
171. Kristensen C, Wiberg FC, Schäffer L et al. Expression and characterization of a 70-kDa fragment of the insulin receptor that binds insulin. Minimizing ligand binding domain of the insulin receptor. J Biol Chem 1998; 273:17780-17786.
172. Marino-Buslje C, Martin-Martinez M, Mizuguchi K et al. The insulin receptor: From protein sequence to structure. Biochem Soc Trans 1999; 27:715-726.
173. Molina L, Marino-Buslje C, Quinn DR et al. Structural domains of the insulin receptor and IGF receptor required for dimerisation and ligand binding. FEBS Lett 2000; 467:226-230.
174. Brandt J, Andersen AS, Kristensen C. Dimeric fragment of the insulin receptor alpha-subunit binds insulin with full holoreceptor affinity. J Biol Chem 2001; 276:12378-12384.
175. Surinya KH, Molina L, Soos MA et al. Role of insulin receptor dimerization domains in ligand binding, cooperativity, and modulation by anti-receptor antibodies. J Biol Chem 2002; 277:16718-16725.
176. Kristensen C, Andersen AS, Ostergaard S et al. Functional reconstitution of insulin receptor binding site from nonbinding receptor fragments. J Biol Chem 2002; 277:18340-18345.
177. Kristensen C, Wiberg FC, Andersen AS. Specificity of insulin and insulin-like growth factor I receptors investigated using chimeric mini-receptors. Role of C-terminal of receptor alpha subunit. J Biol Chem 1999; 274:37351-37356.
178. Schaefer EM, Erickson HP, Federwisch M et al. Structural organization of the human insulin receptor ectodomain. J Biol Chem 1992; 267:23393-23402.
179. Tulloch PA, Lawrence LJ, McKern NM et al. Single-molecule imaging of human insulin receptor ectodomain and its Fab complexes. J Struct Biol 1999; 125:11-18.
180. Woldin CN, Hing FS, Lee J et al. Structural studies of the detergent-solubilized and vesicle-reconstituted insulin receptor. J Biol Chem 1999; 274:34981-34992.
181. Christiansen K, Tranum-Jensen J, Carlsen J et al. A model for the quaternary structure of human placental insulin receptor deduced from electron microscopy. Proc Natl Acad Sci USA 1991; 88:249-252.
182. Tranum-Jensen J, Christiansen K, Carlsen J et al. Membrane topology of insulin receptors reconstituted into lipid vesicles. J Membr Biol 1994; 140:215-223.
183. Luo RZ, Beniac DR, Fernandes A et al. Quaternary structure of the insulin-insulin receptor complex. Science 1999; 285:1077-1080.
184. Ottensmeyer FP, Beniac DR, Luo RZ et al. Mechanism of transmembrane signaling: Insulin binding and the insulin receptor. Biochemistry 2000; 39:12103-12112.
185. Yip CC, Ottensmeyer P. Three-dimensional structural interactions of insulin and its receptor. J Biol Chem 2003; 278:27329-27332.
186. Blundell TL, Dodson GG, Hodgkin DC et al. Insulin: The structure in the crystal and its reflection in chemistry and biology. Adv Prot Chem 1972; 26:279-402.

187. Baker E, Blundell TL, Cutfield JF et al. The structure of 2Zn pig insulin at 1.5 Å resolution. Phil Trans R Soc Lond 1988; B19:369-456.
188. Pullen RA, Lindsay DG, Wood SP et al. Receptor-binding region of insulin. Nature 1976; 259:369-373.
189. Nanjo K, Sanke T, Miyano M et al. Diabetes due to secretion of a structurally abnormal insulin (insulin Wakayama). Clinical and functional characteristics of [LeuA3] insulin. J Clin Invest 1986; 77:514-519.
190. Hua XH, Shoelson SE, Kochoyan M et al. Receptor binding redefined by a structural switch in a mutant human insulin. Nature 1991; 354:238-241.
191. Ludvigsen S, Olsen HB, Kaarsholm NC. A structural switch in a mutant insulin exposes key residues for receptor binding. J Mol Biol 1998; 279:1-7.
192. Conlon JM, Bondareva V, Rusakov Y et al. Characterization of insulin, glucagon, and somatostatin from the river lamprey, Lampetra fluviatilis. Gen Comp Endocrinol 1995; 100:96-105.
193. Kristensen C, Kjeldsen T, Wiberg FC et al. Alanine scanning mutagenesis of insulin. J Biol Chem 1997; 272:12978-12983.
194. Nakagawa SH, Tager HS. Importance of aliphatic side-chain structure at positions 2 and 3 of the insulin A chain in insulin-receptor interactions. Biochemistry 1992; 31:3204-3214.
195. Nakagawa SH, Tager HS, Steiner DF. Mutational analysis of invariant valine B12 in insulin: Implications for receptor binding. Biochemistry 2000; 39:15826-15835.
196. Jensen A-M. Analysis of structure-activity relationships of the insulin molecule by alanine-scanning mutagenesis. Master's thesis Copenhagen University 2000.
197. Weiss MA, Hua Q-X, Jia W et al. Activities of insulin analogues at position A8 are uncorrelated with thermodynamic stability. In: Federwisch M, Dieken ML, De Meyts P, eds. Insulin and Related Proteins - Structure to Function and Pharmacology. Dordrecht, Boston, London: Kluwer Academic Publishers, 2002:103-119.
198. Palsgaard J. Receptor-binding properties of 25 different hystricomorph-like human insulin analogs. Copenhagen, Denmark: Master's thesis Copenhagen University, 2003.
199. Wells JA. Binding in the growth hormone receptor complex. Proc Natl Acad Sci USA 1996; 93:1-6.
200. Van den Brande JL. Structure of the human insulin-like growth factors: Relationship to function. In: Schofield PN, ed. The Insulin-Like Growth Factors. Structure and Biological Functions. Oxford, New York, Tokyo: Oxford University Press, 1992:12-44.
201. Brzozowski AM, Dodson EJ, Dodson GG et al. Structural origins of the functional divergence of human insulin-like growth factor-I and insulin. Biochemistry 2002; 41:9389-9397.
202. Vajdos FF, Ultsch M, Schaffer ML et al. Crystal structure of human insulin-like growth factor-1: Detergent binding inhibits binding protein interactions. Biochemistry 2001; 40:11022-11029.
203. Schaffer ML, Deshayes K, Nakamura G et al. Complex with a phage display-derived peptide provides insight into the function of insulin-like growth factor I. Biochemistry 2003; 42:9324-9334.
204. Zhang W, Gustafson TA, Rutter WJ et al. Positively charged side chains in the insulin-like growth factor-I C- and D- regions determine receptor binding specificity. J Biol Chem 1994; 269:10609-10613.
205. Woods KA, Camacho-Hubner C, Bergman RN et al. Effects of insulin-like growth factor I (IGF-I) therapy on body composition and insulin resistance in IGF-I gene deletion. J Clin Endo Metab 2000; 85:1407-1411.
206. Gill R, Wallach B, Verma C et al. Engineering the C-region of human insulin-like growth factor-1: Implications for receptor binding. Protein Eng 1996; 9:1011-1019.
207. Geddes S, Holst P, Grotzinger J et al. Structure-function studies of an IGF-I analogue that can be chemically cleaved to a two-chain mini-IGF-I. Protein Eng 2001; 14:61-65.
208. Kristensen C, Andersen AS, Hach M et al. A single-chain insulin-like growth factor I/insulin hybrid binds with high affinity to the insulin receptor. Biochem J 1995; 305:981-986.
209. de Vos AM, Ultsch M, Kossiakoff AA. Human growth hormone and extracellular domain of its receptor: Crystal structure of the complex. Science 1992; 255:306-312.
210. DeLisi C, Blumenthal R. Physical chemical aspects of cell surface events in cellular regulation. New York, Amsterdam, Oxford: Elsevier North Holland, 1979:116.
211. Lee J, O'Hare T, Pilch P et al. Insulin receptor autophosphorylation occurs asymmetrically. J Biol Chem 1993; 268:4092-4098.
212. Yip CC. The insulin-binding domain of the insulin receptor is encoded by exons 2 and 3. J Cell Biochem 1992; 48:19-25.
213. Hammond BJ, Tikerpae J, Smith GD. An evaluation of the cross-linking model for the interaction of insulin with its receptor. Am J Physiol 1997; 272:E1136-E1144.

214. Ilondo MM, Damholt A, Cunningham B et al. Receptor dimerization determines the effects of growth hormone in primary rat adipocytes and cultured IM-9 lymphocytes. Endocrinology 1994; 134:2397-2403.
215. Yeh JI, Biemann HP, Prive GG et al. High-resolution structures of the ligand binding domain of the wild-type bacterial aspartate receptor. J Mol Biol 1996; 262:186-201.
216. Remy I, Wilson IA, Michnick SW. Erythropoietin receptor activation by a ligand-induced conformation change. Science 1999; 283:990-993.
217. Livnah O, Stura EA, Middleton SA et al. Crystallographic evidence for preformed dimers of erythropoietin receptor before ligand activation. Science 1999; 283:987-990.
218. Fuh G, Cunningham BC, Fukunaga R et al. Rational design of potent antagonists to the human growth hormone receptor. Science 1992; 256:1677-1680.
219. Zhang B, Roth RA. A region of the insulin receptor important for ligand binding (residues 450-601) is recognized by patients autoimmune antibodies and inhibitory monoclonal antibodies. Proc Natl Acad Sci USA 1991; 88:9858-9862.
220. Muggeo M, Flier JS, Abrams RA et al. Treatment by plasma exchange of a patient with autoantibodies to the insulin receptor. N Engl J Med 1979; 300:477-480.
221. Pillutla RC, Hsiao KC, Beasley JR et al. Peptides identify the critical hotspots involved in the biological activation of the insulin receptor. J Biol Chem 2002; 277:22590-22594.
222. Schäffer L, Brissette RE, Spetzler JC et al. Assembly of high affinity insulin receptor agonists and antagonists from peptide building blocks. Proc Natl Acad Sci 2003; 100:4435-4439.
223. Avruch J. Receptor tyrosine kinases. In: De Groot L, Jameson LJ, eds. Endocrinology. 4th ed. Philadelphia: WB Saunders Co, 2001:25-47.
224. Loren CE, Englund C, Grabbe C et al. A crucial role for the Anaplastic lymphoma kinase receptor tyrosine kinase in gut development in Drosophila melanogaster. EMBO Rep 2003; 4:781-786.
225. Robinson DR, Wu YM, Lin SF. The protein tyrosine kinase family of the human genome. Oncogene 2000; 19:5548-5557.
226. Ogiso H, Ishitani R, Nureki O et al. Crystal structure of the complex of human epidermal growth factor and receptor extracellular domains. Cell 2002; 110:775-787.
227. De Meyts P. My favorite molecule. Insulin and its receptor: Structure, function and evolution. BioEssays 2004; 26:1351-1362.
228. Siddle K. The insulin receptor and downstream signalling. In: Kumar S, O'Rahilly S, eds. Insulin Resistance: Insulin Action and Its Disturbances in Disease. John Wiley and Sons, Ltd, 2005:1-62.
229. Sørensen H, Whittaker L, Hinrichsen J et al. Mapping of the insulin-like growth factor II binding site of the Type 1 insulin-like growth factor receptor by alanine scanning mutagenesis. FEBS Letters 2004; 565:19-22.
230. Whittaker J, Whittaker L. Characterization of the functional insulin binding epitopes of the full length insulin receptor. J Biol Chem 2005; 280:20932-20936.
231. Chakravarty A, Hinrichsen J, Whittaker L et al. Rescue of ligand binding of a mutant IGF-I receptor by complementation. Biochem Biophys Res Commun 2005; 331:74-77.
232. Wan Z, Xu B, Chu YC et al. Enhancing the activity of insulin at the receptor interface: Crystal structures and photo-cross-linking of A8 analogues. Biochemistry 2004; 43:16119-16133.
233. Wan ZL, Huang K, Xu B et al. Diabetes associated mutations in human insulin: Crystal structure and photo-crosslinking studies of A-chain variant insulin Wakayama. Biochemistry 2005; 44:5000-5016.
234. Huang K, Xu B, Chu YC et al. How insulin binds: The B-chain α-helix contacts the L1 β-helix of the insulin receptor. J Mol Biol 2004; 341:529-550.
235. Xu B, Hu SQ, Chu YC et al. Diabetes-associated mutations in insulin: Consecutive residues in the B-chain contact distinct domains of the insulin receptor. Biochemistry 2004; 43:8356-8372.
236. Denley A, Bonython ER, Booker GW et al. Structural determinants for high affinity binding of insulin-like growth factor II to insulin receptor (IR)-A, the exon 11 minus isoform of the IR. Mol Endo 2004; 18:2502-2512.
237. Gauguin L. Structural basis for the lower affinity of the insulin-like growth factors for the insulin receptor. Master's thesis, University of Copenhagen, 2005.
238. Walenkamp MJ, Karperien M, Pereira AM et al. Homozygous and heterozygous expression of a novel insulin-like growth factor mutation. J Clin Endo Metab 2005; 90:2855-2864.
239. Denley A, Wang CC, McNeil KA et al. Structural and functional characteristics of the Val44Met insulin-like growth factor-1 missense mutation: Correlation with effects on growth and development. Mol Endo 2005; 19:711-721.

CHAPTER 2

# Subcellular Compartmentalization of Insulin Signaling Processes and GLUT4 Trafficking Events

Robert T. Watson, Alan R. Saltiel, Jeffrey E. Pessin and Makoto Kanzaki*

## Abstract

Skeletal muscle and adipose tissue are the major sites of postprandial glucose disposal. The insulin-regulated transport of glucose into these tissues is a multi-step process that begins with the binding of insulin to its cell surface receptor. Once activated, the insulin receptor generates multiple intracellular signaling cascades, some of which induce the rapid redistribution of the GLUT4 facilitative glucose transporter from intracellular compartments to the plasma membrane. Although probably best known for its role in glucose homeostasis, insulin regulates a variety of metabolic, mitogenic, and anti-apoptotic processes in specific tissues. Moreover, in addition to insulin several other hormones and growth factors can also activate signaling targets that function downstream of the insulin receptor. For example, phosphatidylinositol-3'-kinase (PI3K), a key enzyme in the signaling pathway leading to insulin-stimulated glucose uptake, can be activated by many extracellular signals. However, only insulin and highly related hormones such as IGF-I efficiently stimulate acute glucose transport. These observations suggest that cellular mechanisms have evolved for maintaining specificity among signaling pathways mediated by various hormones and growth factors. Elucidating the underlying mechanisms for this specificity is a current challenge engaging the attention of many researchers, and will require a thorough understanding of the signaling pathways responsible for each cellular response. To this end, recent work suggests that the coordination of multiple pathways occurs in part through the intracellular compartmentalization of key signaling molecules. Indeed, subcellular compartmentalization plays a critical role in maintaining the specificity of insulin signaling and the fidelity of GLUT4 vesicle trafficking.

## Introduction

The elevated levels of blood sugar and amino acids that occur following a meal signal pancreatic beta cells to release insulin into the bloodstream. Once in the vascular system, circulating insulin markedly enhances glucose transport into skeletal muscle and adipose tissue, the peripheral sites responsible for the majority of postprandial glucose disposal. In response to insulin, glucose enters muscle and fat cells through aqueous pores formed by the glucose transporter 4 (GLUT4) protein. GLUT4 is the fourth of 13 members of a family of facilitative sugar transporters and is the only isoform that is widely accepted as being

*Corresponding Author: Makoto Kanzaki—Tohoku University Biomedical Engineering Research Organization (TUBERO), 2-1 Seiryo-machi, Aoba-ku, Sendai City, Miyagi, Japan, 980-8575. Email: kanzaki@tubero.tohoku.ac.jp

*Mechanisms of Insulin Action*, edited by Alan R. Saltiel and Jeffrey E. Pessin.
©2007 Landes Bioscience and Springer Science+Business Media.

insulin-responsive. Like other GLUT family members, GLUT4 is a 12 transmembrane protein; unlike most other isofoms, GLUT4 is predominantly localized to intracellular compartments in the basal state.[1-4] However, far from being static, GLUT4 slowly cycles between intracellular compartments and the cell surface, even in the absence of insulin.[5,6] Since the rate of GLUT4 vesicle endocytosis is significantly higher than its rate of exocytosis, the vast majority of GLUT4 is excluded from the cell surface under basal conditions. In contrast, activation of the insulin receptor triggers a large increase in the rate of GLUT4 vesicle exocytosis and a concomitant decrease in the rate of endocytosis. This insulin-dependent shift in GLUT4 vesicle trafficking results in a net increase of GLUT4 protein at the cell surface, thus allowing glucose to enter target cells.[6] Once circulating insulin returns to basal levels, GLUT4 is rapidly internalized through clathrin-coated pits and recycled back to its intracellular storage compartments. This regulated transition in the relative rates of GLUT4 vesicle endo- versus exocytosis is orchestrated by a series of vesicular trafficking processes, beginning with the selection of GLUT4 molecules in donor membrane compartments. Subsequent steps include vesicle budding, trafficking, tethering, docking, and fusion with acceptor membrane compartments.[7,8] Importantly, these dynamic processes are regulated by intracellular signals that originate with the activated insulin receptor.

## The Insulin Receptor and Its Immediate Downstream Substrate Proteins

Insulin regulates biological responsiveness by activating the intrinsic tyrosine kinase of its receptor.[9] Unlike most other receptor tyrosine kinases, the insulin receptor is composed of two extracellular $\alpha$-subunits and two transmembrane $\beta$-subunits disulfide linked into a $\alpha_2\beta_2$ heterotetrameric complex.[10] Insulin binds to the $\alpha$-subunits and induces a conformational change that activates the intrinsic tyrosine kinase domain of its intracellular $\beta$-subunits. This results in a series of trans-autophosphorylation reactions on specific tyrosine residues within the two cytoplasmic tails of the receptor.[11,12] Following activation, most receptor tyrosine kinases directly recruit downstream effectors to the phosphotyrosine residues present in their intracellular tails. In contrast, the insulin receptor primarily recruits scaffolding proteins that lack intrinsic enzymatic activity. These substrates are phosphorylated at multiple tyrosine residues, thus providing a range of potential docking sites for downstream signaling molecules. Since these proximal substrate proteins are released from the activated insulin receptor, each can recruit a distinct subset of signaling molecules to the cell surface, including certain microdomains of the plasma membrane, as well as intracellular vesicular or cytoplasmic compartments.

The activated insulin receptor phosphorylates a variety of downstream targets, including members of the insulin receptor substrate (IRS1,2,3,4), Gab1 (Grb2-associated binder 1), p53/58-IRS, APS (adaptor protein containing PH domain and SH2 domain), Cbl, Shc, and SIRPs (signal regulatory proteins) families.[13] Among these various substrates, the IRS proteins are the most extensively characterized.[9,14] The tyrosine phosphorylation of IRS proteins provides binding sites for a range of proteins that contain SH2 domains, including the p85 regulatory subunit of the type 1A PI 3-kinase, the protein tyrosine phosphatase SHP2, the nonreceptor tyrosine kinases Fyn and Csk, and the adaptor proteins Grb2 and Nck. These SH2 adaptor proteins often contain SH3 domains that bind proline-rich sequences harboring the consensus sequence PXXP, and provide further protein-protein interactions with additional downstream effectors. This combinatorial assembly of multiple signaling molecules on the IRS scaffolding proteins may impart one level of specificity to insulin receptor signaling cascades.

## The PI3-Kinase Is Necessary for Insulin-Stimulated GLUT4 Translocation

Among the downstream signaling molecules recruited to tyrosine-phosphorylated IRS proteins, it has been well established that the type 1A PI3-kinase is crucial for a wide variety of insulin's metabolic actions. This enzyme catalyzes the phosphorylation at the D3 position of the inositol ring of PI 4,5P$_2$, producing PI 3,4,5P$_3$ in the plasma membrane. The formation of 3' phosphoinositides (PI3,4,5P$_3$ and PI3,4P$_2$) creates recognition sites for a number of proteins containing pleckstrin homology (PH) domains. These include the serine/threonine kinases phosphoinositide-dependent protein kinase (PDK1) and protein kinase B (PKB/Akt),[15] as well as the guanine nucleotide exchange factor ARNO.[16] Once PI 3,4,5P$_3$ is generated at the cell surface by insulin stimulation, these PH domain-containing proteins are efficiently recruited to the inner surface of the plasma membrane, where they function in the propagation of additional signals. Not surprisingly, the experimental inhibition of PI 3-kinase activity, and thus the formation of PI3,4,5P$_3$, completely abrogates glucose uptake and GLUT4 translocation in muscle and fat cells.[17,18] Following the termination of insulin receptor activity, PI 3,4,5P$_3$ is rapidly dephosphorylated by the 3' phosphoinositide phosphatase PTEN (phosphatase and tensin homolog), which releases the PH-domain containing proteins back to the cytoplasm.[19,20] This regulated transition in the spatial compartmentalization of PH-domain proteins between the cytoplasm and the plasma membrane provides a key regulatory mechanism for the PI 3-kinase signaling pathway.

As mentioned above, ARNO is a PH-domain containing protein that is recruited to the cell surface through interactions with PI 3,4,5P$_3$. ARNO serves as an exchange factor for a subset of ARF (ADP ribosylation factor) proteins, including ARF6.[21] ARFs are members of the Ras superfamily of small GTP binding proteins that function during the early stages of vesicle formation.[22] In the GDP-bound state ARFs are cytosolic or only weakly associated with membranes. However, upon binding GTP they form strong interactions with target membranes and recruit vesicle coat proteins to the donor compartment. The coat proteins interact with cargo molecules and deform the membrane surface into a nascent vesicle. ARF6 is structurally divergent from other ARF isoforms and has been implicated in several vesicular trafficking processes, including the recycling of endosomal compartments at the cell surface,[23] the calcium-regulated exocytosis of dense-core granules,[24] and insulin-regulated GLUT4 translocation.[25] In addition, ARF6 is insensitive to Brefeldin A, a fungal metabolite that inhibits a subset of ARF isoforms.[26] This is significant because GLUT4 translocation is not blocked by BFA treatment. Moreover, phospholipase D (PLD), a downstream target of ARF6, was shown to colocalize with GLUT4 vesicles and to potentiate the effects of insulin on GLUT4 translocation.[27-29] PLD catalyzes the hydrolysis of phosphatidylcholine, producing phosphatidic acid (PA). PA is involved in a variety of signal transduction and membrane trafficking processes.[30] In addition, a recent study demonstrated the insulin-dependent activation of PLD and ARF6 through ARNO in insulin receptor-expressing Rat1 fibroblasts.[31] Thus, the insulin-stimulated recruitment of ARNO to the cell surface could lead to ARF6 and PLD activation and the subsequent production of PA. Although a precise functional role for these molecules in insulin-stimulated GLUT4 translocation remains unclear, PLD activity has recently been shown to potentiate the fusion of GLUT4 vesicles with the plasma membrane.[29]

On the other hand, substantial evidence has demonstrated that two serine/threonine kinases, PDK1 and PKB (also called Akt), play indispensable roles in the insulin-stimulated GLUT4 translocation process. PKB isoforms are a subgroup of the AGC family of protein kinases that possesses two critical regulatory phosphorylation sites, the first of which is a threonine residue (Thr308) located within the activation loop of the kinase domain. This residue is phosphorylated by PDK1.[32] The full activation of PKB also requires phosphorylation on serine 473 (Ser473), which may result from the activity of another putative kinase, or perhaps via autophosphorylation.[15] Although the details remain to be elucidated, the PI 3,4,5P$_3$-dependent redistribution of these kinases is a key step for propagating insulin receptor signals leading to a wide array of insulin actions, including GLUT4 translocation. Indeed, when constitutively targeted to the cell surface through the incorporation of an N-terminal myristoylation signal, PKB was found to induce GLUT4 translocation.[33-36] In contrast, the experimental inhibition of PKB activity, by introducing blocking antibodies or by expressing dominant-interfering mutants, prevented insulin-stimulated GLUT4 translocation.[34,37] In other studies, the insulin-dependent association of PKB with GLUT4-containing membrane compartments was observed.[38-40] More recently, the use of siRNA to reduce the expression levels of PKB isoforms showed that knockdown of PKBβ (Akt2) prevented insulin-stimulated GLUT4 translocation,[41,42] whereas reduction of PKBα (Akt1) had no significant effect. Similarly, genetic ablation of PKBβ in mice caused insulin-resistance by preventing glucose uptake and GLUT4 translocation.[43,44] In contrast, loss of PKBα resulted in growth retardation but no obvious defects in glucose homeostasis.[45,46]

PKB was first implicated in insulin-stimulated GLUT4 translocation in 1996,[33] however the identification of downstream substrates involved in glucose uptake has proven challenging. Recently, a screen was conducted based upon the consensus PKB phosphorylation motif, RXRXXS/T, where X is any amino acid. Using an antibody directed against the phosphorylated form of this motif, Kane et al (2002) were able to immunoprecipitate potential PKB substrates from insulin-stimulated adipocytes.[47] One interesting candidate resulting from this screen was AS160, which contains six consensus PKB phosphorylation sites as well as a Rab GAP domain.[48] Rab proteins comprise the largest branch of the Ras superfamily of small GTP-binding proteins.[49] Rab proteins regulate membrane transport between organelles and may contribute to the specificity of membrane trafficking processes.[50] Like other small GTPases, Rabs oscillate between a GDP-bound ("off") and a GTP-bound ("on") state. In the GTP-bound state, Rabs interact with effector proteins and are thought to regulate several steps of membrane transport, including vesicle budding, motility, tethering, and fusion.[51] Following a round of vesicle fusion, the Rab protein is returned to its GDP-bound state. Insulin-responsive cells express several Rab isoforms, and to date Rabs 4, 5, and 11 have been implicated in GLUT4 trafficking processes.

Of the six consensus PKB phosphorylation sites present in AS160, five were phosphorylated in response to insulin stimulation: Ser318, Ser570, Ser588, Thr642, and Thr571.[48] In addition, when 4 of these five sites were mutated to alanine, the resulting construct (designated AS160-4P) behaved in a dominant-interfering manner and significantly inhibited GLUT4 translocation when over-expressed in adipocytes. As mentioned above, a myristoylated form of PKB induces GLUT4 translocation in the absence of insulin. This provides a method to directly test whether AS160 functions downstream of PKB, and Zeigerer et al (2004) recently demonstrated that expression of AS160-4P blocks the ability of myristoylated PKB to induce GLUT4 translocation.[52] These results support the hypothesis that AS160 functions downstream of PKB, although its precise role remains to be elucidated. Indeed, it is not yet known which Rab isoforms are regulated by the GAP domain of AS160. Nevertheless, these results suggest that insulin-stimulated GLUT4 translocation may require an active, GTP-bound Rab. A model for how AS160 could function in GLUT4 translocation is as follows: In the basal state, the GAP activity of AS160 could maintain a Rab protein in the inactive, GDP-bound form. Insulin stimulation could then result in the phosphorylation and inhibition of AS160 GAP activity, thus allowing the conversion of the Rab protein to the active GTP-bound form. Once in the active GTP-bound state, the Rab protein can then participates in one or more steps of GLUT4 vesicle exocytosis. Although many questions remain regarding the function of AS160, the identification of an insulin-stimulated PKB substrate that has the potential to regulate Rab protein function could provide a possible mechanism for linking insulin signaling to GLUT4 vesicle trafficking.

Another recently identified PKB substrate involved in GLUT4 translocation is the SNARE (soluble N-ethylmaleimide-sensitive fusion (NSF) attachment protein (SNAP) receptors) associated protein synip.[53] Synip was originally identified in a yeast two-hybrid screen using the cytosolic domain of syntaxin 4 as bait.[54] A multidomain protein, synip has an N-terminal PDZ domain, central EF and coiled-coiled domains, and a C-terminal WW motif. In the basal state, synip binds syntaxin 4 and blocks the ability of VAMP2 to interact with syntaxin 4. This may prevent promiscuous fusions between GLUT4 vesicles and the cell surface. However, insulin causes synip to dissociate from syntaxin 4, thus allowing productive SNARE pairing between syntaxin 4 and VAMP2. In addition, recent work has identified an unusual potential PKBβ phosphorylation site within synip (RxKxRS$^{97}$xS$^{99}$).[53] Interestingly, serine 99 appears to be a specific substrate of PKBβ, but not PKBα or PKBγ. Insulin stimulation resulted in the phosphorylation of synip at serine 99, and this lead to the dissociation of the synip-syntaxin 4 complex. Moreover, mutation of serine 99 to phenylalanine prevented the dissociation of synip from syntaxin 4 and also inhibited GLUT4 translocation in a dominant interfering manner. Thus, the insulin-dependent phosphorylation of synip by PKBβ may provide a mechanism for insulin to regulate the docking/fusion of GLUT4 vesicles with the cell surface.

In addition to PKB, a number of studies have suggested that atypical PKCs (PKCλ/ζ) may also function as downstream targets for the IRS-PI 3-kinase signaling pathway leading to GLUT4 translocation.[55-57] Although atypical PKC isoforms lack PH domains, they are recruited to the plasma membrane where they undergo PDK1-dependent activation in response to insulin stimulation.[58,59] Numerous reports have suggested the involvement of aPKCs in various metabolic actions of insulin and their possible dysfunction in insulin-resistant states.[60] For example, the expression of constitutively active PKCλ/ζ mutants increase, whereas dominant-interfering mutants and blocking antibodies inhibit, insulin-induced GLUT4 translocation.[56,57] Moreover, recent work has provided a functional link between PKCλ, Rab4, and the microtubule motor protein KIF3.[61] Imamura et al (2003) found that insulin caused GTP loading of Rab4, an effect efficiently blocked by PI3K inhibitors and a dominant-interfering PKCλ mutant.[61] Furthermore, Rab4 was found to interact with KIF3, and blocking antibodies against KIF3 significantly inhibited GLUT4 translocation. In addition, insulin enhanced the association between KIF3 and microtubules whereas this response was blocked by PI3K inhibitors and a dominant-negative PKCλ mutant. However, other recent work has implicated the microtubule motor KIF5B in GLUT4 translocation.[62] Using DNA microarrays, these authors found that KIF5B is the predominant kinesin expressed in 3T3L1 adipocytes, and that dominant-interfering mutants of this isoform effectively blocked insulin-stimulated GLUT4 translocation.

## Is There a Second Signaling Pathway Required for Insulin-Stimulated Glucose Uptake?

While it has been well established that activation of PI3-kinase and subsequent generation of 3'-phosphoinositides through the IRS signaling pathway are essential for insulin-stimulated GLUT4 translocation, an important open question is the basis for the remarkable specificity of insulin action. For example, the PI 3-kinase signaling pathway is widely activated by a range of hormones and growth factors, yet only insulin evokes specific metabolic activities. For example, activation of integrin receptor signaling induces PI 3-kinase and $PI3,4,5P_3$ production in adipocytes, but does not stimulated glucose uptake or GLUT4 translocation.[63,64] Moreover, a cell-permeable analog of $PI3,4,5P_3$ did not stimulate glucose uptake when added alone to cells.[65] However, in the presence of wortmannin, which completely blocks endogenous PI 3-kinase activity, the coadministration of the $PI3,4,5P_3$ analog together with insulin resulted in enhanced glucose uptake by adipocytes. Subsequent studies revealed that the $PI3,4,5P_3$ analog induced GLUT4 translocation, although the transporter was not functional, suggesting the presence of a PI 3-kinase independent glucose transport activation pathway.[66,67] Together, these data provide compelling evidence that the IRS-PI 3-kinase pathway is necessary, but not sufficient for insulin-dependent glucose transport.

Biological responses often reflect the integration of outputs from multiple downstream pathways activated by a given receptor. Indeed, even when downstream signaling molecules are shared by several different receptors, specificity can result from unique wiring circuitries or distinct combinations of signaling intermediates. In the context of insulin-induced GLUT4 translocation in adipocytes, Cbl, a substrate for insulin receptor tyrosine kinase, triggers a second insulin receptor pathway that functions in concert with the IRS-PI 3-kinase signaling pathway.[6] In addition to multiple tyrosine residues that can be phosphorylated by the insulin receptor kinase, the Cbl protein also contains SH2, zinc ring finger, proline-rich, and leucine zipper domains. Importantly, the stimulation of Cbl phosphorylation by insulin is observed in fully differentiated adipocytes that can display 10-20 fold increases in glucose uptake in response to insulin. In contrast, fibroblasts overexpressing the insulin receptor are incapable of stimulating Cbl phosphorylation.[68]

Fully differentiated adipocytes express two Cbl adaptor proteins, the adaptor protein containing a PH and SH2 domain (APS) and the Cbl associated protein (CAP); both adaptors are required for insulin-induced tyrosine phosphorylation of Cbl.[69-72] The activated insulin receptor recruits APS via its SH2 domain. Structural analysis of the insulin receptor kinase domain/APS SH2 domain complex revealed that the APS SH2 domain forms a homodimer, with each partner binding to a different subunit of the insulin receptor heterodimer.[73] The recruitment of APS to the insulin receptor then results in phosphorylation of a carboxyl-terminal tyrosine residue, which in turn serves as a binding site for the SH2 domain of Cbl.[74] Cbl also interacts with CAP, and is recruited to APS together with CAP.[69,75] Intriguingly, CAP is under the transcriptional control of the nuclear receptor PPARγ, and its expression is stimulated by a class of insulin-sensitizing PPARγ agonists, the thiazolidinediones (TZDs). CAP expression correlates well with both insulin stimulation of Cbl phosphorylation and insulin sensitivity in 3T3L1 adipocytes and in mice.[76,77] Consistent with these findings, dominant negative forms of APS, Cbl, and CAP inhibited insulin-stimulated glucose uptake and GLUT4 translocation.[74,78-80] In addition, the use of RNAi to reduce the protein levels of Cbl or APS in 3T3L1 adipocytes also abrogated glucose uptake,[79] although some data appear to be at variance with these results.[81]

At present, the preponderance of evidence supports the existence of a second, tissue-specific insulin signaling pathway that is necessary for GLUT4 translocation. This pathway, defined by APS-CAP-Cbl, provides an important conceptual framework for our understanding of insulin's metabolic actions, especially with regard to its highly restricted tissue specificity. Thus, insulin-induced GLUT4 translocation in adipocytes results from two independent insulin receptor signals, one being the CAP-Cbl signaling pathway that is activated only in fully differentiated adipocytes, and the other being the IRS-PI 3-kinase signaling pathway that is relatively broadly activated by a number of hormones and growth factors (Fig. 1).

## The APS-CAP-Cbl Pathway Is Compartmentalized Within Plasma Membrane Microdomains

In addition to an activation mechanism that relies on signaling molecules expressed exclusively in adipocytes, the second insulin signaling pathway is also compartmentalized within specialized lipid raft/caveolae microdomains of the plasma membrane. Lipid raft microdomains are defined by their distinct lipid compositions. They are highly enriched in cholesterol and sphingolipids. A subset of these domains contains the structural protein caveolin, which forms small invaginations of the membrane termed caveolae. Caveolae are abundant in fat, muscle, and endothelial cells. Lipid rafts/caveolae are enriched in a number of proteins involved in signaling, including heterotrimeric Gαq, H-Ras, endothelial Nitric Oxide Synthase (eNOS), growth factor receptors and the src family tyrosine kinases Fyn and Lyn.[82] The insulin receptor has also been reported to segregate into these domains, and to catalyze the tyrosine phosphorylation of caveolin.[83-85]

The stimulation of Cbl tyrosine phosphorylation by insulin appears to occur primarily in lipid raft subdomains, due mainly to the associated protein CAP. CAP is a multifunctional adaptor protein with three adjacent SH3 domains in the carboxyl terminus, which directly binds the proline-rich domain of Cbl, and a Sorbin Homology (SoHo) domain in the amino terminus, which binds flotillin, a lipid raft/caveolar protein.[78,85] This latter interaction stabilizes the CAP/Cbl complex in rafts. Once tyrosine-phosphorylated, Cbl recruits a signaling complex containing the adaptor protein CrkII and the guanine nucleotide exchange factor C3G to lipid rafts. C3G acts on TC10, a Rho family small GTP-binding protein. The activation of TC10 by insulin via the CAP-Cbl-CrkII-C3G signaling complex appears to be necessary for GLUT4 translocation, since prevention of its proper activation by over-expression of either

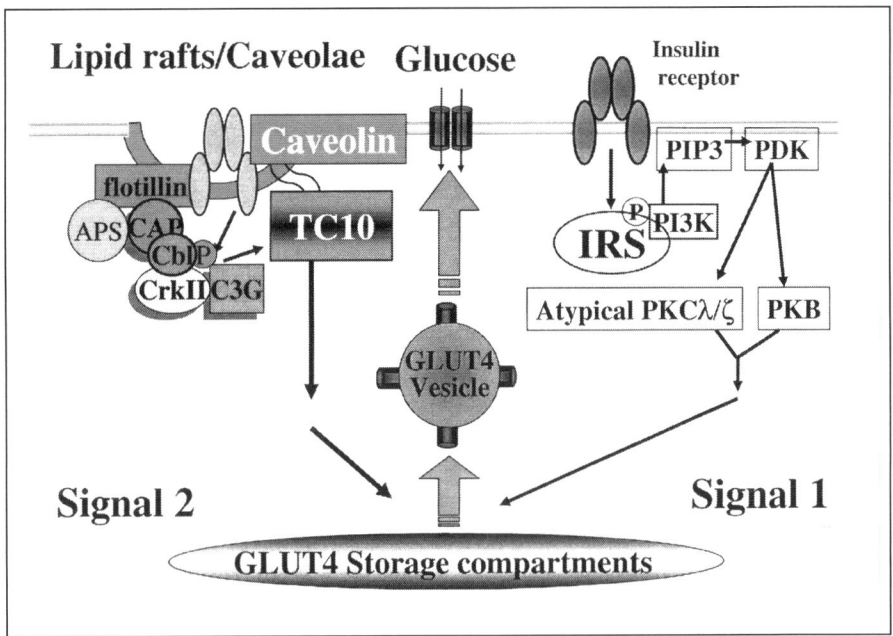

Figure 1. Insulin stimulation of GLUT4 translocation requires two independent signaling pathways; one is dependent on the IRS-PI 3-kinase pathway (signal 1) and the other is dependent on the CAP/Cbl/TC10 pathway (signal 2). Signal 1 involves insulin-induced IRS phosphorylation and PI 3-kinase recruitment, resulting in the PIP3 formation, which subsequently activates PDK1. Signal 2 involves insulin-induced Cbl phosphorylation and the recruitment of CrkII-C3G complex to the caveolae/lipid raft microdomain where TC10 is localized. The lipid-raft-recruited C3G serves as a guanine nucleotide exchange factor for TC10, resulting in the activation of this small GTPase. These two insulin receptor signals act together to induce the translocation of GLUT4.

TC10 or CAP mutants significantly inhibits insulin-induced GLUT4 translocation in adipocytes.[78,86] Moreover, this signaling pathway must occur in lipid rafts, since cholesterol removal from the plasma membrane by methyl-β-cyclodextrin or by overexpression of mutant forms of caveolin3 also markedly inhibits insulin activation of TC10 mediated through the CAP-Cbl signaling pathway.[87]

## TC10 Generates Spatially Compartmentalized Signals That Contribute to the Specificity of Insulin Action

TC10 is a member of the Rho family of GTP-binding proteins and has a high degree of sequence similarity with Rac, Rho, and Cdc42, well known actin regulators in various cell types.[88-91] Like other Rho family members, TC10 operates as a molecular switch cycling between inactive GDP-bound and active GTP-bound conformational states.[89] The active GTP-bound TC10 can bind numerous potential effector molecules possessing a Cdc42/Rac interactive binding (CRIB) domain such as p21-activated protein kinase (PAK), the Borg family of interacting proteins, the mammalian partition-defective homologue Par6, and the N-WASP isoform of the Wiscott-Aldrich syndrome protein,[88,92-94] as well as others without such domains, including Exo70, CIP4, and PIST.[95-97]

One interesting morphological feature of the adipocyte plasma membrane is the presence of both multiple individual caveolae and clusters of caveolae organized into large ring-shape structures (caveolae-rosettes) that are visualized at both the electron and light microscopic levels.[87,98,99]

Intriguingly, TC10 also colocalizes with these caveolin-positive caveolae-rosette structures. Most Rho members contain a single carboxyl-terminal cysteine residue in the appropriate sequence context for geranylgeranylation, and interact with guanine nucleotide dissociation inhibitors.[90] In contrast, TC10 contains a sequence similar to that of H-Ras, encoding for both farnesylation and palmitoylation. These post-translational modifications target TC10 persistently to lipid raft/caveolae microdomains in adipocytes.[87,100] Furthermore, the localization of TC10 in lipid rafts is necessary for its activation by insulin. Indeed, when experimentally mis-targeted to nonraft regions of the plasma membrane, TC10 failed to undergo insulin-dependent activation.

Recent results have demonstrated that the caveolae-rosette domains are directly involved in the organization of a unique filamentous actin structure in 3T3L1 adipocytes.[99] In most cell types, F-actin forms stress fibers, lamellipodia and fillopodia. However, the actin cytoskeleton is dramatically changed during the differentiation of fibroblast-like preadipocytes into adipocytes. Although preadipocytes contain well defined stress fibers following differentiation into adipocytes, this F-actin converts to cortical actin lining the inner face of the cell surface membrane.[91,101] At the same time, the levels of caveolin mRNA and protein expression increase 20 fold, and the number of caveolae increases 10 fold.[98,102] This marked induction of caveolin is accompanied by the clustering of individual caveolae (50-80nm) into higher-order caveolae-rosettes structures that can be visualized by fluorescent microscopy.[87,98,99] Intriguingly, fully differentiated 3T3L1 adipocytes display patches of punctate F-actin that emanate from the organized caveolae-rosette structure. This unique arrangement of caveolae and actin has been designated caveolin-associated F-actin (Cav-actin).[99]

Currently, the molecular basis underlying the conversion from stress fiber type F-actin to the Cav-actin structures in adipocytes remains unknown. However, this dramatic structural change suggests that actin modulators and their regulatory mechanisms are essential for adipocyte function. In this regard, treatment of adipocytes with a variety of agents that perturb actin turnover, including cytochalasin D, latrunculin A or B, jasplakinolide, or swinholide all inhibit insulin-induced GLUT4 translocation.[101,103,104] Furthermore, insulin stimulates dynamic actin remodeling at both the inner surface of the plasma membrane and in the perinuclear region that is sensitive to *C. difficile* toxin B.[101] Moreover, over-expression of TC10 mutants completely disrupts the cortical F-actin including Cav-actin structures, and abolishes insulin-induced actin remodeling that results in an inhibition of GLUT4 translocation.[99,101] Interestingly, recent work has suggested that the blockade of GLUT4 translocation caused by actin-disrupting agents can be overcome by the expression of an active, myristoylated form of PKB.[105] These results suggest that actin may be critical for maintaining the intracellular organization of signaling complexes, rather than being required for the GLUT4 exocytotic process per se.

The observation that Cdc42, a very close relative of TC10, has no impact on either Cav-actin or GLUT4 translocation clearly indicates that the compartmentalization of TC10 to the caveolae-rosette structures is necessary to maintain and control the adipocyte actin structure. In addition, the dynamic remodeling of actin by TC10 is required for insulin-stimulated GLUT4 translocation. Although the actin regulatory mechanism and effector molecules governed by TC10 in adipocytes are still unclear, available evidence indicates an involvement of N-WASP in the dynamic actin remodeling and GLUT4 translocation process, since the dominant-interfering N-WASP mutant, partially but significantly inhibits both events.[106,107]

## Downstream Targets of TC10

The data described above suggest that lipid raft/caveolae microdomains function as signaling platforms for the regulation of specific biological actions of insulin such as GLUT4 translocation, via the localization of TC10 effectors (Fig. 2). The exocyst complex consists of eight proteins: Sec3, Sec5, Sec6, Sec8, Sec10, Sec15, Exo70 and Exo84 that were originally identified in yeast.[108] This octameric complex is involved in the tethering or docking of exocytotic vesicles.[108] In 3T3L1 adipocytes, it was found that the exocyst component Exo70 can bind active GTP-bound

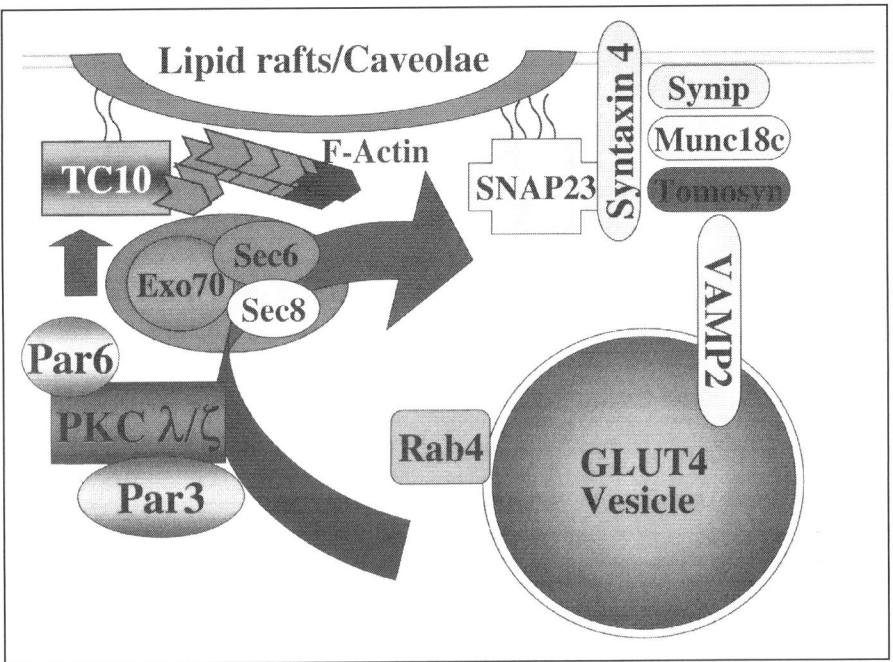

Figure 2. The lipid-raft-resident TC10 recruits multiple downstream effectors including the exocyst protein complex and the Par6-Par3-atypical PKC protein complex to the plasma membrane. Unique F-actin structure (Cav-actin) is also associated with the caveolae-rosette structure where TC10 exists. Since the exocyst protein complex is implicated in the regulation of tethering/docking step of exocytic vesicles, these microdomains may function as important platforms for regulating the GLUT4-vesicle exocytosis mediated by SNARE proteins (SNAP-23, syntaxin4, VAMP2).

TC10, in the process mediating translocation or assembly of the complex at the plasma membrane.[97]

Overexpression of full length Exo70 in 3T3L1 adipocytes resulted in the potentiation of insulin stimulated glucose uptake. In contrast, a carboxyl-terminal-deletion mutant of Exo70 (C-truncated Exo70) markedly inhibited glucose uptake without any obvious inhibition of GLUT4 translocation to the plasma membrane. By using an exofacial Myc epitope-tagged GLUT4 reporter protein, the C-truncated-Exo70 mutant was found to specifically prevent the tethering and/or docking of GLUT4-containing vesicles, a necessary prerequisite for the final plasma membrane fusion. However, the C-truncated-Exo70 mutant did not affect the intracellular trafficking of GLUT4 from storage compartments to the plasma membrane, as GLUT4 vesicles were observed to accumulate underneath the plasma membrane.[97] Together, these findings suggest that the recruitment or assembly of the exocyst complex to the lipid raft microdomain through the interaction between TC10 and Exo70 is required for the final steps in the GLUT4 exocytosis process.

At the final stage of exocytosis, the lipid bilayers of the plasma membrane and vesicle membrane fuse in a reaction catalyzed by the interactions between integral membrane proteins that are present in the target membrane (t-SNAREs) and the exocytotic vesicle membrane (v-SNAREs).[109] The formation of a stable ternary complex between these SNARE proteins brings the exocytotic vesicle and target membrane into close proximity, and eventually leads to their fusion. In adipocytes, the v-SNARE VAMP-2 (or synaptobrevin 2) in the GLUT4-containing vesicles, and the t-SNAREs syntaxin4 and SNAP23 in the plasma membrane play a crucial role in the final fusion step of GLUT4-containing vesicles.[110] Among these SNARE proteins, SNAP23 is known to be palmitoylated and is targeted to the detergent-insoluble region of the plasma membrane.[111] Interestingly, the exocyst complex has been reported to be required for exocytotic vesicle targeting and docking at specific areas of the plasma membrane such as sites of polarized exocytosis.[108] In the context of insulin-induced GLUT4 translocation in adipocytes, this might be equivalent to the spatially restricted microdomains of the plasma membrane. In any case, it should also be recognized that in addition to these interactions, there are several accessory proteins involved in the exocytosis of GLUT4-containing vesicles including Munc18c, Tomosyn, and Synip.[54,112,113]

Another downstream effector of TC10, TCGAP, was recently identified in a yeast two-hybrid screen.[114] TCGAP is a multi-domain protein with N-terminal PX and SH3 domains, a central Rho-GAP domain, and several C-terminal proline-rich regions. TCGAP interacts with Cdc42 and TC10β via its GAP domain, although GAP activity towards these proteins was not detected in vivo. Nevertheless, TCGAP translocated to the plasma membrane in response to insulin, and overexpression of either the full-length or the C-terminal region of TCGAP inhibited insulin-stimulated GLUT4 translocation and glucose uptake. Although the PX domain was found to interact specifically with PI4,5P2, the binding partners for the SH3 and proline-rich domains of TCGAP remain to be determined. Thus, TCGAP appears to function downstream of TC10 in the insulin-regulated translocation of GLUT4, although its precise functional role remains to be elucidated and is an active area of current investigation.

Recent studies have also suggested that the atypical protein kinase C isoforms lambda and zeta (PKCλ/ζ) are downstream targets for the IRS-PI 3-kinase signaling pathway, since these enzymes are direct substrate for PDK1.[115,116] However, PKCλ/ζ also form a quaternary complex with Par6 and Par3/ASIP.[93,117] The Par proteins were originally identified as molecules involved in asymmetric cell division and polarized growth in *C. elegans* development.[118] Par6 contains a PDZ (PSD-95/Dlg/ZO-1) domain downstream of a motif that is similar to a CRIB domain, and both are apparently required for the association of Par6 and Cdc42, a Rho family member structurally close to TC10, with exception of its carboxy-terminal CAAX motif.[87,100] Par6 and PKCλ/ζ both contain PB1 (Phox and Bem1) domains that are responsible for forming heterodimeric complexes.[119] Par3, also termed ASIP, contains three PDZ domains and specifically binds to both Par6 and PKCλ/ζ.[120] Thus Par6 and Par3 proteins appear to serve as scaffolding molecules, linking PKCλ/ζ and the Rho family small GTP-binding proteins.[93,121]

In fully differentiated 3T3L1 adipocytes, expression of either constitutively active TC10 mutants or C3G, which activates endogenous TC10, results in the recruitment of PKCλ/ζ to the plasma membrane through the Par6-Par3 complex. Furthermore, PKCλ/ζ translocates to the plasma membrane with Par proteins in a manner dependent on insulin activation of TC10.[94] More specifically, PKCλ/ζ appears to translocate to caveolae-rosette structures and undergoes the phosphorylation of Thr402/410 at the activation loop. Because insulin activates PDK1 and induces Thr402/Thr410 phosphorylation, it has been assumed that PKCλ/ζ is recruited to the plasma membrane by PDK1. However, the activation of the PI 3-kinase pathway by expression of the constitutively active membrane targeted p110 (catalytic subunit of the PI 3-kinase) failed to recruit PKCλ/ζ. In contrast, insulin-induced recruitment of PKCλ/ζ was prevented by *C. difficile* toxin B and methyl-β-cyclodextrin pretreatments. Moreover, PKCλ/ζ recruitment was also prevented by overexpression of the dominant-interfering Par6 mutant lacking CRIB domain. Thus, these data provide an important connection between two independent insulin receptor signaling pathways by demonstrating that PKCλ/ζ is a convergent downstream target of both the IRS-PI 3-kinase and Cbl-TC10 signaling cascades. Importantly, only the TC10 pathway results in the recruitment of PKCλ/ζ to caveolae in fully differentiated 3T3L1 adipocytes.

Although the physiological role of the recruited PKCλ/ζ remains to be determined, emerging evidence suggests that the proper compartmentalization of PKCλ/ζ to caveolae may provide an important clue to explain the basis for signaling specificity. For example, PKC is a relatively promiscuous serine/threonine kinase and PKCλ/ζ-dependent phosphorylation sites are highly degenerate (RXS, RXXS, or RXXSXS), compared to consensus site for the related kinase PKB/Akt (RXRXXS).[122] Thus, many proteins are potentially PKC substrates and are phosphorylated in vitro, however, only a subset of these is actually phosphorylated in vivo. The spatial compartmentalization of PKCλ/ζ to the caveolae may therefore allow the enzyme to phosphorylate substrate proteins only in the restricted region that could lead to signaling specificity.

## Sorting GLUT4 Into and Out of the Insulin-Responsive Storage Compartment

### *Exit of GLUT4 from the Insulin-Responsive Compartment*

Regulated exocytosis requires that target proteins be compartmentalized in a way that renders them highly responsive to specific extracellular signals.[123] The most studied examples of this process include the secretion of soluble proteins such as hormones and neurotransmitters. In these cases, the membrane compartment itself is mobilized to the cell surface and the vesicular contents are released from the cell following the fusion event. In contrast, GLUT4 is an integral membrane protein that continually cycles between the cell surface and intracellular compartments, even in the absence of insulin.[6] A key question concerns the mechanism by which muscle and fat cells efficiently skew the distribution of GLUT4 toward intracellular compartments under basal conditions. In one current model, GLUT4-containing vesicles are sequestered intracellularly through direct interactions between the transporter and a retention receptor. Using a novel functional screening approach, Bogan et al (2003) have recently identified a candidate GLUT4-interacting partner that may serve as an intracellular tether for this transporter.[124] This interacting partner, termed TUG (Tether, containing a UBX domain, for GLUT4), interacts specifically with GLUT4, but not with GLUT1. In addition, insulin appears to disassemble the TUG-GLUT4 complex, as detected by an insulin-stimulated reduction in the amount of TUG coimmunoprecipitated with GLUT4 in 3T3L1 adipocytes. The C-terminal region of TUG, from residues 463-550, was necessary for the efficient sequestration of GLUT4, and expression of a C-terminal fragment that encompassed this region inhibited the fold-stimulation of GLUT4 translocation in response to insulin. In part, this occurred through an increase in the basal translocation of GLUT4. In contrast, expression of full-length TUG resulted in a more rapid response to insulin together with a greater extent of translocation. These results suggest that TUG may function as a tether that efficiently sequesters GLUT4 within intracellular compartment(s) in the absence of insulin. Presumably TUG also interacts with other, as yet unidentified proteins, but in any event insulin stimulation appears to release GLUT4 from TUG, allowing the transporter to traffic to the cell surface. In the future, it will be interesting to reduce the levels of the TUG protein, perhaps by RNA interference or targeted gene disruption, as this would be predicted to cause increased basal GLUT4 translocation.

In addition to the 'retention receptor' model described above for retaining GLUT4 intracellularly, another mechanism, termed 'dynamic retention' has also been proposed.[125,126] In this latter model, GLUT4 continually undergoes budding and fusion processes between the GLUT4 storage compartments, endosomes, and the plasma membrane. This model is based in part on the observation that GLUT4 is about five times more likely to fuse with an endosome compartment than with the plasma membrane in the basal state. Indeed, using a GLUT4 reporter construct with an engineered exofacial myc-epitope, Karylowski et al (2004) showed that GLUT4 was equally distributed between endosomes and the insulin-responsive storage compartment under basal conditions.[125] In contrast, the transferrin receptor (TfR) was confined to the general endosome population and was excluded from the specialized GLUT4 storage compartment. Based on these and other observations, the authors propose that GLUT4 is in dynamic equilibrium with the endosome compartments and that specific sorting processes allow the continuous retrieval of GLUT4, but not general endosome markers such as TfR, back to the insulin-responsive compartment. The authors further propose that the large increase in GLUT4 exocytosis induced by insulin results from the mobilization of GLUT4 from both the endosomal and the specialized insulin-responsive storage compartments. Although the details remain to be elucidated, it should be noted that the 'retention receptor' and 'dynamic retention' models are not mutually exclusive.

## *Entry of Newly-Synthesized GLUT4 into the Insulin-Responsive Compartment*

GLUT4 is a fairly stable protein with a half-life of ~50 h in 3T3L1 adipocytes.[127] Since GLUT4 continually cycles to and from the cell surface, there must be a sorting process within endosomes such that GLUT4 is returned, directly or indirectly, to the insulin-responsive compartment.[5,128] Indeed, using various exofacial labeling techniques, the internalization of GLUT4 from the cell surface has been studied by several groups.[5] Following endocytosis, GLUT4 is routed through endosomes and equilibrates with the unlabeled intracellular GLUT4 population. Importantly, this internalized GLUT4 was able to undergo subsequent rounds of insulin-stimulated translocation, indicating that it was returned to the insulin-responsive compartment.[2,129-132] Although numerous studies have examined the trafficking dynamics of GLUT4 following endocytosis from the cell surface, much less attention has been given to the sorting events responsible for targeting newly-synthesized GLUT4 into the insulin-responsive compartment.

Like other membrane proteins, GLUT4 is inserted into membranes of the endoplasmic reticulum following its initial biosynthesis. GLUT4 then traffics to the Golgi complex, presumably en mass with other membrane proteins. At the trans-Golgi network (TGN), many proteins undergo sorting processes that allow them to traffic to the cell surface or to specific intracellular compartments.[133] In the case of newly-synthesized GLUT4, it is possible that this transporter is sorted directly into the insulin-responsive compartment from the TGN, or alternatively that GLUT4 first traffics to the cell surface and is then routed through endosomes before entering the insulin responsive compartment.

To distinguish between these possibilities, we recently examined the sorting of GLUT4 immediately following its initial biosynthesis.[134] By monitoring the localization of GLUT4 at short time intervals following transient transfection, we have been able to track the GLUT4 protein as it makes its way out of the endoplasmic reticulum and traffics through the Golgi to the insulin-responsive compartment. Although the precise identity of the insulin-responsive compartment remains unknown, it can be operationally defined because GLUT4 becomes insulin-responsive when it is therein sequestered. In contrast, general membrane proteins such as VSV-G and GLUT1 localized to the plasma membrane in as little as 2-3 h post-transfection, whereas GLUT4 was retained within the peri-nuclear region and required 9-12 h to display the full extent of insulin-stimulated translocation.[134] Furthermore, there was no apparent requirement for newly synthesized GLUT4 to traffic to the cell surface before acquiring the ability to respond to insulin. These results suggest that newly-synthesized GLUT4 undergoes a time-dependent sorting process that results in its compartmentalization within insulin-responsive vesicular structures.

Golgi-localized, γ-ear-containing, Arf-binding (GGA) proteins are a relatively new family of monomeric clathrin adaptors that function at the TGN to regulate the exit of certain cargo molecules, including the two mannose-6-phosphate receptors and sortilin.[135,136] Expression of a dominant-interfering GGA mutant comprising the VHS-GAT domains, completely inhibited the insulin-stimulated translocation of the newly synthesized GLUT4 protein.[134] In addition, this mutant had no measurable effect on the insulin-stimulated translocation of endogenous GLUT4. These results suggest that the entry of newly-made GLUT4 into the insulin-responsive compartment is GGA-dependent, whereas insulin-stimulated exit from this compartment is GGA-independent. Furthermore, GGA has recently been shown to be involved in the endocytotic recycling from the plasma membrane back into the insulin-responsive compartment.[137] Consistent with these data, Shewan et al (2003) observed that the endocytotic recycling of GLUT4 traffics through the TGN.[138] Together, these results are consistent with the involvement of GGA in a specific, GLUT4-selective budding event, most likely at the TGN.

GGA proteins interact with cargo molecules that harbor a specific type of acidic cluster-dileucine motif, such as is present in the mannose-6-phosphate receptors.[135] Although GLUT4 contains a region rich in acidic amino acids and a pair of leucines in its C-terminal cytosolic domain, they are not in the correct context for interactions with GGA proteins. Indeed, the GLUT4 protein does not directly bind to GGA. However, GLUT4-containing vesicles do coprecipitate with GGA,[137] indicating the likely existence of an as yet unidentified intermediary binding partner. Taken together, the above data suggest that newly synthesized GLUT4 is transported to the Golgi and arrives at the TGN. At the TGN GLUT4 undergoes a specific sorting event that excludes other proteins such as GLUT1 and the TfR, through a GGA-dependent selection process that results in the compartmentalization of GLUT4 within insulin-responsive storage vesicles. This latter sorting step could occur directly from the TGN or through an intermediate endosome compartment. Once in the insulin-responsive compartment, GLUT4 slowly leaks to the cell surface under basal conditions. Following insulin stimulation, exit from this compartment is markedly increased. Although initially identified for newly-synthesized GLUT4, recent work indicates that GLUT4 molecules that have been retrieved from the cell surface may also undergo a GGA-dependent sorting step during their subsequent entry into the insulin-responsive compartment.[137] In any event, the ability to functionally isolate the processes responsible for sorting GLUT4 into the insulin-responsive compartment has allowed for a more detailed analysis of the mechanisms and molecules responsible for this critical sorting event. In addition, we now have the means for biochemically isolating and characterizing the GLUT4 storage compartment, an avenue of research currently under vigorous investigation.

## Does Insulin Regulate the Intrinsic Transport Activity of GLUT4?

Although insulin clearly induces the translocation of GLUT4 from intracellular storage sites to the cell surface, it remains possible that the intrinsic transport activity of GLUT4 is also regulated by insulin. Indeed, discrepancies between the fold-increase in glucose uptake and the fold-increase in GLUT4 translocation have been reported by several investigators.[139] In general, the fold-increase in glucose uptake exceeded the fold-increase in GLUT4 translocation, although results varied considerably. In addition, time course experiments using a GLUT4 reporter construct engineered with an exofacial myc-epitope have shown that GLUT4 translocates to the plasma membrane prior to a measurable increase in glucose uptake. Together, these data raise the possibility that insulin may enhance the intrinsic glucose transport activity of GLUT4 transporters at the plasma membrane.

Over the past several years, evidence implicating the p38 mitogen activated protein kinases (MAPKs) in GLUT4 activation has accumulated. For example, it was shown that inhibitors of p38 MAPK, such as SB203580, prevent glucose uptake without affecting the ability of GLUT4 to translocate to the cell surface in response to insulin.[140-142] However, subsequent work showed that the SB203580 compound interacts directly with the endofacial surface of GLUT4 and competitively inhibits glucose transport.[143] Nevertheless, recent work has demonstrated that exogenously delivered PI3,4,5P$_3$ can stimulate GLUT4 translocation without causing a concomitant increase in glucose uptake.[144] Moreover, it has been reported that a cell-permeable phosphoinositide-binding peptide (PBP10) can also induce the translocation of GLUT4 in the absence of insulin, without an increase in glucose transport.[145] The PBP10 peptide is derived from the N-terminus of gelsolin and binds phosphoinositides at the D3 and D4 positions. Although its mechanism of action with regard to GLUT4 translocation remains unclear, cells pretreated with PBP10 were able to take up glucose when subsequently stimulated with insulin. Thus, insulin's ability to regulate the intrinsic transport activity of GLUT4 remains an open and active area of research. The general acceptance of this hypothesis will probably require the experimental identification of a specific mechanism by which insulin could potentially regulate the intrinsic transport activity of GLUT4.

## Conclusions and Future Directions

The past several years have seen many advances on two separate investigative fronts: Insulin signaling and GLUT4 vesicle trafficking, two distinct fields united by the common theme of subcellular compartmentalization. At some point the information carried by the insulin receptor signaling cascade must be translated into the language of vesicular trafficking. How this is accomplished remains unclear, however the recent discovery of AS160 as a downstream substrate of PKB provides a potential mechanism. AS160 contains a Rab GAP domain, and Rab proteins are thought to function at several steps of the vesicular transport process. However, the AS160 target Rab remains unidentified and this is an area that clearly is of great interest. In addition to AS160, the recent identification of Exo70 as a binding partner for activated TC10 provides another link between signaling and trafficking processes. In this case Exo70 appears to act at a relatively late stage in the translocation process, during the docking/fusion step with the plasma membrane. Similarly, the identification of Synip as a downstream substrate of PKB also provides a link between insulin signaling and GLUT4 trafficking, again occurring during the docking/fusion step of vesicular transport. Thus, although several important connections have been made between signaling processes and trafficking events, it still remains unclear how the precise mobilization of GLUT4 storage compartments in response to insulin is accomplished.

GLUT4 traffics through several internal compartments during its exocytsosis and subsequent retrieval from the cell surface. Although GGA appears to play a role in selecting GLUT4 molecules at the TGN for delivery to the insulin responsive compartment, there almost certainly are analogous sorting processes occurring in endosomes such that GLUT4 is efficiently retrieved back to its storage compartment. Indeed, following its internalization at the cell surface, GLUT4 is routed through a series of endosome compartments before returning to the insulin-responsive compartment. Although the precise trafficking itinerary remains unclear, for example whether GLUT4 is routed back through the TGN before entering its storage compartment, it nevertheless seem apparent that many sorting decisions are made in the process of correctly routing GLUT4 back to its storage depot. Some of these sorting decisions likely employ general factors, such as AP2 at the cell surface. However, other sorting events may be specifically selective for GLUT4, and deciphering these cargo-specific steps represents an important area for current and future work.

## References

1. Jhun BH, Rampal AL, Liu H et al. Effects of insulin on steady state kinetics of GLUT4 subcellular distribution in rat adipocytes. J Biol Chem 1992; 268:17710-5.
2. Satoh S, Nishimura H, Clark AE et al. Use of bismannose photolabel to elucidate insulin-regulated GLUT4 subcellular trafficking kinetics in rat adipose cells. Evidence that exocytosis is a critical site of hormone action. J Biol Chem 1993; 268:17820-9.
3. Yang J, Holman GD. Comparison of GLUT4 and GLUT1 subcellular trafficking in basal and insulin-stimulated 3T3-L1 cells. J Biol Chem 1993; 268:4600-3
4. Bryant NJ, Govers R, James DE. Regulated transport of the glucose transporter GLUT4. Nat Rev Mol Cell Biol 2002; 3:267-77.
5. Rea S, James DE. Moving GLUT4: The biogenesis and trafficking of GLUT4 storage vesicles. Diabetes 1997; 46:1667-77.
6. Watson RT, Kanzaki K, Pessin JE. Regulated membrane trafficking of the insulin-responsive glucose transporter 4 in adipocytes. Endocr Rev 2004; 25:177-204
7. Thurmond DC, Pessin JE. Molecular machinery involved in the insulin-regulated fusion of GLUT4-containing vesicles with the plasma membrane. Mol Membr Biol 2001; 18:237-45.

8. Kanzaki M, Pessin JE. Insulin signaling: GLUT4 vesicles exit via the exocyst. Curr Biol 2003; 13:R574-R6.
9. Saltiel AR. New perspectives into the molecular pathogenesis and treatment of type 2 diabetes. Cell 2001; 104:517-29.
10. Czech MP. The nature and regulation of the insulin receptor: structure and function. Ann Rev Physiol 1985; 47:357-81.
11. Frattali AL, Pessin JE. Relationship between a subunit ligand occupancy and b subunit autophosphorylation in insulin/insulin-like growth factor-1 hybrid receptors. J Biol Chem 1993; 268:7393-400.
12. Lee J, O'Hare T, Pilch PF et al. Insulin receptor autophosphorylation occurs asymmetrically. J Biol Chem 1993; 268:4092-8.
13. Pessin JE, Saltiel AR. Signaling pathways in insulin action: Molecular targets of insulin resistance. J Clin Invest 2000; 106:165-9.
14. White MF, Yenush L. The IRS-signaling system: A network of docking proteins that mediate insulin and cytokine action. Curr Top Microbiol Immunol 1998; 228:179-208.
15. Toker A, Newton AC. Akt/protein kinase B is regulated by autophosphorylation at the hypothetical PDK-2 site. J Biol Chem 2000; 275:8271-4.
16. Mossessova E, Gulbis JM, Goldberg J. Structure of the guanine nucleotide exchange factor Sec7 domain of human ARNO and analysis of the interaction with Arf GTPase. Cell 1998; 92:415-23.
17. Cheatham B, Vlahos CJ, Cheatham L et al. Phosphatidylinositol 3-kinase activation is required for insulin stimulation of pp70 S6 kinase, DNA synthesis, and glucose transporter translocation. Mol Cell Biol 1994; 14:4902-11.
18. Okada T, Kawano Y, Sakakibara R et al. Essential role of phophatidylinositol 3-kinase in insulin-induced glucose transport and antilypolysis in rat adipocytes. Studies with a selective inhibitor wortmannin. J Biol Chem 1994; 269:3568-73.
19. Maehama T, Dixon JE. The tumor suppressor, PTEN/MMAC1, dephosphorylates the lipid second messenger, phosphatidylinositol 3,4,5-trisphosphate. J Biol Chem 1998; 273:13375-8.
20. Nakashima N, Sharma PM, Imamura T et al. The tumor suppressor PTEN negatively regulates insulin signaling in 3T3-L1 adipocytes. J Biol Chem 2000; 275:12889-95.
21. Frank S, Upender S, Hansen SH et al. ARNO is a guanine nucleotide exchange factor for ADP-ribosylation factor 6. J Biol Chem 1998; 273:23-7.
22. Chavrier P, Goud B. The role of Arf and Rab GTPases in membrane transport. Curr Opin Cell Biol 1999; 11:466-75.
23. D'Souza-Schorey C, van Donselaar E, Hsu VW et al. Arf6 targets recycling vesicles to the plasma membrane: Insights from an ultrastructural investigation. J Cell Biol 1998; 140:603-16.
24. Vitale N, Chasserot-Golaz S, Bailly Yet al. Calcium-regulated exocytosis of dense-core vesicles requires the activation of ADP-ribosylation factor (Arf)6 by arf nucleotide binding site opener at the plasma membrane. J Cell Biol 2002; 159:79-89.
25. Bose A, Cherniack AD, Langille SE et al. G(alpha)11 signaling through Arf6 regulates F-actin mobilization and GLUT4 glucose transporter translocation to the plasma membrane. Mol Cell Biol 2001; 21:5262-75.
26. Jackson CL, Casanova JE. Turning on arf: The Sec7 family of guanine-nucleotide-exchange factors. Trends Cell Biol 2000; 10:60-7.
27. Emoto M, Langille SE, Czech MP. A role for kinesin in insulin-stimulated GLUT4 glucose transporter translocation in 3T3-L1 adipocytes. J Biol Chem 2001; 276:10677-82.
28. Kristiansen S, Richter EA. GLUT4-containing vesicles are released from membranes by phospholipase D cleavage of a GPI anchor. Am J Physiol Endocrinol Metab 2002; 283:E374-E82.
29. Huang P, Altshuller YM, Hou JC et al. Insulin-stimulated plasma membrane fusion of GLUT4 glucose transporter-containing vesicles is regulated by phospholipase D1. Mol Biol Cell 2005; 16(6):2614-23.
30. Liscovitch M, Czarny M, Fiucci Get al. Phospholipase d: Molecular and cell biology of a novel gene family. Biochem J 2000; 345:401-15.
31. Li HS, Shome K, Rojas R et al. The guanine nucleotide exchange factor ARNO mediates the activation of Arf and phospholipase D by insulin. BMC Cell Biol 2003; 4:13.
32. Stokoe D, Stephens LR, Copeland T et al. Dual role of phosphatidylinositol-3,4,5-trisphosphate in the activation of protein kinase B. Science 1997; 277:567-70.
33. Kohn AD, Summers SA, Birnbaum MJ et al. Expression of a constitutively active Akt Ser/Thr kinase in 3T3-L1 adipocytes stimulates glucose uptake and glucose transporter 4 translocation. J Biol Chem 1996; 271:31372-8.
34. Cong LN, Chen H, Li Y et al. Physiological role of Akt in insulin-stimulated translocation of GLUT4 in transfected rat adipose cells. Mol Endocrinol 1997; 11:1881-90.

35. Hajduch E, Alessi DR, Hemmings BA et al. Constitutive activation of protein kinase B alpha by membrane targeting promotes glucose and system a amino acid transport, protein synthesis, and inactivation of glycogen synthase kinase 3 in L6 muscle cells. Diabetes 1998; 47:1006-13.
36. Kohn AD, Barthel A, Kovacina KS et al. Construction and characterization of a conditionally active version of the Serine/Threonine kinase Akt. J Biol Chem 1998; 273:11937-43.
37. Wang Q, Somwar R, Bilan PJ et al. Protein kinase B/Akt participates in GLUT4 translocation by insulin in L6 myoblasts. Mol Cell Biol 1999; 19:4008-18
38. Heller-Harrison RA, Morin M, Guilherme A et al. Insulin-mediated targeting of phosphatidylinositol 3-kinase to GLUT4-containing vesicles. J Biol Chem 1996; 271:10200-4.
39. Kupriyanova TA, Kandror KV. Akt-2 binds to GLUT4-containing vesicles and phosphorylates their component proteins in response to insulin. J Biol Chem 1999; 274:1458-64.
40. Calera MR, Martinez C, Liu H et al. Insulin increases the association of Akt-2 with GLUT4-containing vesicles. J Biol Chem 1998; 273:7201-4.
41. Katome T, Obata T, Matsushima R et al. Use of RNA interference-mediated gene silencing and adenoviral overexpression to elucidate the roles of Akt/protein kinase B isoforms in insulin actions. J Biol Chem 2003; 278:28312-23.
42. Jiang ZY, Zhou QL, Coleman KA et al. Insulin signaling through Akt/protein kinase B analyzed by small interfering RNA-mediated gene silencing. Proc Natl Acad Sci USA 2003; 100:7569-74.
43. Bae SS, Cho H, Mu J et al. Isoform-specific regulation of insulin-dependent glucose uptake by Akt/protein kinase B. J Biol Chem 2003; 278:49530-6.
44. Cho H, Mu J, Kim JK et al. Insulin resistance and a diabetes mellitus-like syndrome in mice lacking the protein kinase Akt2 (PKB beta). Science 2001; 292:1728-31.
45. Cho H, Thorvaldsen JL, Chu Q et al. Akt1/PKB alpha is required for normal growth but dispensable for maintenance of glucose homeostasis in mice. J Biol Chem 2001; 276:38349-52.
46. Verdu J, Buratovich MA, Wilder EL et al. Cell-autonomous regulation of cell and organ growth in drosophila by Akt/PKB. Nat Cell Biol 1999; 1:500-6.
47. Kane S, Sano H, Liu SC et al. A method to identify serine kinase substrates. Akt phosphorylates a novel adipocyte protein with a Rab GTPase-activating protein (GAP) domain. J Biol Chem 2002; 277:22115-8.
48. Sano H, Kane S, Sano E et al. Insulin-stimulated phosphorylation of a Rab GTPase-activating protein regulates GLUT4 translocation. J Biol Chem 2003; 278:14599-602.
49. Colicelli J. Human Ras superfamily proteins and related GTPases. Sci STKE 2004.; 250:RE13.
50. Pfeffer SR. Structural clues to Rab GTPase functional diversity. J Biol Chem 2005; 280(16):15485-8.
51. Maxfield FR, McGraw TE. Endocytic recycling. Nat Rev Mol Cell Biol 2004; 2:121-32.
52. Zeigerer A, McBayer MK, McGraw TE. Insulin stimulation of GLUT4 exocytosis, but not its inhibition of endocytosis, is dependent on RabGAP as160. Mol Biol Cell 2004; 10:4406-15.
53. Yamada E, Okada S, Saito T et al. Akt2 phosphorylates synip to regulate docking and fusion of GLUT4-containing vesicles. J Cell Biol 2005; 168:921-8.
54. Min J, Okada S, Coker K et al. Synip: A novel insulin-regulated syntaxin 4 binding protein mediating GLUT4 translocation in adipocytes. Mol Cell 1999; 3:751-60.
55. Standaert ML, Galloway L, Karnam P et al. Protein kinase C-zeta as a downstream effector of phosphatidylinositol 3-kinase during insulin stimulation in rat adipocytes. Potential role in glucose transport. J Biol Chem 1997; 272:30075-82.
56. Kotani K, Ogawa W, Matsumoto M et al. Requirement of atypical protein kinase C lambda for insulin stimulation of glucose uptake but not for Akt activation in 3T3-L1 adipocytes. Mol Cell Biol 1998; 18:6971-82.
57. Bandyopadhyay G, Standaert ML, Kikkawa U et al. Effects of transiently expressed atypical (zeta, lambda), conventional (alpha, beta) and novel (delta, epsilon) protein kinase C isoforms on insulin-stimulated translocation of epitope-tagged GLUT4 glucose transporters in rat adipocytes: Specific interchangeable effects of protein kinases C-zeta and C-lambda. Biochem J 1999; 337:461-70.
58. Standaert ML, Bandyopadhyay G, Perez L et al. Insulin activates protein kinases C-zeta and C-lambda by an autophosphorylation-dependent mechanism and stimulates their translocation to GLUT4 vesicles and other membrane fractions in rat adipocytes. J Biol Chem 1999; 274:25308-16.
59. Bandyopadhyay G, Standaert ML, Sajan MP et al. Dependence of insulin-stimulated glucose transporter 4 translocation on 3-phosphoinositide-dependent protein kinase-1 and its target Threonine-410 in the activation loop of protein kinase C-zeta. Mol Endocrinol 1999; 13:1766-72.
60. Farese RV. Function and dysfunction of aPKC isoforms for glucose transport in insulin-sensitive and insulin-resistant states. Am J Physiol Endocrinol Metab 2002; 283:E1-E11.
61. Imamura T, Huang J, Usui I et al. Insulin-induced GLUT4 translocation involves protein kinase C-lambda-mediated functional coupling between Rab4 and the motor protein kinesin. Mol Cell Biol 2003; 23:4892-900.
62. Semiz S, Park JG, Nicoloro SM et al. Conventional kinesin KIF5B mediates insulin-stimulated GLUT4 movements on microtubules. EMBO J 2003; 22:2387-9239.

63. Isakoff SJ, Taha C, Rose E et al. The inability of phosphatidylinositol 3-kinase activation to stimulate GLUT4 translocation indicates additional signaling pathways are required for insulin-stimulated glucose uptake. Proc Natl Acad Sci USA 1995; 92:10247-51.
64. Guilherme A, Czech MP. Stimulation of IRS-1-associated phosphatidylinositol 3-kinase and Akt/protein kinase B but not glucose transport by beta1-integrin signaling in rat adipocytes. J Biol Chem 1998; 273:33119-22.
65. Jiang T, Sweeney G, Rudolf MT et al. Membrane-permeant esters of phosphatidylinositol 3,4,5-trisphosphate. J Biol Chem 1998; 273:11017-24.
66. Somwar R, Koterski S, Sweeney G et al. A dominant-negative p38 MAPK mutant and novel selective inhibitors of p38 MAPK reduce insulin-stimulated glucose uptake in 3T3-L1 adipocytes without affecting GLUT4 translocation. J Biol Chem 2002; 277:50386-95.
67. Konrad D, Bilan PJ, Nawaz Z et al. Need for GLUT4 activation to reach maximum effect of insulin-mediated glucose uptake in brown adipocytes isolated from GLUT4myc-expressing mice. Diabetes 2002; 51:2719-26.
68. Ribon V, Saltiel AR. Insulin stimulates tyrosine phosphorylation of the proto-oncogene product of c-Cbl in 3T3-L1 adipocytes. Biochem J 1997; 324:839-45.
69. Ribon V, Printen JA, Hoffman NG et al. A novel, multifuntional c-Cbl binding protein in insulin receptor signaling in 3T3-L1 adipocytes. Mol Cell Biol 1998; 18:872-9.
70. Ribon V, Herrera R, Kay BK et al. A role for CAP, a novel, multifunctional Src homology 3 domain-containing protein in formation of actin stress fibers and focal adhesions. J Biol Chem 1998; 273:4073-80.
71. Ahmed Z, Smith BJ, Pillay TS. The APS adapter protein couples the insulin receptor to the phosphorylation of c-Cbl and facilitates ligand-stimulated ubiquitination of the insulin receptor. FEBS Lett 2000; 475:31-4.
72. Moodie SA, Alleman-Sposeto J, Gustafson TA. Identification of the APS protein as a novel insulin receptor substrate. J Biol Chem 1999; 274:11186-93.
73. Hu J, Liu J, Ghirlando R et al. Structural basis for recruitment of the adaptor protein APS to the activated insulin receptor. Mol Cell 2003; 12:1379-89.
74. Liu J, Kimura A, Baumann CA et al. Aps facilitates c-Cbl tyrosine phosphorylation and GLUT4 translocation in response to insulin in 3T3-L1 adipocytes. Mol Cell Biol 2002; 22:3599-609.
75. Ahmed Z, Smith BJ, Kotani K et al. APS, an adapter protein with a PH and SH2 domain, is a substrate for the insulin receptor kinase. Biochem J 1999; 341:665-8.
76. Ribon V, Johnson JH, Camp HS et al. Thiazolidinediones and insulin resistance: Peroxisome proliferatoractivated receptor gamma activation stimulates expression of the CAP gene. Proc Natl Acad Sci USA 1998; 95:14751-6.
77. Baumann CA, Chokshi N, Saltiel AR et al. Cloning and characterization of a functional peroxisome proliferator activator receptor-gamma-responsive element in the promoter of the CAP gene. J Biol Chem 2000; 75:9131-5.
78. Baumann CA, Ribon V, Kanzaki M et al. CAP defines a second signalling pathway required for insulin-stimulated glucose transport. Nature 2000; 407:202-7.
79. Ahn MY, Katsanakis KD, Bheda F et al. Primary and essential role of the adaptor protein APS for recruitment of both c-Cbl and its associated protein CAP in insulin signaling. J Biol Chem 2004; 279:21526-32.
80. Alcazar O, Ho RC, Fujii N et al. cDNA cloning and functional characterization of a novel splice variant of c-Cbl-associated protein from mouse skeletal muscle. Biochem Biophys Res Commun 2004; 317:285-93.
81. Mitra P, Zheng X, Czech MP. RNAi-based analysis of CAP, Cbl, and CrkII function in the regulation of GLUT4 by insulin. J Biol Chem 2004; 279:37431-5.
82. Mastick CC, Saltiel AR. Insulin-stimulated tyrosine phosphorylation of caveolin is specific for the differentiated adipocyte phenotype in 3T3-L1 cells. J Biol Chem 1997; 272:20706-14.
83. Gustavsson J, Parpal S, Karlsson M et al. Localization of the insulin receptor in caveolae of adipocyte plasma membrane. FASEB J 1999; 13:1961-71.
84. Parpal S, Karlsson M, Thorn H et al. Cholesterol depletion disrupts caveolae and insulin receptor signaling for metabolic control via insulin receptor substrate-1, but not for mitogen-activated protein kinase control. J Biol Chem 2001; 276:9670-8.
85. Kimura A, Baumann CA, Chiang SH et al. The Sorbin homology domain: A motif for the targeting of proteins to lipid rafts. Proc Natl Acad Sci USA 2001; 98:9098-103.
86. Chiang S-H, Baumann CA, Kanzaki M et al. Insulin-stimulated GLUT4 translocation requires the CAP-dependent activation of the small GTP binding protein TC10. Nature 2001; 410:944-8.
87. Watson RT, Shigematsu S, Chiang SH et al. Lipid raft microdomain compartmentalization of TC10 is required for insulin signaling and GLUT4 translocation. J Cell Biol 2001; 154:829-40.
88. Neudauer CL, Joberty G, Tatsis N, Macara IG et al. Distinct cellular effects and interactions of the Rho-family GTPase TC10. Curr Biol 1998; 8:1151-60.

89. Murphy GA, Solski PA, Jillian SA et al. Cellular functions of TC10, a Rho family GTPase: Regulation of morphology, signal transduction and cell growth. Oncogene 1999; 18:3831-45.
90. Etienne-Manneville S, Hall A. Rho GTPases in cell biology. Nature 2002; 420:629-35.
91. Kanzaki M, Watson RT, Hou JC et al. Small GTP-binding protein TC10 differentially regulates two distinct populations of filamentous actin in 3T3L1 adipocytes. Mol Biol Cell 2002; 13:2334-46.
92. Joberty G, Perlungher RR, Macara IG. The Borgs, a new family of Cdc42 and TC10 GTPase-interacting proteins. Mol Cell Biol 1999; 19:6585-97.
93. Joberty G, Petersen C, Gao L et al. The cell-polarity protein Par6 links Par3 and atypical protein kinase C to Cdc42 . Nat Cell Biol 2000; 2:531-9.
94. Kanzaki M, Furukawa M, Raab W et al. Phosphatidylinositol-4, 5-bisphosphate (PI4,5P2) regulates adipocyte actin dynamics and GLUT4 vesicle recycling. J Biol Chem 2004; 279:30622-33.
95. Neudauer CL, Joberty G, Macara IG. PIST: A novel PDZ/coiled-coil domain binding partner for the Rho-family GTPase TC10. Biochem Biophys Res Commun 2001; 280:541-7.
96. Chang L, Adams RD, Saltiel AR. The TC10-interacting protein CIP4/2 is required for insulin-stimulated GLUT4 translocation in 3T3L1 adipocytes. Proc Natl Acad Sci USA 2002; 99:12835-40.
97. Inoue M, Chang L, Hwang J et al. The exocyst complex is required for targeting of GLUT4 to the plasma membrane by insulin. Nature 2003; 422:629-33.
98. Scherer PE, Lisanti MP, Baldini G et al. Induction of caveolin during adipogenesis and association of GLUT4 with caveolin-rich vesicles. J. Cell Biol 1994; 127:1233-43.
99. Kanzaki M, Pessin JE. Caveolin-associated filamentous actin (Cav-actin) defines a novel F-actin structure in adipocytes. J Biol Chem 2002; 277:25867-9.
100. Watson RT, Furukawa M, Chiang SH et al. The exocytotic trafficking of TC10 occurs through both classical and nonclassical secretory transport pathways in 3T3L1 adipocytes. Mol Cell Biol 2003; 23:961-74.
101. Kanzaki M, Pessin JE. Insulin-stimulated GLUT4 translocation in adipocytes is dependent upon cortical actin remodeling. J Biol Chem 2001; 276(45):42436-44.
102. Rothberg KG, Heuser JE, Donzell WC et al. Caveolin, a protein component of caveolae membrane coats. Cell 1992; 68:673-82.
103. Tsakiridis T, Vranic M, Klip A. Disassembly of the actin network inhibits insulin-dependent stimulation of glucose transport and prevents recruitment of glucose transporters to the plasma membrane. J Biol Chem 1994; 269:29934-42.
104. Omata W, Shibata H, Li L et al. Actin filaments play a critical role in insulin-induced exocytotic recruitment but not in endocytosis of GLUT4 in isolated rat adipocytes. Biochem J 2000; 346 Pt 2:321-8.
105. Eyster CA, Duggins QS, Olson AL. Expression of a constitutively active Akt/PKB signals GLUT4 translocation in the absence of an intact actin cytoskeleton. J Biol Chem 2005; 280(18):17978-85.
106. Kanzaki M, Watson RT, Khan A et al. Insulin stimulates actin comet tails on intracellular GLUT4-containing compartments in differentiated 3t3l1 adipocytes. J Biol Chem 2001; 276(52):49331-6.
107. Jiang ZY, Chawla A, Bose A et al. A phosphatidylinositol 3-kinase-independent insulin signaling pathway to N-WASP/Arp2/3/F-actin required for GLUT4 glucose transporter recycling. J Biol Chem 2002; 277:509-15.
108. Lipschutz JH, Mostov KE. Exocytosis: The many masters of the exocyst. Curr Biol 2002; 12:R212-4.
109. Rothman JE, Warren G. Implications of the SNARE hypothesis for intracellular membrane topology and dynamics. Curr Biol 1994; 4:220-33.
110. Pessin JE, Thurmond DC, Elmendorf JS et al. Molecular basis of insulin-stimulated GLUT4 vesicle trafficking. Location! Location! Location! J Biol Chem 1999; 274:2593-6.
111. Chamberlain LH, Gould GW. The vesicle- and target-SNARE proteins that mediate GLUT4 vesicle fusion are localized in detergent-insoluble lipid rafts present on distinct intracellular membranes. J Biol Chem 2002; 277:49750-4.
112. Thurmond DC, Ceresa BP, Okada S et al. Regulation of insulin-stimulated GLUT4 translocation by Munc18c in 3T3L1 adipocytes. J Biol Chem 1998; 273:33876-83.
113. Widberg CH, Bryant NJ, Girotti M et al. Tomosyn interacts with the t-SNAREs syntaxin4 and SNAP23 and plays a role in insulin-stimulated GLUT4 translocation. J Biol Chem 2003; 278:35093-101.
114. Chiang S-H, Hwang J, Legendre M et al. TCGAP, a multidomain Rho GTPase-activating protein involved in insulin-stimulated glucose transport. EMBO J 2003; 22:2679-91.
115. Chou MM, Hou W, Johnson J et al. Regulation of protein kinase C zeta by PI 3-kinase and PDK-1. Curr Biol 1998; 8:1069-77.
116. Le Good JA, Ziegler WH, Parekh DB et al. Protein kinase C isotypes controlled by phosphoinositide 3-kinase through the protein kinase PDK1. Science 1998; 281:2042-5.
117. Lin D, Edwards AS, Fawcett JP et al. A mammalian Par-3-Par-6 complex implicated in Cdc4 2/Rac1 and aPKC signalling and cell polarity . Nat Cell Biol 2000; 2:540-7.

118. Etemad-Moghadam B, Guo S, Kemphues KJ. Asymmetrically distributed Par-3 protein contributes to cell polarity and spindle alignment in early C. elegans embryos. Cell 1995; 83:743-52.
119. Ponting CP, Ito T, Moscat J et al. OPR, PC and AID: All in the PB1 family. Trends Biochem Sci 2002; 27:10.
120. Izumi Y, Hirose T, Tamai Y et al. An atypical PKC directly associates and colocalizes at the epithelial tight junction with ASIP, a mammalian homologue of caenorhabditis elegans polarity protein Par-3. J Cell Biol 1998; 143:95-106.
121. Noda Y, Takeya R, Ohno S et al. Human homologues of the caenorhabditis elegans cell polarity protein Par6 as an adaptor that links the small GTPases Rac and Cdc42 to atypical protein kinase C. Genes Cells 2001; 6:107-19.
122. Nishikawa K, Toker A, Johannes FJ et al. Determination of the specific substrate sequence motifs of protein kinase C isozymes. J Biol Chem 1997; 272:952-60.
123. Chieregatti E, Meldolesi J.. Regulated exocytosis: New organelles for non-secretory purposes. Nat Rev Mol Cell Biol 2005; 6:181-7.
124. Bogan JS, Hendon N, McKee AE et al. Functional cloning of TUG as a regulator of GLUT4 glucose transporter trafficking. Nature 2003; 425:727-33.
125. Karylowski O, Zeigerer A, Cohen A et al. GLUT4 is retained by an intracellular cycle of vesicle formation and fusion with endosomes. Mol Biol Cell 2004; 15:870-82.
126. Zeigerer A, Lampson MA, Karylowski O et al. GLUT4 retention in adipocytes requires two intracellular insulin-regulated transport steps. Mol Biol Cell 2002; 13:2421-35.
127. Sargeant RJ, Paquet MR. Effect of insulin on the rates of synthesis and degradation of GLUT1 and GLUT4 glucose transporters in 3T3-L1 adipocytes. Bioch J 1993; 290:913-9.
128. Holman GD, Sandoval IV. Moving the insulin-regulated glucose transporter GLUT4 into and out of storage. Trends Cell Biol 2001; 11:173-9.
129. Palacios S, Lalioti V, Martinez-Arca S et al. Recycling of the insulin-sensitive glucose transporter GLUT4. Access of surface internalized GLUT4 molecules to the perinuclear storage compartment is mediated by the Phe5-Gln6-Gln7-Ile8 motif. J Biol Chem 2001; 276:3371-83.
130. Foster LJ, Li D, Randhawa VK et al. Insulin accelerates inter-endosomal GLUT4 traffic via phosphatidylinositol 3-kinase and protein kinase B. J Biol Chem 2001; 276:44212-21.
131. Lampson MA, Racz A, Cushman SW et al. Demonstration of insulin-responsive trafficking of GLUT4 and vpTR in fibroblasts. J Cell Sci 2000; 113:4065-76.
132. Shigematsu S, Khan AH, Kanzaki M et al. Intracellular insulin-responsive glucose transporter (GLUT4) distribution but not insulin-stimulated GLUT4 exocytosis and recycling are microtubule dependent. Mol Endocrinol 2002; 16:1060-8.
133. Gleeson PA, Lock JG, Luke MR et al. Domains of the tgn: Coats, tethers and G proteins. Traffic 2004; 5:315-26.
134. Watson RT, Khan AH, Furukawa M et al. Entry of newly synthesized GLUT4 into the insulin-responsive storage compartment is GGA-dependent. EMBO J 2004; 23:2059-70.
135. Bonifacino JS. The GGA proteins: Adaptors on the move. Nat Rev Mol Cell Biol 2004; 5:23-32.
136. Ghosh P, Kornfeld S. The GGA proteins: Key players in protein sorting at the trans-Golgi network. Eur J Cell Biol 2004; 83:257-62.
137. Li LV, Kandror KV. GGA adaptors mediate insulin-responsive trafficking of GLUT4 in 3T3-L1 adipocytes. Mol Endocrinol 2005; 19:2145-53.
138. Shewan AM, Van Dam EM, Martin S et al. GLUT4 recycles via a trans-Golgi network (TGN) subdomain enriched in syntaxins 6 and 16 but not TGN38: Involvement of an acidic targeting motif. Mol Biol Cell 2003;14:973-86.
139. Furtado LM, Somwar R, Sweeney Get al. Activation of the glucose transporter GLUT4 by insulin. Biochem Cell Biol 2002; 80:569-78.
140. Sweeney G, Somwar R, Ramlal T et al. An inhibitor of p38 mitogen-activated protein kinase prevents insulin-stimulated glucose transport but not glucose transporter translocation in 3T3-L1 adipocytes and L6 myotubes. J Biol Chem 1999; 274:10071-8.
141. Somwar R, Kim DY, Sweeney G et al. GLUT4 translocation precedes the stimulation of glucose uptake by insulin in muscle cells: Potential activation of GLUT4 via p38 mitogen-activated protein kinase. Biochem J 2001; 359:639-49.
142. Bazuine M, Carlotti F, Rabelink MJ et al. The p38 mitogen-activated protein kinase inhibitor SB203580 reduces glucose turnover by the glucose transporter-4 of 3T3-L1 adipocytes in the insulin-stimulated state. Endocrinology 2005; 146:1818-24.
143. Ribe D, Yang J, Patel S, Koumanov F, Cushman SW, Holman GD et al. Endofacial competitive inhibition of glucose transporter-4 intrinsic activity by the mitogen-activated protein kinase inhibitor SB203580. Endocrinology 2005; 146:1713-7.
144. Sweeney G, Garg RR, Ceddia RB et al. Intracellular delivery of phosphatidylinositol (3,4,5)-trisphosphate causes incorporation of glucose transporter 4 into the plasma membrane of muscle and fat cells without increasing glucose uptake. J Biol Chem 2004; 279:32233-42.
145. Funaki M, Randhawa P, Janmey PA. Separation of insulin signaling into distinct GLUT4 translocation and activation steps. Mol Cell Biol 2004; 24:7567-77.

CHAPTER 3

# Regulation of Insulin Action and Insulin Secretion by SNARE-Mediated Vesicle Exocytosis

Debbie C. Thurmond*

## Summary

Maintenance of glucose homeostasis requires 'cross-talk' between pancreatic insulin secretion and insulin signaling in the peripheral tissues. Both insulin secretion and glucose uptake are regulated exocytotic processes mediated by SNARE protein complexes. SNARE core complexes are heterotrimeric, composed of syntaxin, SNAP-25/23 and VAMP2 proteins in a 1:1:1 ratio. It has become clear that we must thoroughly define the precise mechanisms underlying insulin-stimulated GLUT4 vesicle translocation by skeletal muscle and adipose cells as well as glucose-stimulated insulin secretion by pancreatic beta cells in order to develop therapeutic strategies to better treat and eventually cure diabetic patients. In this chapter, I will examine the molecular interactions responsible for these SNARE dependent events. Similarities and differences in molecular machinery, exocytosis mechanisms as well as the impact of accessory proteins such as Munc18 and Rab GTPases will be discussed. New findings on alterations of particular SNARE proteins upon glucose homeostasis are described in terms of potential for pharmacological targets for intervention.

In 1957 Lacy and colleagues published immunofluorescent and later electron microscopy images of islet beta cells filled with darkened organelles, or insulin granules. These granules were later shown to fuse with the plasma membrane to release insulin.[1-3] Analogously, in 1990 the localization of the insulin-responsive glucose transporter 'GLUT4' to intracellular vesicles of adipocytes was revealed, and these GLUT4-storage vesicles were later shown to fuse with the plasma membrane upon stimulation with insulin, to ultimately facilitate glucose disposal.[4,5] However, it was not until 1993 that the molecular basis for vesicle fusion and secretion, termed 'vesicle exocytosis', was elucidated by the discovery of SNARE proteins (soluble N-ethylmaleimide sensitive factor attachment protein receptor).[6,7]

### Vesicle Exocytosis

Vesicle exocytosis entails the pairing of a vesicle associated membrane protein (v)-SNARE (VAMP) with a binary cognate receptor complex at the target membrane composed of SNAP-25/ 23 and Syntaxin proteins (t-SNAREs)[8-14] to form the SNARE core complex (Fig. 1). The v-SNARE VAMP2 (also called synaptobrevin) is an 18 kDa protein oriented with its N-terminus towards the cytoplasm and the carboxyl-terminus spanning the membrane facing the vesicle

*Debbie C. Thurmond—Department of Biochemistry and Molecular Biology Center for Diabetes Research, Indiana University School of Medicine, Indianapolis, Indiana, U.S.A.
Email: dthurmon@iupui.edu

*Mechanisms of Insulin Action*, edited by Alan R. Saltiel and Jeffrey E. Pessin.
©2007 Landes Bioscience and Springer Science+Business Media.

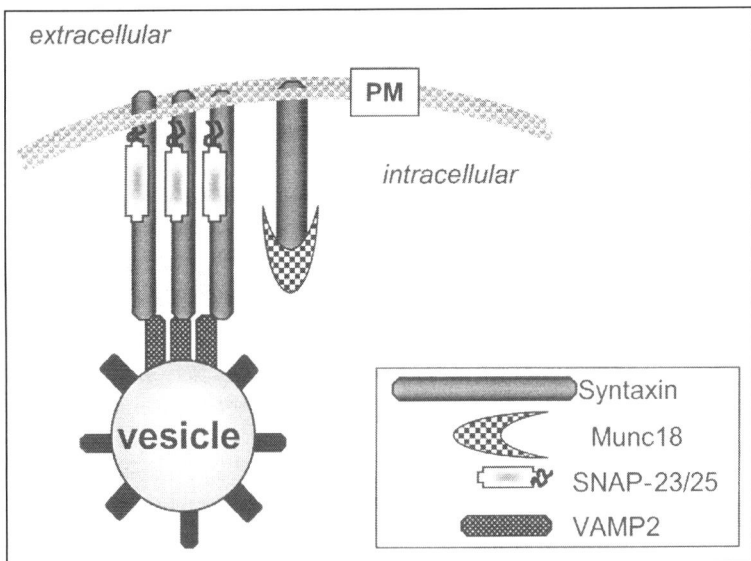

Figure 1. Schematic model of SNARE core complexes and Munc18-Syntaxin complexes important in insulin-stimulated GLUT4 translocation and insulin secretion. Syn4 is integrated into the plasma membrane (PM, depicted as a variegated vertical line); SNAP-23/25 is associated with the PM via palmitoylated moieties; VAMP2 is integrated into the vesicle membrane; Munc18 associates with the 'closed' form of Syntaxin at the PM.

lumen.[15-17] The t-SNARE Syntaxin is a 35 kDa protein oriented with its amino terminal domain towards the cytoplasm and carboxyl-terminus spanning the plasma membrane facing the extracellular matrix.[8,18] The other t-SNARE, SNAP-23/25 (a 23 or 25 kDa synaptosome associated protein) is tethered to the target membrane via palmitoylated cysteine residues.[19] A combination of electron microscopic, spectroscopic and X-ray crystallographic evidence shows that these SNAREs form a rod shaped complex that is a coiled-coil of four helices: one from VAMP, one from Syntaxin, and the remaining two from SNAP-25.[20-23] Upon forming the core SNARE complex the syntaxin protein increases it helical content whereas VAMP and SNAP-25 convert from unstructured forms into α helices.[20] These conformational changes result in the formation of the heterotrimeric complex that is extremely stable, melting only at temperatures greater than 95°C.[6,10,12]

Multiple isoforms of each SNARE protein were found to exist and it is hypothesized that by the particular pairing and compartmentalization of these proteins, specificity of vesicle targeting could be achieved.[24-32] On the basis of these studies in neuroendocrine cells, investigators from diabetes-related fields discovered that SNARE protein core complexes were also responsible for regulated secretion of insulin from islet beta cells in response to increased blood glucose,[33-40] as well as facilitating the downstream action of insulin on peripheral glucose disposal via the translocation to and integration of intracellular glucose transporter (GLUT4) vesicles into the plasma membrane of adipocytes and skeletal muscle transverse tubules/sarcolemma.[41-50]

## Regulation of Exocytosis by Munc18 Proteins

Concurrent with the discovery of SNARE proteins came evidence from yeast genetics suggesting that other proteins also participate in exocytosis and play a role in the regulation of the SNARE complex,[51] leading to the discovery of Sec1 secretory proteins. Yeast Sec1 was found to interact directly with the t-SNARE Syntaxin leading to the quick identification of Sec1 homologues in *C. elegans* (unc18), *D. melanogaster* (Rop)[52-61] and mammals.[53,54,59,62,63]

**Table 1. Comparison of SNARE protein composition utilized for GLUT4 translocation and insulin exocytosis**

| Protein | GLUT4 Translocation | Functional? | Insulin Exocytosis | Functional? |
|---|---|---|---|---|
| Syntaxin 1 | Absent | N.D. (49,50) | Present | Yes (34,39,40, 158,164,165) |
| Syntaxin 2 | Present | No (46,50,71) | Present | No (34,164) |
| Syntaxin 3 | Present | No (46,50) | Present | Yes (34,164) |
| Syntaxin 4 | Present | Yes (44,46,50,70,72,109,120) | Present | Yes (34,162-164) |
| VAMP2 | Present | Yes (42,43,45,46,48,49) | Present | Yes (34,37,39) |
| VAMP3 | Present | Yes/No (42,46,48,49,110) | Present | Yes/No (34,37) |
| SNAP-23 | Present | Yes (41,44,47) | Present | Yes (160) |
| SNAP-25 | Absent | N.D. (49,72) | Present | Yes (38) |
| Munc18a | Absent | N.D. (70) | Present | Yes (172) |
| Munc18b | Present | No (64,71,73) | Present | No (172) |
| Munc18c | Present | Yes (41,64,70,73,84,116-119) | Present | Yes (39,117,118) |

N.D.: not determined

Three homologues were identified in mammalian plasma membranes, and named Munc18 (for Mammalian unc18) proteins. Munc18 proteins are ~66-68 kDa in size and are soluble factors with no transmembrane domain,[54] although they are found localized to the plasma membrane through their high-affinity for binding to their cognate Syntaxin.[64] Munc18-1 (also called Munc18a/n-Sec1/rbSec1) was demonstrated to interact with Syntaxin 1 in a manner mutually exclusive of the other SNARE core complex proteins,[52,54,65] as depicted as a separate binary complex in Figure 1. By 1995 Munc18b and Munc18c were identified but found to be ubiquitously expressed, unlike Munc18a which was expressed only in neurons and islet cells.[56,66-71] In addition, it was shown that Munc18a and Munc18b shared binding preferences for Syntaxin isoforms 1-3, while only Munc18c bound Syntaxin 4 (Syn4) with high affinity.[67,68,70,71] Then it was in 1997 that Syntaxin 4 was demonstrated to be the functional isoform in insulin-stimulated GLUT4 vesicle translocation,[46,72] and in 1998 we and others found that Munc18c over-expression inhibited this process[64,73] (Table 1). Collectively, the Sec1 and Munc18 protein family are now referred to as 'SM' proteins (Sec1 and Munc18).

The evidence cited above strongly suggests that Munc18 proteins are essential regulators of SNARE mediated exocytosis. The most recent crystallographic and NMR structural analyses support and extend this concept, showing that Munc18-1 holds Syntaxin 1 in a "closed" conformational state in a 1:1 Munc18:Syntaxin molar ratio and unable to interact with VAMP2. Munc18-1 is proposed to orchestrate the conversion of Syntaxin 1 to the "open" conformational state to facilitate the interaction between Syntaxin 1, VAMP2 and SNAP-25.[54,74-80] Structural studies have led to the general conclusion that Munc18 proteins share a similar overall structure in which a small folded N-terminal domain mediates the interaction with Syntaxins,[81,82] whereas the remainder of the C-terminal domain carries out the poorly understood effector function that appears essential for fusion. Dulubova and colleagues have furthermore speculated that a particular loop2/3 domain of Munc18 proteins may be critical for this effector function,[81,83] which is consistent with studies showing that an inhibitory peptide directed at this region or a single point mutation within it alters Syntaxin 4-Munc18c interaction in 3T3L1 adipocytes.[84,85]

One common theme in studying SNARE protein complexes and accessory proteins is that over-expression of one member of the complex can alter that stoichiometry which is optimal

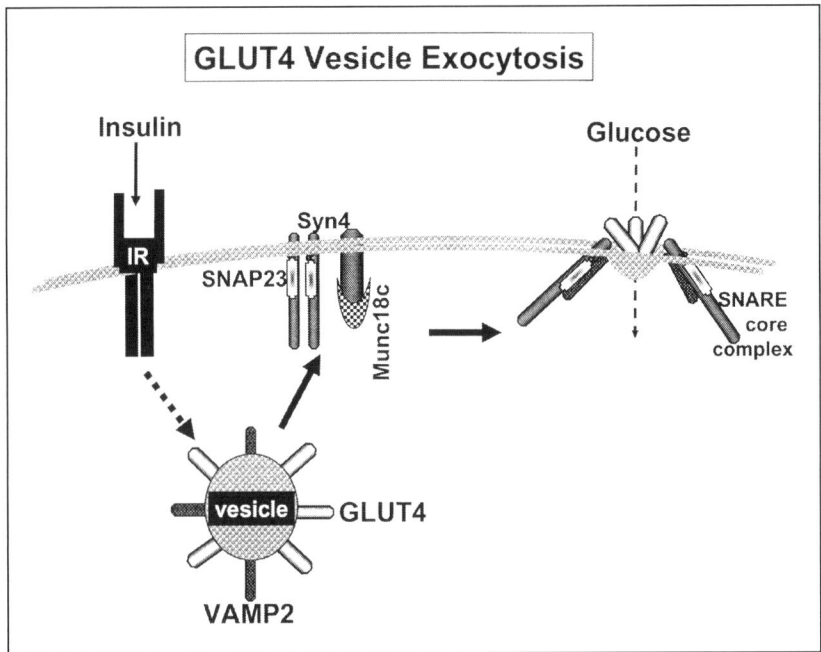

Figure 2. Schematic model depicting the cascade of events occurring in the process of insulin-stimulated GLUT4 translocation. Insulin binds to the insulin receptor (IR) present on the cell surface of insulin-responsive adipose and skeletal muscle cells, triggering a cascade of signaling events leading to the movement of GLUT4 vesicles from an intracellular location to the plasma membrane. At the plasma membrane the VAMP2-bound vesicles dock and tether to t-SNARE proteins Syntaxin 4 (Syn4) and SNAP-23 which leads to eventual vesicle fusion. This results in the intercalation of GLUT4 protein at the cell surface which facilitates the uptake of glucose from the extracellular milieu.

for the exocytotic event, resulting in loss/inhibition of function. Throughout the chapter this will become evident, as most initial studies of SNARE and SNARE accessory proteins were conducted using over-expression. Now as more data on SNARE knockout mice and siRNA-depletion evolves we will be able to disseminate the essentiality of these proteins in the various exocytotic processes.

## Insulin Action: GLUT4 Vesicle Translocation

The insulin-responsive glucose transporter GLUT4 is predominantly expressed in striated muscle and adipose tissue and is responsible for the majority of insulin-stimulated glucose uptake.[86] In the basal noninsulin stimulated state, GLUT4 localizes to tubulovesicular elements and small intracellular vesicles throughout the cell cytoplasm.[87,88] Upon stimulation with insulin these GLUT4-containing compartments undergo a series of regulated steps leading to their eventual fusion with the plasma membrane.[86,89-95] This ultimately results in a large increase in the number of functional glucose transporters on the cell surface (a process termed translocation), which accounts for nearly all of the insulin-stimulated glucose uptake.

The insulin-stimulated translocation of GLUT4-containing vesicles is a complex multi-step process (Fig. 2). Several specific molecular events and signaling molecules have been identified. Initially, insulin binding to the insulin receptor activates the intrinsic protein kinase of the receptor β subunit resulting in its autophosphorylation and tyrosine phosphorylation of several proteins, most notably the IRS family of insulin receptor substrate proteins.[96] The tyrosine phosphorylation of IRS results in the association, activation and targeting of the phosphatidylinositol kinase (PI3-kinase).[97-100] The active PI3-kinase can then generate phosphatidylinositol-3,4,5-trisphosphate, which is necessary for the stimulation of both protein kinase B (PKB/Akt) and atypical protein kinase C isoforms through activation of the phosphoinositide-dependent protein kinases (PDK1 and PDK2) and/or through engagement of the PKB pleckstrin homology (PH) domain.[101-103] In 2002, the Akt substrate AS160 was identified.[104] AS160, a Rab GTPase-activating protein, becomes phosphorylated in response to insulin and is required for insulin-stimulated GLUT4 translocation.[105-107] Unlike neuronal vesicles, the GLUT4 vesicles are mostly intracellularly localized and physically separated from the plasma membrane, and following insulin receptor signaling, these intracellular GLUT4 storage vesicles traffic towards the plasma membrane, afterwards becoming docked just below the plasma membrane. Docked vesicles undergo priming mediated by binding between syntaxin 4, SNAP-23 and VAMP2 and ultimately the primed vesicles are functionally incorporated into the plasma membrane through the physical mixing of the two lipid bilayers by membrane fusion events, leaving the GLUT4 protein positioned to facilitate the uptake of glucose.[92,94]

Substantial advances have been made in our understanding of the priming and fusion steps of GLUT4 translocation. GLUT4 vesicles copurify with the VAMP2 v-SNARE, and specific proteolytic cleavage of VAMP2, expression of a dominant-negative VAMP2 mutant or inhibitory peptides all impair insulin-stimulated GLUT4 translocation.[42,43,45,46,48,49] Although VAMP2 (-/-) knockout mice have been generated, no reports on their glucose homeostasis are available at this time as they fail to thrive upon birth.[108] Using dominant-interfering mutants and inhibitory peptides, Syntaxin 4 and SNAP-23 have been identified as the t-SNARE proteins required for insulin-stimulated GLUT4 translocation.[41,44,46,47,72] Syntaxin 4 (-/+) mice show deficiencies in skeletal muscle GLUT4 translocation and glucose uptake, although adipocyte glucose uptake is normal.[109] While VAMP3 was initially found to be part of this process in 3T3L1 adipocytes, it was later determined to be nonessential since VAMP3 (-/-) knockout mice showed no defects in skeletal muscle or adipocyte insulin-stimulated GLUT4 translocation or alterations in whole-body glucose homeostasis.[110] In all, these data provide compelling evidence that VAMP2 functions by directing the association of the GLUT4 containing vesicles with Syntaxin4 and SNAP-23 at the plasma membrane.

## *Impact of Syntaxin4 Binding Proteins Upon GLUT4 Vesicle Translocation*

### Munc18c

We and others have demonstrated that the Munc18c-Syntaxin4 complex is functional in insulin-stimulated GLUT4 translocation in 3T3L1 adipocytes. Over-expression of Munc18c resulted in the inhibition of insulin-stimulated glucose uptake and GLUT4 translocation, while having no effect upon GLUT1 translocation or the trafficking of mannose-6-phosphate receptor or transferrin receptor.[64,73,111] Furthermore, competition for the association of endogenous Munc18c-Syntaxin4 complexes using an interfering peptide of Munc18c demonstrated a specific requirement of this complex for insulin-stimulated GLUT4 vesicle fusion,[84] but was not necessary for proximal trafficking steps.[85] This functional profile is consistent with the role of the homologous Sec1 protein in yeast.[112,113]

Post-translational modification of Munc18c or alteration of Munc18c abundance has been related to insulin resistance. For example, glucosamine-induced insulin resistance in adipocytes has recently been correlated with increased O-linked glycosylation of Munc18c and a reduced association of VAMP2 with Syntaxin4.[114,115] In vivo, localized over-expression of Munc18c in skeletal muscle via adenoviral particle injection markedly impaired insulin-stimulated GLUT4 translocation to sarcolemmal and transverse tubule membranes.[116] Moreover, over-expression of a tetracycline-repressible (tet-off) CMV-driven Munc18c transgene in skeletal muscle, adipose tissue and pancreas of mice significantly impaired whole body glucose tolerance, skeletal muscle glucose uptake and GLUT4 translocation.[117] Most recently we have shown that Munc18c (-/+) knockout mice are insulin resistant and have severely impaired insulin-stimulated GLUT4 translocation in skeletal muscle.[118] This contrasts however with

*Table 2. Transgenic and knockout strains of mice with genetic alterations in SNARE proteins*

| Mouse Strain | Glucose Homeostasis | Glucose Uptake | Insulin Secretion | References |
|---|---|---|---|---|
| Syntaxin 1A Tg | Insulin resistant | N.D. | Decreased | 158 |
| Syntaxin 4 Tg | Insulin sensitive | Increased | No effect | 120 |
| Munc18c Tg | Insulin resistant | Decreased | Decreased | 117 |
| Syntaxin 4 (-/+) | Insulin resistant | Decreased | Decreased | 109,163 |
| VAMP3 (-/-) | No effect | No effect | N.D. | 110 |
| VAMP2 (-/-) | N.D. | N.D. | N.D. | 108 |
| Munc18c (-/+) | Insulin resistant | Decreased | Decreased | 118 |
| Munc18c (-/-) | N.D. | Increased | N.D. | 119 |
| Munc13-1 (-/-) | N.D. | N.D. | N.D. | 190 |

N.D.: not determined

the response of adipocytes differentiated from mouse embryonic fibroblasts (MEF) derived from an independent line of Munc18c (-/-) knockout mice: the MEF-derived adipocytes showed enhanced appearance of GLUT4 at the cell surface in response to insulin.[119] These seemingly conflicting data are actually quite consistent with the emerging data from other genetically-altered SNARE mouse models of insulin resistance, showing defects in GLUT4 translocation and glucose uptake in skeletal muscle but not in adipocytes (Table 2).[109,117,118,120] An alternate explanation for these differences might be that different genes encoding Munc18c were targeted, a concept which is supported by differences in exon-intron structure depicted in the two reports.[117,118]

## Rabs

In yeast, the SM protein Sly1p interacts with the small GTPase protein Ypt1p.[83] However, an analogous Munc18c interacting protein has remained elusive. Members of the small GTPase Rab protein family play crucial regulatory roles in vesicular trafficking. In neurotransmitter release, Rab3A is involved in regulating the efficiency of vesicle priming and membrane fusion.[121] Also, in the yeast trafficking pathway from Golgi to vacuole, the Rab effector protein Vac1 has been shown to interact with the Sec1 family member Vps45.[122] Vps45 binds to the yeast syntaxin Pep12,[123] connecting the functions of a Rab, a soluble Rab effector and the target-membrane-localized Sec1/syntaxin family. To date, there are more than 30 different Rab GTPases, most with distinct cellular localizations and individual roles in secretory and exocytic pathways. Rabs cycle between an active GTP-bound form and the inactive GDP-bound form such that vesicle-associated Rabs are found in the GTP-bound state and following membrane fusion are hydrolyzed to the inactive form.[124] Rab4 has been shown to be associated with GLUT4-containing vesicles and implicated in the insulin action on glucose transport in rat adipocytes.[125] Upon insulin stimulation Rab4 is depleted from the GLUT4 containing microsomal fraction of adipocytes and is redistributed to the cytosolic fraction, where it specifically interacts with the GDP-dissociation inhibitor GDI-1.[126-128] In addition, introduction of a carboxyl-terminal peptide, expression of a carboxyl-terminal Rab4 deletion mutant, or increased expression of wild type Rab4 all inhibit insulin-stimulated GLUT4 translocation,[129-131] consistent with a vital functional role for Rab4 in GLUT4 trafficking.

## Tomosyn and Synip

In 1998 the Syntaxin 1 binding protein Tomosyn was identified in neuronal cells and found to dissociate Munc18a from Syntaxin 1.[132] Tomosyn has a VAMP2-like region that is necessary for its interaction with Syntaxin 1 and for complex formation with Syntaxin 1 and

SNAP-25.[133,134] Tomosyn has further been shown to exist as multiple isoforms sharing a conserved structure.[134,135] Since tomosyn over-expression results in down-regulated exocytosis,[132,133] and it dissociates the Munc18a-Syntaxin 1 complex known to be important for exocytosis, tomosyn has been characterized as a negative-regulator of neuronal exocytosis. More specifically, it is proposed to inhibit the priming step.[136] In an analogous manner, the b-tomosyn isoform is found to bind to Syntaxin 4 through a VAMP2-like domain, form a ternary complex with Syntaxin 4 and SNAP-23, and inhibit the priming of GLUT4 vesicles in 3T3L1 adipocytes when over-expressed.[137] Thus tomosyn and Munc18c appear to function with Syntaxin 4 at the priming step of GLUT4 vesicle fusion, although the details of their roles remain unclear.

The Syntaxin 4 binding protein Synip was isolated and characterized in 1999 in 3T3L1 adipocytes, and the syntaxin 4-synip complex was found to be dissociated upon stimulation with insulin.[138] Moreover, deletion analyses suggested that the amino terminal domain may provide a regulatory role in modulating the interaction of the carboxyl terminal domain of synip with syntaxin 4.[138] More recently, synip was shown to contain an unusual dual Akt/PKB consensus phosphorylation motif, where serine 99 is a substrate for Akt2 but not Akt1 or Akt3.[139] The expression of a synip-ser99 mutant inhibited insulin-stimulated GLUT4 vesicle docking/fusion, concurrent with its inability to undergo insulin-stimulated dissociation from syntaxin 4. It is thus postulated that insulin activation of Akt2 in adipocytes regulates the docking/fusion step of GLUT4 vesicle translocation through regulation of synip phosphorylation.

## Exocyst Complex

In the yeast *S. cerevisiae* a complex of seven proteins were found to function together at sites on the plasma membrane to promote exocytosis, and were so named collectively the Exocyst.[140] Mammalian homologues were isolated shortly thereafter and one Exocyst protein, rSec6, was specifically immunolocalized to sites of granule exocytosis in neurosecretory PC12 cell processes.[141] Another exocyst protein, Exo70, has been identified in 3T3L1 adipocytes and found to associate with other exocyst proteins Sec6 and Sec8.[142] Exo70 translocates to the plasma membrane in response to insulin through the activation of TC10, which is suggested to function in the insulin signaling cascade as part of a PI-3-kinase independent pathway convergent at stimulation of GLUT4 trafficking.[94,143] Most recently, Sec6 and Sec8 have also been shown to redistribute to the plasma membrane in response to insulin in 3T3L1 adipocytes,[144] altogether supporting a model whereby the exocyst complex functions in insulin-stimulated GLUT4 vesicle targeting to the docking/fusion sites at the plasma membrane.

## Insulin Exocytosis in Pancreatic Beta Cells

Greater than 99% of insulin secreted from the pancreatic beta cell proceeds via a regulated secretory pathway.[145] Insulin is contained within dense core granules located inside the pancreatic beta cell. These granules form in the trans-golgi network (TGN), at which time they are loaded with proinsulin. In these immature beta cell granules the proinsulin is cleaved to insulin and C-peptide. Glucose stimulates these granules to move from their intracellular location to the cell surface, and also from more interior pools to the readily releasable pools located just beneath the cell surface. The amount of insulin contained in beta cells is very constant, as the balance between insulin secretion, proinsulin biosynthesis and insulin granule degradation is carefully maintained.[146,147] Based upon the amount of stored insulin in normal beta cells, it is generally accepted that at least for the initial release of insulin, insulin content is not rate-limiting.[148]

As depicted in Figure 3, when extracellular glucose rises, the glucose transporters (GLUT2) at the beta cell surface facilitate the uptake of glucose into the cell, resulting in an increase in ATP through metabolism of the glucose. This ATP upsets the basal ATP/ADP ratio causing the $K_{ATP}$-channels to close and cell depolarization,[149,150] leading to the opening of $Ca^{2+}$-channels, overall increasing cytoplasmic $Ca^{2+}$ concentration.[151] In response to the rise in intracellular $Ca^{2+}$, some docked and primed insulin storage granules fuse and release insulin, while storage granules traffic to join the readily releasable pool at the cell surface (see review, ref. 152). Thus,

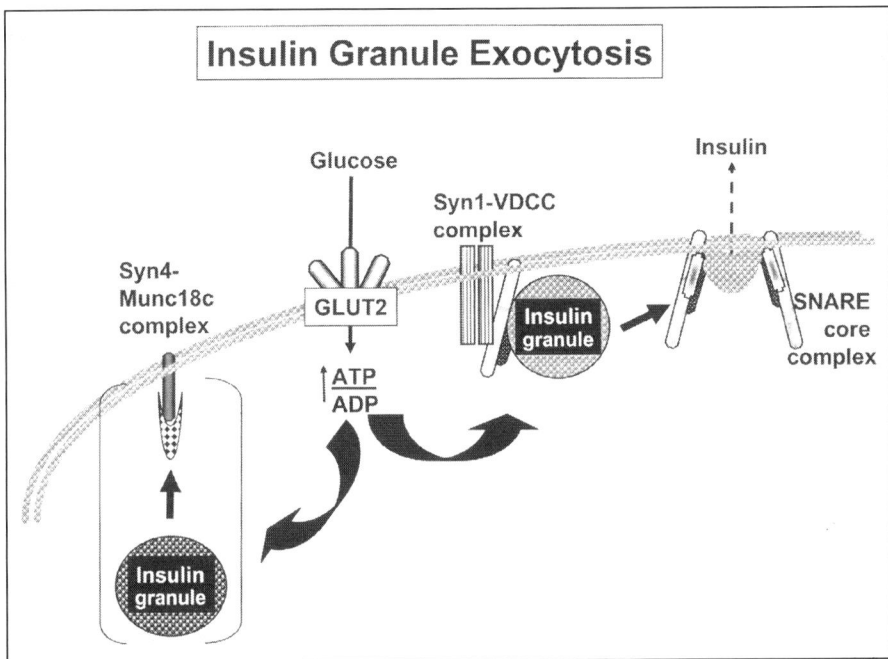

Figure 3. Model depicting the hypothetical pathways of glucose-stimulated insulin release from insulin secretory granules in islet beta cells regulated by Syntaxin 1 (Syn1) and Syntaxin 4 (Syn4) based SNARE complexes. Primed granules are often found colocalized with the voltage-dependent calcium channels (VDCC) and Syn1-SNAP25 t-SNARE complexes at the plasma membrane. This pool of granules is thought to account for the initial/first phase of insulin secretion. Syntaxin 4-Munc18c complexes are also localized to the plasma membrane and mediate both first and second phase insulin secretion in islets isolated from Syntaxin 4 and Munc18c (-/+) knockout mouse models.

insulin secretion in islets is biphasic,[153,154] with the first phase resulting from the release of insulin granules from an immediate releasable pool preprimed at the plasma membrane.[155] First phase secretion peaks within 5-10 min, and without returning to baseline insulin secretion gradually rises over time after this if glucose is present (representing a second phase of secretion).[153] The second phase only occurs with nutrient stimulation[156] and is currently thought to represent secretion from those granules which are mobilized to refill the readily and immediate releasable pools of granules.[157]

## SNARE Proteins in Insulin Secretion

To date, insulin secretion is known to involve the same SNARE isoforms as those utilized in synaptic vesicle exocytosis and neurotransmitter release, namely Syntaxin 1, VAMP2, and SNAP-25.[33-40] Similar to neurosecretory cell over-expression of Syntaxin 1, over-expression of Syntaxin 1 in insulin-secreting beta cells as well as in transgenic mice results in decreased insulin release.[158,159] Cleavage of VAMP2 by botulinum toxin severely reduces insulin release from islets.[37,38] Cleavage of SNAP-25 only reduces insulin secretion by 50%, but can be substituted for by the abundant SNAP-23 isoform in Syntaxin 1 based SNARE complexes.[38,160] Cleavage of Syntaxin 1 inhibited 95% of $K^+$-induced insulin secretion, but only 25% of glucose induced secretion was inhibited.[40,161] Taken together, these data suggest that although the Syntaxin 1-based SNARE complexes are responsible for $K^+$ induced secretion, glucose-stimulated secretion likely involves a second VAMP2-dependent SNARE complex.

In actuality, pancreatic beta cells contain numerous other Syntaxin isoforms (Table 1), suggesting that additional SNARE complexes participate in the regulation of insulin secretion.[34,39,160] Recently we and others have determined that Syntaxin 4 plays a positive functional role in glucose-stimulated insulin secretion. This was initially shown indirectly in over-expression studies whereby expression of a dominant-negative form of Syntaxin 4 resulted in inhibition of insulin secretion in βHC9 cultured beta cells.[162] However this has now been substantiated by detection of a 50% reduction of insulin secretion in islets isolated from Syntaxin 4 (-/+) knockout mice.[163] Furthermore, we have shown that Syntaxin 4 is specifically required by the beta cells of the islet by using siRNA-mediated depletion and immunodepletion of Syntaxin 4 from MIN6 beta cells in culture.[163] While over-expression studies have provided some evidence for a role for Syntaxin 3 but not Syntaxin 2 in beta cells,[164] further studies using depletion of these Syntaxin isoforms will be required to determine their essential nature to insulin secretion.

In addition to isoform specificity, the abundance of SNARE proteins in pancreatic beta cells has been shown to impinge upon insulin secretion in two different rodent models of diabetes. For example, in islets isolated from diabetic GK rats, a nonobese rodent model of Type II diabetes, reduced levels of SNAP-25, Syntaxin 1, Syntaxin 2 and VAMP2 SNARE isoform proteins were correlated with impaired insulin secretion, and exogenous replacement of these particular SNARE proteins improved insulin secretion.[165] The Type II diabetic Zucker fa/fa rat also has reduced levels of these SNARE proteins in beta cells.[166] These findings illustrate the importance of SNARE protein stoichiometry to overall islet function.

## SNARE Binding Proteins in Insulin Secretion

### Calcium and Potassium Channels

The insulin granules found primed at the plasma membrane release insulin rapidly in response to stimuli, and thus this group of granules is referred to as the 'readily releasable pool'. It was found that granules in this pool are tightly associated with the voltage-dependent $\alpha_{1C}$ $Ca^{2+}$ channels (VDCC)[155] (Fig. 3). This occurs through an interaction between a region in $\alpha_{1C}$ subunit of the L-type channel, corresponding to the 'synprint' site which binds to Syntaxin 1 and SNAP-25.[167] Expression of this peptide inhibits exocytosis evoked by voltage clamp depolarization, suggesting that this interaction facilitated first phase insulin secretion. Moreover, over-expression of Syntaxin 1A in beta cell lines inhibits L-type channel activity.[164] In addition, Syntaxin 1A also binds to the Kv2.1 channel, the dominant membrane-repolarizing voltage gated $K^+$ channel in pancreatic beta cells,[168] and the open form of Syntaxin 1A was found to inhibit the channel, which was proposed to limit $K^+$ efflux during exocytosis to optimize insulin release.[169] SNAP-25 also associates with the voltage-gated $\alpha_{1C}$ $Ca^{2+}$ channels and Kv2.1 channels.[170,171] Thus, Syntaxin 1A and SNAP-25 mediate secretion by regulating calcium entry and membrane potential though their associations with calcium and potassium channels in pancreatic beta cells.

### Munc18a and Mint1

The Syntaxin 1 binding protein Munc18a (n-Sec1/Munc18-1/rbSec1) was initially immunolocalized in rat pancreatic islet cells and shown to interact with Syntaxin 1 in extracts of HIT-T15 cultured beta cells.[172] Functionally Munc18a is described as a negative regulator of exocytosis in the beta cell, as evidenced by an increase in insulin secretion in streptolysin-O treated HIT-T15 cells administered antibody to Munc18 or a Munc18 peptide.[172] Munc18b is also expressed in islet cells, although its function is unknown. Mint1, a Munc18a interacting protein, is expressed in rat pancreatic islets, localized primarily to the cell periphery.[173] In addition to Mint1, RT-PCR analysis of rat islets and two culture beta cells lines (HIT-T15 and RINm5F) showed the presence of mRNA for Mint2 and Mint3. Mint2 but not Mint3 can bind to Munc18a or Munc18b (Munc18-2),[174] thus leaving the importance of Mint3 expression in beta cells unknown.

## Munc18c

Munc18c has been immunolocalized in rat islets and multiple cultured insulin secreting cell lines.[39] Munc18c transgenic mice, which have over-expression of Munc18c in pancreas, showed inhibited insulin secretion in isolated islets,[117] indicating for the first time, that Munc18c played a role in the islet beta cell (Table 2). However, just as in GLUT4 translocation, while over-expression of Munc18c resulted in inhibition, so did depletion of Munc18c in the skeletal muscle of the Munc18c (-/+) knockout mouse.[118] Consistent with this, we have now found that islets isolated from Munc18c (-/+) knockout mice have a 50% reduction in glucose-stimulated insulin secretion.[118] This requirement for Munc18c is further supported by siRNA-mediated depletion and antibody immunodepletion of endogenous Munc18c in MIN6 beta cells (D.C. Thurmond, unpublished results). Thus, since the islet beta cells were thought previously to use the same t-SNAREs as neurons, namely Syntaxin 1 and Munc18a, our data now suggests that Syntaxin 4 and its binding partner Munc18c regulates insulin secretion as well.

## Rab Proteins and Effectors

Rab3 has been suggested to participate in the dissociation of the Munc18a-Syntaxin 1 complex.[175] In fact, the number of synaptic vesicles that fuse with the plasma membrane in response to stimuli is increased in Rab3A (-/-) knockout mice.[176] Rab3 was originally identified in homogenates of rat pancreatic islets and in HIT-T15 and RINm5F insulin secreting cell lines,[177,178] and later was immunolocalized to the insulin secretory granules.[179] Munc18a is also associated with Rab3 in pancreatic beta cells,[180] and over-expression of dominant active mutants of Rab3 decreases exocytosis.[179,181] Numerous Rab3 effectors have since been identified in pancreatic beta cells, such as RIM, granuphilin and Noc2. RIM was localized on the plasma membrane of rat islet beta cells, INS-1E and HIT-T15 beta cell lines.[182] Competitive over-expression of the Rab3 binding domain of RIM resulted in enhanced glucose-stimulated insulin secretion, implicating RIM in the regulation of insulin exocytosis. Granuphilin interacts with Rab3 in its GTP-bound form and Munc18a, and over-expression of granuphilin inhibits stimulated insulin secretion.[180,183] Granuphilin also binds to Rab27 on the insulin granule,[184] which forms a heterotrimeric complex with Syntaxin 1 at the plasma membrane, effectively tethering granules to the fusion sites at the plasma membrane.[185] Noc2 was cloned and characterized as a Rab3 effector protein expressed in the MIN6 insulin-secreting cell line.[186] RNA interference was used to silence Noc2 in INS-1E beta cells, which strongly impaired secretagogue-induced insulin secretion.[187] Furthermore, the defect was particularly detrimental during the sustained release phase, suggesting that Noc2 may be involved in the recruitment of secretory granules. Noc2 also binds to Munc13-1,[187] which has been implicated in insulin granule priming[188] and is also a RIM binding protein[189] and is functional in synaptic vesicle priming. Munc13-1 (-/-) knockout mice have been generated but have yet to be examined for defects in insulin secretion.[190] This is consistent with data showing Noc2 interaction with the cytoskeletal-associated protein zyxin, and thus Noc2 is postulated to regulate exocytosis by interacting with the actin cytoskeleton.[186]

## *Insulin Granule Trafficking and Actin Remodeling*

Filamentous actin (F-actin) is known to be important to the process of insulin secretion.[191-193] Our recent studies in MIN6 beta cells and isolated rat islets demonstrate that glucose transiently modulates cortical actin organization and disrupts the interaction of F-actin with the t-SNARE complex at the plasma membrane to facilitate glucose-stimulated insulin secretion.[194] Further evidence indicates that the cortical actin reorganization induced by glucose occurs at a proximal step in the stimulus-secretion pathway, perhaps at a step concurrent with the $K_{ATP}$ channel closure.[195] However, our understanding of the coupling of SNARE mediated exocytosis to F-actin reorganization remains incomplete.

Multiple lines of evidence suggest that the F-actin effector protein Cdc42, a Rho family small GTPase, is a downstream target of glucose signaling in beta cells. For example, we have recently shown that stimulation with glucose results in alterations in the cycling of Cdc42 between the GTP-bound activated and GDP-bound inactivated states, which correlate with a transient depolymerization of cortical F-actin.[195] Cdc42 has also been demonstrated to colocalize with VAMP2-containing insulin secretory granules in pancreatic beta cells.[196,197] Moreover, Cdc42 has been shown to interact indirectly with the t-SNARE Syntaxin 1, linking Cdc42 and the actin cytoskeleton to the plasma membrane exocytotic machinery.[198] We have recently demonstrated that VAMP2 bridged the interaction between Cdc42 and Syntaxin 1 and that the interactions were functionally important for SNARE dependent insulin exocytosis.[199]

## Perspectives

Diabetes now strikes 6.2% of the U.S. population, having risen to afflict approximately 1 in every 17 Americans, and is expected to continue to rise. In addition, although it is well-recognized that frank diabetes develops over years of cellular dysfunction in insulin secretion and insulin action, clear markers of this predisposition have remained elusive. Significant progress has been made in defining signaling cascades leading to the secretion of insulin from the islets and uptake of glucose into skeletal muscle and adipose tissues, however each signaling cascade is rate-limited by the distal steps of vesicle exocytosis.[85,148,200] The functional regulation of Syntaxins by the Munc18 proteins is thought to be crucial in these distal steps, although the details remain unclear. There are also numerous animal models of insulin resistance and diabetes which exhibit alterations in SNARE protein abundance with which therapeutic agents are tested.[165,166,201,202] However, given the evidence showing that Syntaxin 4-based SNARE core complexes and Syntaxin 4-Munc18c complexes participate in both insulin secretion and insulin action, therapies which target or even affect secondarily the SNARE protein abundances will have consequences in both insulin secretion and insulin action. This could be beneficial if the regulation occurred in parallel, but hazardous if the complex regulation were opposing (i.e., upregulation of Syntaxin 4 resulted in decreased glucose uptake but increased insulin secretion). Lack of such knowledge represents an important gap in our understanding of the etiology of insulin resistance and diabetes. Progress in therapeutic treatment of diabetes will need to take into account the mechanism(s) underlying the regulation of insulin secretion in islets cells and glucose uptake in adipocytes by these SNARE complexes and SNARE accessory proteins, and relate these tissue-specific effects to the overall control of glucose homeostasis.

## *Acknowledgements*

I would like to thank Dr. Herbert Gaisano for personal communications regarding his Syntaxin 1 transgenic mice, and Angela Nevins for her critical review of the manuscript. This work was supported by research grants from the National Institute of Health (DK067912) and the American Diabetes Association (1-03-CD-10).

## References

1. Lacy PE. Electron microscopy of the beta cell of the pancreas. Am J Med 1961; 31:851.
2. Lacy PE, Davies J. Preliminary studies on the demonstration of insulin in the islets by the fluorescent antibody technic. Diabetes 1957; 6:354.
3. Orci L, Amherdt M, Malaisse-Lagae F et al. Insulin release by emiocytosis: Demonstration with freeze-etching technique. Science 1973; 179(68):82-84.
4. Clancy BM, Czech MP. Hexose transport stimulation and membrane redistribution of glucose transporter isoforms in response to cholera toxin, dibutyryl cyclic AMP, and insulin in 3T3-L1 adipocytes. J Biol Chem 1990; 265(21):12434-12443.
5. Holman GD, Kozka IJ, Clark AE et al. Cell surface labeling of glucose transporter isoform GLUT4 by bis-mannose photolabel. Correlation with stimulation of glucose transport in rat adipose cells by insulin and phorbol ester. J Biol Chem 1990; 265:18172-18179.
6. Sollner T, Bennett MK, Whiteheart SW et al. A protein assembly-disassembly pathway in vitro that may correspond to sequential steps of synaptic vesicle docking, activation, and fusion. Cell 1993; 75(3):409-418.

7. Sollner T, Whiteheart SW, Brunner M et al. SNAP receptors implicated in vesicle targeting and fusion. Nature 1993; 362(6418):318-324.
8. Calakos N, Bennett MK, Peterson KE. Protein-protein interactions contributing to the specificity of intracellular vesicular trafficking. Science 1994; 263:1146-1149.
9. Chapman E, An S, Barton N et al. SNAP-25, a t-SNARE which binds to both syntaxin and synaptobrevin via domains that may form coiled coils. J Biol Chem 1994; 269(44):27427-27432.
10. Fasshauer D, Otto H, Eliason WK et al. Structural changes are associated with soluble N-ethylmaleimide- Sensitive fusion protein attachment protein receptor complex formation. J Biol Chem 1997; 272(44):28036-28041.
11. Hayashi T, McMahon H, Yamasaki S et al. Synaptic vesicle membrane fusion complex: Action of clostridial neurotoxins on assembly. EMBO J 1994; 13:5051-5061.
12. Hayashi T, Yamasaki S, Nauenburg S et al. Disassembly of the reconstituted synaptic vesicle membrane fusion complex in vitro. EMBO J 1995; 14(10):2317-2325.
13. Kee Y, Lin RC, Hsu SC et al. Distinct domains of syntaxin are required for synaptic vesicle fusion complex formation and dissociation. Neuron 1995; 14(5):991-998.
14. Poirier MA, Hao JC, Malkus PN et al. Protease resistance of syntaxin·SNAP-25·VAMP complexes. Implications for assembly and structure. J Biol Chem 1998; 273(18):11370-11377.
15. Baumert M, Mollard GFV, Jahn R et al. Synaptobrevin: An integral membrane protein of 18,000 daltons present in small synaptic vesicles of rat brain. EMBO J 1989; 8:379-384.
16. Sudhof TC, Baumert M, Perin MS et al. A synaptic vesicle membrane protein is conserved from mammals to Drosophila. Neuron 1989; 2:1475-1481.
17. Trimble WS, Cowan DM, Scheller RH. Vamp-1: A synaptic vesicle-associated integral membrane protein. Proc Natl Acad Sci USA 1988; 85:4538-4542.
18. Pevsner J, Hsu S-C, Braun JEA et al. Specificity and regulation of a synaptic vesicle docking complex. Neuron 1994; 13:353-361.
19. Hess DT, Slater TM, Wilson MC et al. The 25 kDa synaptosomal-associated protein SNAP-25 is the major methionine-rich polypeptide in rapid axonal transport and a major substrate for palmitoylation in adult CNS. J Neurosci 1992; 12(12):4634-4641.
20. Canaves JM, Montal M. Assembly of a ternary complex by the predicted minimal coiled-coil-forming domains of syntaxin, SNAP-25, and synaptobrevin. A circular dichroism study. J Biol Chem 1998; 273(51):34214-34221.
21. Lin RC, Scheller RH. Structural organization of the synaptic exocytosis core complex. Neuron 1997; 19(5):1087-1094.
22. Nicholson KL, Munson M, Miller RB et al. Regulation of SNARE complex assembly by an N-terminal domain of the t- SNARE Sso1p. Nat Struct Biol 1998; 5(9):793-802.
23. Sutton RB, Fasshauer D, Jahn R et al. Crystal structure of a SNARE complex involved in synaptic exocytosis at 2.4 A resolution. Nature 1998; 395(6700):347-353.
24. Advani RJ, Bae H-R, Bock JB et al. Seven novel mammalian SNARE proteins localize to distinct membrane compartments. J Biol Chem 1998; 273(17):10317-10324.
25. Bock JB, Matern HT, Peden AA et al. A genomic perspective on membrane compartment organization. Nature 2001; 409(6822):839-841.
26. Chen YA, Scales SJ, Patel SM et al. SNARE complex formation is triggered by Ca2+ and drives membrane fusion. Cell 1999; 97(2):165-174.
27. Chen YA, Scheller RH. SNARE-mediated membrane fusion. Nat Rev Mol Cell Biol 2001; 2(2):98-106.
28. McNew JA, Parlati F, Fukuda R et al. Compartmental specificity of cellular membrane fusion encoded in SNARE proteins. Nature 2000; 407(6801):153-159.
29. McNew JA, Weber T, Parlati F et al. Close is not enough: SNARE-dependent membrane fusion requires an active mechanism that transduces force to membrane anchors. J Cell Biol 2000; 150(1):105-118.
30. Pombo I, Rivera J, Blank U. Munc18-2/syntaxin3 complexes are spatially separated from syntaxin3-containing SNARE complexes. FEBS Letters 2003; 550(1-3):144-148.
31. Scales SJ, Chen YA, Yoo BY et al. SNAREs contribute to the specificity of membrane fusion. Neuron 2000; 26(2):457-464.
32. Watson RT, Pessin JE. Transmembrane domain length determines intracellular membrane compartment localization of syntaxins 3, 4, and 5. Am J Physiol Cell Physiol 2001; 281(1):C215-223.
33. Daniel S, Noda M, Straub SG et al. Identification of the docked granule pool responsible for the first phase of glucose-stimulated insulin secretion. Diabetes 1999; 48(9):1686-1690.
34. Jacobsson G, Bean AJ, Scheller RH et al. Identification of synaptic proteins and their isoform mRNAs in compartments of pancreatic endocrine cells. Proc Natl Acad Sci USA 1994; 91(26):12487-12491.

35. Kiraly-Borri CE, Morgan A, Burgoyne RD et al. Soluble N-ethylmaleimide-sensitive-factor attachment protein and N-ethylmaleimide-insensitive factors are required for Ca2+-stimulated exocytosis of insulin. Biochem J 1996; 314(Pt 1):199-203.
36. Nakamichi Y, Nagamatsu S. Alpha-SNAP functions in insulin exocytosis from mature, but not immature secretory granules in pancreatic beta cells. Biochem Biophys Res Commun 1999; 260(1):127-132.
37. Regazzi R, Wollheim CB, Lang J et al. VAMP-2 and cellubrevin are expressed in pancreatic beta-cells and are essential for Ca(2+)-but not for GTP gamma S-induced insulin secretion. EMBO J 1995; 14(12):2723-2730.
38. Sadoul K, Lang J, Montecucco C et al. SNAP-25 is expressed in islets of Langerhans and is involved in insulin release. J Cell Biol 1995; 128(6):1019-1028.
39. Wheeler MB, Sheu L, Ghai M et al. Characterization of SNARE protein expression in beta cell lines and pancreatic islets. Endocrinology 1996; 137(4):1340-1348.
40. Yang SN, Larsson O, Branstrom R et al. Syntaxin 1 interacts with the L(D) subtype of voltage-gated Ca(2+) channels in pancreatic beta cells. Proc Natl Acad Sci USA 1999; 96(18):10164-10169.
41. Araki S, Tamori Y, Kawanishi M et al. Inhibition of the binding of SNAP-23 to syntaxin 4 by Munc18c. Biochem Biophys Res Commun 1997; 234:257-262.
42. Cain CC, Trimble WS, Lienhard GE. Members of the VAMP family of synaptic vesicle proteins are components of glucose transporter-containing vesicles from rat adipocytes. J Biol Chem 1992; 267:11681-11684.
43. Cheatham B, Volchuk A, Kahn CR et al. Insulin-stimulated translocation of GLUT4 glucose transporters requires SNARE-complex proteins. Proc Natl Acad Sci USA 1996; 93:15169-15173.
44. Kawanishi M, Tamori Y, Okazawa H et al. Role of SNAP23 in insulin-induced translocation of GLUT4 in 3T3-L1 adipocytes. Mediation of complex formation between syntaxin4 and VAMP2. J Biol Chem 2000; 275(11):8240-8247.
45. Martin LB, Shewan A, Millar CA et al. Vesicle-associated membrane protein 2 plays a specific role in the insulin-dependent trafficking of the facilitative glucose transporter GLUT4 in 3T3-L1 adipocytes. J Biol Chem 1998; 273:1444-1452.
46. Olson AL, Knight JB, Pessin JE. Syntaxin 4, VAMP2, and/or VAMP3/cellubrevin are functional target membrane and vesicle SNAP receptors for insulin-stimulated GLUT4 translocation in adipocytes. Mol Cell Biol 1997; 17:2425-2435.
47. Rea S, Martin LB, McIntosh S et al. Syndet, an adipocyte target SNARE involved in the insulin-induced translocation of GLUT4 to the cell surface. J Biol Chem 1998; 273:18784-18792.
48. Tamori Y, Hashiramoto M, Araki S et al. Cleavage of vesicle-associated membrane protein (VAMP)-2 and cellubrevin on GLUT4-containing vesicles inhibits the translocation of GLUT4 in 3T3-L1 adipocytes. Biochem Biophys Res Commun 1996; 220:740-745.
49. Volchuk A, Mitsumoto Y, He L et al. Expression of vesicle-associated membrane protein 2 (VAMP-2)/synaptobrevin II and cellubrevin in rat skeletal muscle and in a muscle cell line. Biochem J 1994; 304(Pt 1):139-145.
50. Timmers KI, Clark AE, Omatsu-Kanbe M et al. Identification of SNAP receptors in rat adipose cell membrane fractions and in SNARE complexes coimmunoprecipitated with epitope-tagged N-ethylmaleimide-sensitive fusion protein. Biochem J 1996; 320:429-436.
51. Ferro-Novick S, Jahn R. Vesicle fusion from yeast to man. Nature 1994; 370(6486):191-193.
52. Garcia EP, Gatti E, Butler M et al. A rat brain Sec1 homologue related to Rop and UNC18 interacts with syntaxin. Proc Natl Acad Sci USA 1994; 91:2003-2007.
53. Harrison SD, Broadie K, van de Goor J et al. Mutations in the Drosophila Rop gene suggest a function in general secretion and synaptic transmission. Neuron 1994; 13:555-566.
54. Hata Y, Slaughter CA, Sudhof TC. Synaptic vesicle fusion complex contains unc-18 homologue bound to syntaxin. Nature 1993; 366:347-351.
55. Hosono R, Hekimi S, Kamuya Y et al. The unc-18 gene encodes a novel protein affecting the kinetics of acetylcholine metabolism in the nematode Caenorhabditis elegans. J Neurochem 1992; 58:1517-1525.
56. Katagiri H, Terasaki J, Murata T et al. A novel isoform of syntaxin-binding protein homologous to yeast Sec1 expressed ubiquitously in mammalian cells. J Biol Chem 1995; 270:4963-4966.
57. Novick P, Schekman R. Secretion and cell-surface growth are blocked in a temperature-sensitive mutant of Saccharomyces cerevisiae. Proc Natl Acad Sci USA 1979; 76:1858-1862.
58. Ogawa H, Harada S, Sassa T et al. Functional properties of the unc-64 gene encoding a Caenorhabditis elegans syntaxin. J Biol Chem 1998; 273(4):2192-2198.
59. Pevsner J, Hsu SC, Scheller RH. n-Sec1: A neural-specific syntaxin-binding protein. Proc Natl Acad Sci USA 1994; 91(4):1445-1449.

60. Salzberg A, Cohen N, Halachmi N et al. The Drosophila Ras2 and Rop gene pair: A dual homology with a yeast Ras-like gene and a suppressor of its loss-of-function phenotype. Development 1993; 117(4):1309-1319.
61. Schulze KL, Littleton JT, Salzberg A et al. Rop, a Drosophila homolog of yeast Sec1 and vertebrate n-Sec1/Munc-18 proteins, is a negative regulator of neurotransmitter release in vivo. Neuron 1994; 13:1099-1108.
62. Gengyo-Ando K, Kamiya Y, Yamakawa A et al. The C. elegans unc-18 gene encodes a protein expressed in motor neurons. Neuron 1993; 11(4):703-711.
63. Verhage M, Maia AS, Plomp JJ et al. Synaptic assembly of the brain in the absence of neurotransmitter secretion: Dynamics of munc18-1 phosphorylation/dephosphorylation in rat brain nerve terminals. Science 2000; 287(5454):864-869.
64. Thurmond DC, Ceresa BP, Okada S et al. Regulation of insulin-stimulated GLUT4 translocation by Munc18c in 3T3L1 adipocytes. J Biol Chem 1998; 273:33876-33883.
65. Wu MN, Fergestad T, Lloyd TE et al. Syntaxin 1A interacts with multiple exocytic proteins to regulate neurotransmitter release in vivo. Neuron 1999; 23(3):593-605.
66. Fujita Y, Sasaki T, Fukui K et al. Phosphorylation of Munc-18/n-Sec1/rbSec1 by protein kinase C: Its implication in regulating the interaction of Munc-18/n-Sec1/rbSec1 with syntaxin. J Biol Chem 1996; 271(13):7265-7268.
67. Garcia EP, McPherson PS, Chilcote TJ et al. rbSec1A and B colocalize with syntaxin 1 and SNAP-25 throughout the axon, but are not in a stable complex with syntaxin. J Cell Biol 1995; 129:105-120.
68. Halachmi N, Lev Z. The Sec1 family: A novel family of proteins involved in synaptic transmission and general secretion. J Neurochem 1996; 66(3):889-897.
69. Hata Y, Sudhof TC. A novel ubiquitous form of Munc-18 interacts with multiple syntaxins. J Biol Chem 1995; 270:13022-13028.
70. Tellam JT, Macaulay SL, McIntosh S et al. Characterization of Munc-18c and syntaxin-4 in 3T3-L1 adipocytes. Putative role in insulin-dependent movement of GLUT-4. J Biol Chem 1997; 272:6179-6186.
71. Tellam JT, McIntosh S, James DE. Molecular identification of two novel Munc-18 isoforms expressed in nonneuronal tissues. J Biol Chem 1995; 270:5857-5863.
72. Volchuk A, Wang Q, Ewart HS et al. Syntaxin 4 in 3T3-L1 adipocytes: Regulation by insulin and participation in insulin-dependent glucose transport. Mol Biol Cell 1996; 7:1075-1082.
73. Tamori Y, Kawanishi M, Niki T et al. Inhibition of insulin-induced GLUT4 translocation by Munc18c through interaction with syntaxin4 in 3T3-L1 adipocytes. J Biol Chem 1998; 273:19740-19746.
74. Bracher A, Perrakis A, Dresbach T et al. The X-ray crystal structure of neuronal Sec1 from squid sheds new light on the role of this protein in exocytosis. Structure Fold Des 2000; 8(7):685-694.
75. Dulubova I, Sugita S, Hill S et al. A conformational switch in syntaxin during exocytosis: Role of munc18. EMBO J 1999; 18(16):4372-4382.
76. Fernandez I, Ubach J, Dulubova I et al. Three-dimensional structure of an evolutionarily conserved N-terminal domain of syntaxin 1A. Cell 1998; 94(6):841-849.
77. Lerman JC, Robblee J, Fairman R et al. Structural analysis of the neuronal SNARE protein syntaxin-1A. Biochemistry 2000; 39(29):8470-8479.
78. Misura KM, Scheller RH, Weis WI. Three-dimensional structure of the neuronal-Sec1-syntaxin 1a complex. Nature 2000; 404(6776):355-362.
79. Rowe J, Corradi N, Malosio ML et al. Blockade of membrane transport and disassembly of the Golgi complex by expression of syntaxin 1A in neurosecretion-incompetent cells: Prevention by rbSEC1. J Cell Sci 1999; 112(Pt 12):1865-1877.
80. Yang B, Steegmaier M, Gonzalez Jr LC et al. nSec1 binds a closed conformation of syntaxin1A. J Cell Biol 2000; 148(2):247-252.
81. Dulubova I, Yamaguchi T, Arac D et al. Convergence and divergence in the mechanism of SNARE binding by Sec1/Munc18-like proteins. PNAS 2003; 100(1):32-37.
82. Grusovin J, Stoichevska V, Gough KH et al. Definition of a minimal munc18c domain that interacts with syntaxin 4. Biochem J 2000; 350(Pt 3):741-746.
83. Dascher C, Ossig R, Gallwitz D et al. Identification and structure of four yeast genes (SLY) that are able to suppress the functional loss of YPT1, a member of the RAS superfamily. Mol Cell Biol 1991; 11(2):872-885.
84. Thurmond DC, Kanzaki M, Khan AH et al. Munc18c function is required for insulin-stimulated plasma membrane fusion of GLUT4 and insulin-responsive amino peptidase storage vesicles. Mol Cell Biol 2000; 20(1):379-388.
85. Thurmond DC, Pessin JE. Discrimination of GLUT4 vesicle trafficking from fusion using a temperature-sensitive Munc18c mutant. EMBO J 2000; 19(14):3565-3575.

86. Olson AL, Pessin JE. Structure, function and regulation of the mammalian facilitative glucose transporter gene family. Ann Rev Nutr 1996; 16:235-256.
87. Slot JW, Geuze HJ, Gigengack S et al. Translocation of the glucose transporter GLUT4 in cardiac myocytes of the rat. Proc Natl Acad Sci USA 1991; 88:7815-7819.
88. Slot JW, Geuze HJ, Gigengack S et al. Immuno-localization of the insulin regulatable glucose transporter in brown adipose tissue of the rat. J Cell Biol 1991; 113:123-135.
89. Czech MP. Molecular actions of insulin on glucose transport. Annu Rev Nutr 1995; 15:441-471.
90. Kandror KV, Pilch PF. Compartmentalization of protein traffic in insulin-sensitive cells. Am J Physiol 1996; 271:E1-E14.
91. Klip A, Tsakiridis T, Marette A et al. Regulation of expression of glucose transporters by glucose: A review of studies in vivo and in cell cultures. FASEB J 1994; 8:43-53.
92. Pessin JE, Thurmond DC, Elmendorf JS et al. Molecular basis of insulin-stimulated GLUT4 vesicle trafficking. Location! location! location! J Biol Chem 1999; 274:2593-2596.
93. Rea S, James DE. Moving GLUT4: The biogenesis and trafficking of GLUT4 storage vesicles. Diabetes 1997; 46:1667-1677.
94. Thurmond DC, Pessin JE. Molecular machinery involved in the insulin-regulated fusion of GLUT4-containing vesicles with the plasma membrane. Mol Membr Biol 2001; 18(4):237-245.
95. Kahn BB. Facilitative glucose transporters: Regulatory mechanisms and dysregulation in diabetes. J Clin Invest 1992; 89:1367-1374.
96. White MF. The IRS-signalling system: A network of docking proteins that mediate insulin action. Mol Cell Biochem 1998; 182:3-11.
97. Chen KS, Friel JC, Ruderman NB. Regulation of phosphatidylinositol 3-kinase by insulin in rat skeletal muscle. Am J Physiol 1993; 265:E736-E742.
98. Folli F, Saad MJ, Backer JM et al. Regulation of phosphatidylinositol 3-kinase activity in liver and muscle of animal models of insulin-resistant and insulin-deficient diabetes mellitus. J Clin Invest 1993; 92(4):1787-1794.
99. Kelly KL, Ruderman NB, Chen KS. Phosphatidylinositol-3-kinase in isolated rat adipocytes. Activation by insulin and subcellular distribution. J Biol Chem 1992; 267(5):3423-3428.
100. Ruderman NB, Kapeller R, White MF et al. Activation of phosphatidylinositol 3-kinase by insulin. Proc Natl Acad Sci USA 1990; 87:1411-1415.
101. Bandyopadhyay G, Standaert ML, Zhao L et al. Activation of protein kinase C (alpha, beta, and zeta) by insulin in 3T3/L1 cells. Transfection studies suggest a role for PKC-zeta in glucose transport. J Biol Chem 1997; 272:2551-2558.
102. Kotani K, Ogawa W, Matsumoto M et al. Requirement of atypical protein kinase clambda for insulin stimulation of glucose uptake but not for Akt activation in 3T3-L1 adipocytes. Mol Cell Biol 1998; 18:6971-6982.
103. Vollenweider P, Clodi M, Martin SS et al. An SH2 domain-containing 5' inositolphosphatase inhibits insulin-induced GLUT4 translocation and growth factor-induced actin filament rearrangement. Mol Cell Biol 1999; 19:1081-1091.
104. Kane S, Sano H, Liu SC et al. A method to identify serine kinase substrates. Akt phosphorylates a novel adipocyte protein with a Rab GTPase-activating protein (GAP) domain. J Biol Chem 2002; 277(25):22115-22118.
105. Bruss MD, Arias EB, Lienhard GE et al. Increased phosphorylation of Akt substrate of 160 kDa (AS160) in rat skeletal muscle in response to insulin or contractile activity. Diabetes 2005; 54(1):41-50.
106. Sano H, Kane S, Sano E et al. Insulin-stimulated phosphorylation of a Rab GTPase-activating protein regulates GLUT4 translocation. J Biol Chem 2003; 278(17):14599-14602.
107. Zeigerer A, McBrayer MK, McGraw TE. Insulin stimulation of GLUT4 exocytosis, but not its inhibition of endocytosis, is dependent on RabGAP AS160. Mol Biol Cell 2004; 15(10):4406-4415.
108. Schoch S, Deak F, Konigstorfer A et al. SNARE function analyzed in synaptobrevin/VAMP knockout mice. Science 2001; 294(5544):1117-1122.
109. Yang C, Coker KJ, Kim JK et al. Syntaxin 4 heterozygous knockout mice develop muscle insulin resistance. J Clin Invest 2001; 107(10):1311-1318.
110. Yang C, Mora S, Ryder JW et al. VAMP3 null mice display normal constitutive, insulin- and exercise-regulated vesicle trafficking. Mol Cell Biol 2001; 21(5):1573-1580.
111. Macaulay SL, Grusovin J, Stoichevska V et al. Cellular munc18c levels can modulate glucose transport rate and GLUT4 translocation in 3T3L1 cells. FEBS Letters 2002; 528(1-3):154-160.
112. Cao X, Ballew N, Barlowe C. Initial docking of ER-derived vesicles requires Uso1p and Ypt1p but is independent of SNARE proteins. EMBO J 1998; 17:2156-2165.
113. VanRheenen SM, Cao X, Sapperstein SK et al. Sec34p, a protein required for vesicle tethering to the yeast Golgi apparatus, is in a complex with Sec35p. J Cell Biol 1999; 147(4):729-742.

114. Chen G, Liu P, Thurmond DC et al. Glucosamine-induced insulin resistance is coupled to O-linked glycosylation of Munc18c. FEBS Lett 2003; 534(1-3):54-60.
115. Nelson BA, Robinson KA, Buse MG. Insulin acutely regulates Munc18-c subcellular trafficking: Altered response in insulin-resistant 3T3-L1 adipocytes. J Biol Chem 2002; 277(6):3809-3812.
116. Khan AH, Thurmond DC, Yang C et al. Munc18c regulates insulin-stimulated GLUT4 translocation to the transverse tubules in skeletal muscle. J Biol Chem 2001; 276(6):4063-4069.
117. Spurlin BA, Thomas RM, Nevins AK et al. Insulin resistance in tetracycline-repressible Munc18c transgenic mice. Diabetes 2003; 52(8):1910-1917.
118. Oh E, Spurlin BA, Pessin JE et al. Munc18c heterozygous knockout mice display increased susceptibility for severe glucose intolerance. Diabetes 2005; 54(3):638-647.
119. Kanda H, Tamori Y, Shinoda H et al. Adipocytes from Munc18c-null mice show increased sensitivity to insulin-stimulated GLUT4 externalization. J Clin Invest 2005; 115(2):291-301.
120. Spurlin BA, Park SY, Nevins AK et al. Syntaxin 4 transgenic mice exhibit enhanced insulin-mediated glucose uptake in skeletal muscle. Diabetes 2004; 53(9):2223-2231.
121. Geppert M, Bolshakov VY, Siegelbaum SA et al. The role of Rab3A in neurotransmitter release. Nature 1994; 369(6480):493-497.
122. Tall GG, Hama H, DeWald DB et al. The phosphatidylinositol 3-phosphate binding protein Vac1p interacts with a Rab GTPase and a Sec1p homologue to facilitate vesicle-mediated vacuolar protein sorting. Mol Biol Cell 1999; 10(6):1873-1889.
123. Webb GC, Hoedt M, Poole LJ et al. Genetic interactions between a pep7 mutation and the PEP12 and VPS45 genes: Evidence for a novel SNARE component in transport between the Saccharomyces cerevisiae Golgi complex and endosome. Genetics 1997; 147(2):467-478.
124. Schimmoller F, Simon I, Pfeffer SR. Rab GTPases, directors of vesicle docking. J Biol Chem 1998; 273(35):22161-22164.
125. Cormont M, Tanti JF, Zahraoui A et al. Insulin and okadaic acid induce Rab4 redistribution in adipocytes. J Biol Chem 1993; 268(26):19491-19497.
126. Li L, Omata W, Kojima I et al. Direct interaction of Rab4 with syntaxin 4. J Biol Chem 2000; 276:5265-5273.
127. Shibata H, Omata W, Kojima I. Insulin stimulates guanine nucleotide exchange on Rab4 via a wortmannin-sensitive signaling pathway in rat adipocytes. J Biol Chem 1997; 272(23):14542-14546.
128. Shisheva A, Czech MP. Association of cytosolic Rab4 with GDI isoforms in insulin-sensitive 3T3-L1 adipocytes. Biochemistry 1997; 36(22):6564-6570.
129. Cormont M, Bortoluzzi MN, Gautier N et al. Potential role of Rab4 in the regulation of subcellular localization of Glut4 in adipocytes. Mol Cell Biol 1996; 16:6879-6886.
130. Mora S, Monden I, Zorzano A et al. Heterologous expression of rab4 reduces glucose transport and GLUT4 abundance at the cell surface in oocytes. Biochem J 1997; 324:455-459.
131. Shibata H, Omata W, Suzuki Y et al. A synthetic peptide corresponding to the Rab4 hypervariable carboxyl-terminal domain inhibits insulin action on glucose transport in rat adipocytes. J Biol Chem 1996; 271:9704-9709.
132. Fujita Y, Shirataki H, Sakisaka T et al. Tomosyn: A syntaxin-1-binding protein that forms a novel complex in the neurotransmitter release process. Neuron 1998; 20(5):905-915.
133. Hatsuzawa K, Lang T, Fasshauer D et al. The R-SNARE motif of tomosyn forms SNARE core complexes with syntaxin 1 and SNAP-25 and down-regulates exocytosis. J Biol Chem 2003; 278(33):31159-31166.
134. Yokoyama S, Shirataki H, Sakisaka T et al. Three splicing variants of tomosyn and identification of their syntaxin-binding region. Biochem Biophys Res Commun 1999; 256(1):218-222.
135. Groffen AJ, Jacobsen L, Schut D et al. Two distinct genes drive expression of seven tomosyn isoforms in the mammalian brain, sharing a conserved structure with a unique variable domain. J Neurochem 2005; 92(3):554-568.
136. Yizhar O, Matti U, Melamed R et al. Tomosyn inhibits priming of large dense-core vesicles in a calcium-dependent manner. Proc Natl Acad Sci USA 2004; 101(8):2578-2583.
137. Widberg CH, Bryant NJ, Girotti M et al. Tomosyn interacts with the t-SNAREs syntaxin4 and SNAP23 and plays a role in insulin-stimulated GLUT4 translocation. J Biol Chem 2003; 278(37):35093-35101.
138. Min J, Okada S, Kanzaki M et al. Synip: A novel insulin-regulated syntaxin 4-binding protein mediating GLUT4 translocation in adipocytes. Mol Cell 1999; 3(6):751-760.
139. Yamada E, Okada S, Saito T et al. Akt2 phosphorylates Synip to regulate docking and fusion of GLUT4-containing vesicles. J Cell Biol 2005; 168(6):921-928.
140. TerBush DR, Maurice T, Roth D et al. The Exocyst is a multiprotein complex required for exocytosis in Saccharomyces cerevisiae. EMBO J 1996; 15(23):6483-6494.
141. Kee Y, Yoo JS, Hazuka CD et al. Subunit structure of the mammalian exocyst complex. Proc Natl Acad Sci USA 1997; 94(26):14438-14443.

142. Inoue M, Chang L, Hwang J et al. The exocyst complex is required for targeting of Glut4 to the plasma membrane by insulin. Nature 2003; 422(6932):629-633.
143. Thurmond DC, Pessin JE. Molecular basis for insulin-stimulated GLUT4 translocation. Current Opinion in Endocrinology and Diabetes 2001; 8(2):67-73.
144. Ewart MA, Clarke M, Kane S et al. Evidence for a role of the exocyst in insulin-stimulated Glut4 trafficking in 3T3-L1 adipocytes. J Biol Chem 2005; 280(5):3812-3816.
145. Rhodes CJ. Processing of the insulin molecule. In: LeRoith T, Olefsky, eds. Diabetes Mellitus: A fundamental and clinical text. Philadelphia, PA: Lippincott Williams and Wilkins, 2000:20-38.
146. Halban PA, Renold AE. Influence of glucose on insulin handling by rat islets in culture. A reflection of integrated changes in insulin biosynthesis, release, and intracellular degradation. Diabetes 1983; 32(3):254-261.
147. Halban PA, Wollheim CB. Intracellular degradation of insulin stores by rat pancreatic islets in vitro. An alternative pathway for homeostasis of pancreatic insulin content. J Biol Chem 1980; 255(13):6003-6006.
148. Lang J. Molecular mechanisms and regulation of insulin exocytosis as a paradigm of endocrine secretion. Eur J Biochem 1999; 259(1-2):3-17.
149. Cook DL, Hales CN. Intracellular ATP directly blocks K+ channels in pancreatic B-cells. Nature 1984; 311(5983):271-273.
150. Meglasson MD, Matschinsky FM. Pancreatic islet glucose metabolism and regulation of insulin secretion. Diabetes Metab Rev 1986; 2:163-214.
151. Satin LS, Cook DL. Voltage-gated Ca2+ current in pancreatic B-cells. Pflugers Arch 1985; 404(4):385-387.
152. Rorsman P, Eliasson L, Renstrom E et al. The cell physiology of biphasic insulin secretion. News Physiol Sci 2000; 15:72-77.
153. Curry DL, Bennett LL, Grodsky GM. Dynamics of insulin secretion by the perfused rat pancreas. Endocrinology 1968; 83(3):572-584.
154. Heinemann C, von Ruden L, Chow RH et al. A two-step model of secretion control in neuroendocrine cells. Pflugers Arch 1993; 424(2):105-112.
155. Barg S, Eliasson L, Renstrom E et al. A subset of 50 secretory granules in close contact with L-type Ca(2+) channels accounts for first-phase insulin secretion in mouse beta-cells. Diabetes 2002; 51(Suppl 1):S74-82.
156. Gembal M, Gilon P, Henquin JC. Evidence that glucose can control insulin release independently from its action on ATP-sensitive K+ channels in mouse B cells. J Clin Invest 1992; 89(4):1288-1295.
157. Renstrom E, Eliasson L, Bokvist K et al. Cooling inhibits exocytosis in single mouse pancreatic B-cells by suppression of granule mobilization. J Physiol 1996; 494(Pt 1):41-52.
158. Lam PL, Leung YM, Sheu L et al. Transgenic mouse over-expressing Syntaxin-1A as a diabetes model. Diabetes 2005; 54:2744-54.
159. Watanabe T, Fujiwara T, Komazaki S et al. HPC-1/syntaxin 1A suppresses exocytosis of PC12 cells. J Biochem (Tokyo) 1999; 125(4):685-689.
160. Sadoul K, Berger A, Niemann H et al. SNAP-23 is not cleaved by botulinum neurotoxin E and can replace SNAP-25 in the process of insulin secretion. J Biol Chem 1997; 272(52):33023-33027.
161. Land J, Zhang H, Vaidyanathan VV et al. Transient expression of botulinum neurotoxin C1 light chain differentially inhibits calcium and glucose induced insulin secretion in clonal beta-cells. FEBS Lett 1997; 419(1):13-17.
162. Saito T, Okada S, Yamada E et al. Syntaxin 4 and Synip (Syntaxin 4 Interacting Protein) regulate insulin secretion in the pancreatic {beta} HC-9 cell. J Biol Chem 2003; 278(38):36718-36725.
163. Spurlin BA, Thurmond DC. Syntaxin 4 facilitates biphasic insulin secretion from pancreatic beta cells. Mol Endocrinol 2005; in press.
164. Kang Y, Huang X, Pasyk EA et al. Syntaxin-3 and syntaxin-1A inhibit L-type calcium channel activity, insulin biosynthesis and exocytosis in beta-cell lines. Diabetologia 2002; 45(2):231-241.
165. Nagamatsu S, Nakamichi Y, Yamamura C et al. Decreased expression of t-SNARE, syntaxin 1, and SNAP-25 in pancreatic beta-cells is involved in impaired insulin secretion from diabetic GK rat islets: Restoration of decreased t-SNARE proteins improves impaired insulin secretion. Diabetes 1999; 48(12):2367-2373.
166. Chan CB, MacPhail RM, Sheu L et al. Beta-cell hypertrophy in fa/fa rats is associated with basal glucose hypersensitivity and reduced SNARE protein expression. Diabetes 1999; 48(5):997-1005.
167. Wiser O, Bennett MK, Atlas D. Functional interaction of syntaxin and SNAP-25 with voltage-sensitive L- and N-type Ca2+ channels. EMBO J 1996; 15(16):4100-4110.
168. Leung YM, Kang Y, Gao X et al. Syntaxin 1A binds to the cytoplasmic C terminus of Kv2.1 to regulate channel gating and trafficking. J Biol Chem 2003; 278(19):17532-17538.

169. Leung YM, Kang Y, Xia F et al. Open form syntaxin-1A is a more potent inhibitor than wild type syntaxin-1A of Kv2.1 channels. Biochem J 2005; 387(Pt 1):195-202.
170. Ji J, Yang SN, Huang X et al. Modulation of L-type Ca(2+) channels by distinct domains within SNAP-25. Diabetes 2002; 51(5):1425-1436.
171. MacDonald PE, Wang G, Tsuk S et al. Synaptosome-associated protein of 25 kilodaltons modulates Kv2.1 voltage-dependent K(+) channels in neuroendocrine islet beta-cells through an interaction with the channel N terminus. Mol Endocrinol 2002; 16(11):2452-2461.
172. Zhang W, Efanov A, Yang SN et al. Munc-18 associates with syntaxin and serves as a negative regulator of exocytosis in the pancreatic beta -cell. J Biol Chem 2000; 275(52):41521-41527.
173. Zhang W, Lilja L, Bark C et al. Mint1, a Munc-18-interacting protein, is expressed in insulin-secreting beta-cells. Biochem Biophys Res Commun 2004; 320(3):717-721.
174. Okamoto M, Sudhof TC. Mints, Munc18-interacting proteins in synaptic vesicle exocytosis. J Biol Chem 1997; 272:31459-31464.
175. Lupashin VV, Waters MG. t-SNARE activation through transient interaction with a rab-like guanosine triphosphatase. Science 1997; 276(5316):1255-1258.
176. Geppert M, Goda Y, Stevens CF et al. The small GTP-binding protein Rab3A regulates a late step in synaptic vesicle fusion. Nature 1997; 387(6635):810-814.
177. Regazzi R, Kikuchi A, Takai Y et al. The small GTP-binding proteins in the cytosol of insulin-secreting cells are complexed to GDP dissociation inhibitor proteins. J Biol Chem 1992; 267(25):17512-17519.
178. Regazzi R, Vallar L, Ullrich S et al. Characterization of small-molecular-mass guanine-nucleotide-binding regulatory proteins in insulin-secreting cells and PC12 cells. Eur J Biochem 1992; 208(3):729-737.
179. Regazzi R, Ravazzola M, Iezzi M et al. Expression, localization and functional role of small GTPases of the Rab3 family in insulin-secreting cells. J Cell Sci 1996; 109(Pt 9):2265-2273.
180. Coppola T, Frantz C, Perret-Menoud V et al. Pancreatic beta-cell protein granuphilin binds Rab3 and Munc-18 and controls exocytosis. Mol Biol Cell 2002; 13(6):1906-1915.
181. Iezzi M, Escher G, Meda P et al. Subcellular distribution and function of Rab3A, B, C, and D isoforms in insulin-secreting cells. Mol Endocrinol 1999; 13(2):202-212.
182. Iezzi M, Regazzi R, Wollheim CB. The Rab3-interacting molecule RIM is expressed in pancreatic beta-cells and is implicated in insulin exocytosis. FEBS Lett 2000; 474(1):66-70.
183. Torii S, Zhao S, Yi Z et al. Granuphilin modulates the exocytosis of secretory granules through interaction with syntaxin 1A. Mol Cell Biol 2002; 22(15):5518-5526.
184. Yi Z, Yokota H, Torii S et al. The Rab27a/granuphilin complex regulates the exocytosis of insulin-containing dense-core granules. Mol Cell Biol 2002; 22(6):1858-1867.
185. Torii S, Takeuchi T, Nagamatsu S et al. Rab27 effector granuphilin promotes the plasma membrane targeting of insulin granules via interaction with syntaxin 1A. J Biol Chem 2004; 279(21):22532-22538.
186. Kotake K, Ozaki N, Mizuta M et al. Noc2, a putative zinc finger protein involved in exocytosis in endocrine cells. J Biol Chem 1997; 272(47):29407-29410.
187. Cheviet S, Coppola T, Haynes LP et al. The Rab-binding protein Noc2 is associated with insulin-containing secretory granules and is essential for pancreatic beta-cell exocytosis. Mol Endocrinol 2004; 18(1):117-126.
188. Sheu L, Pasyk EA, Ji J et al. Regulation of Insulin Exocytosis by Munc13-1. J Biol Chem 2003; 278(30):27556-27563.
189. Betz A, Thakur P, Junge HJ et al. Functional interaction of the active zone proteins Munc13-1 and RIM1 in synaptic vesicle priming. Neuron 2001; 30(1):183-196.
190. Augustin I, Rosenmund C, Sudhof TC et al. Munc13-1 is essential for fusion competence of glutamatergic synaptic vesicles. Nature 1999; 400(6743):457-461.
191. Li G, Rungger-Brandle E, Just I et al. Effect of disruption of actin filaments by Clostridium botulinum C2 toxin on insulin secretion in HIT-T15 cells and pancreatic islets. Mol Biol Cell 1994; 5(11):1199-1213.
192. Somers G, Blondel B, Orci L et al. Motile events in pancreatic endocrine cells. Endocrinology 1979; 104(1):255-264.
193. Wilson JR, Ludowyke RI, Biden TJ. A redistribution of actin and myosin IIA accompanies Ca(2+)-dependent insulin secretion. FEBS Lett 2001; 492(1-2):101-106.
194. Thurmond DC, Gonelle-Gispert C, Furukawa M et al. Glucose-stimulated insulin secretion is coupled to the interaction of actin with the t-SNARE (Target Membrane Soluble N-Ethylmaleimide-Sensitive Factor Attachment Protein Receptor Protein) complex. Mol Endocrinol 2003; 17(4):732-742.
195. Nevins AK, Thurmond DC. Glucose regulates the cortical actin network through modulation of Cdc42 cycling to stimulate insulin secretion. Am J Physiol Cell Physiol 2003; 285(3):C698-710.
196. Kowluru A, Metz SA. Regulation of guanine-nucleotide binding proteins in islet subcellular fractions by phospholipase-derived lipid mediators of insulin secretion. Biochim Biophys Acta 1994; 1222(3):360-368.

197. Kowluru A, Seavey SE, Li G et al. Glucose- and GTP-dependent stimulation of the carboxyl methylation of CDC42 in rodent and human pancreatic islets and pure beta cells. Evidence for an essential role of GTP-binding proteins in nutrient-induced insulin secretion. J Clin Invest 1996; 98(2):540-555.
198. Daniel S, Noda M, Cerione RA et al. A link between Cdc42 and syntaxin is involved in mastoparan-stimulated insulin release. Biochemistry 2002; 41(30):9663-9671.
199. Nevins AK, Thurmond DC. A direct interaction between Cdc42 and vesicle-associated membrane protein 2 regulates SNARE-dependent insulin exocytosis. J Biol Chem 2005; 280(3):1944-1952.
200. Parsons TD, Coorssen JR, Horstmann H et al. Docked granules, the exocytic burst, and the need for ATP hydrolysis in endocrine cells. Neuron 1995; 15(5):1085-1096.
201. Maier V, Melvin D, Lister C et al. v- and t-SNARE protein expression in models of insulin resistance: Normalization of glycemia by rosiglitazone treatment corrects overexpression of cellubrevin, vesicle-associated membrane protein-2, and syntaxin 4 in skeletal muscle of Zucker diabetic fatty rats. Diabetes 2000; 49(4):618-625.
202. Miura T, Suzuki W, Ishihara E et al. Impairment of insulin-stimulated GLUT4 translocation in skeletal muscle and adipose tissue in the Tsumura Suzuki obese diabetic mouse: A new genetic animal model of type 2 diabetes. Eur J Endocrinol 2001; 145(6):785-790.

# CHAPTER 4

# Control of Protein Synthesis by Insulin

Joseph F. Christian and John C. Lawrence, Jr.*

## Introduction

The stimulation of protein synthesis is a classic action of insulin.[1-3] Loss of the stimulatory effect of insulin on protein synthesis contributes to the cessation of growth and weight loss, which are hallmarks of untreated Type 1 diabetes mellitus. The effect of insulin on protein metabolism is complex and involves changes in both synthesis and degradation.[1-3] In some cell types an increase in rate of protein synthesis may be detected within minutes of insulin treatment. This response to insulin occurs within a timeframe comparable to that of other acute actions of the hormone, such as the activation of glucose transport and glycogen synthase activation. The rapid effects of insulin on protein synthesis involve increases in mRNA translation, the process through which the genetic code transcribed in the mRNA template is translated into protein.

Translation takes place on ribosomes in a complex series of reactions that can be segregated into three phases— initiation, elongation and termination.[4,5] Each phase involves a select group of translation factors, referred to as initiation factors, elongation factors, and release factors (abbreviated eIF, eEF, and eRF, respectively; where "e" is eukaryotic). Although the effects of insulin on mRNA translation have received less attention than those on carbohydrate and lipid metabolism, recent studies have increased our understanding of the mechanisms involved in the control of protein synthesis.

The activities of several translation factors have been found to be controlled by insulin, explaining at least in part the stimulatory effect of insulin on translation. A newly discovered signaling system based on the Ser/Thr protein kinase, mTOR, has been found to have a key role in the control of mRNA translation. Protein synthesis and the control of carbohydrate metabolism have now been linked in unexpected ways, and many of the same signaling elements utilized by insulin to control glucose metabolism have been found to be involved in the control of protein synthesis. With these discoveries has come an appreciation that signaling molecules identified in studies of protein synthesis may turn out to be important in the control of carbohydrate and lipid metabolism. This chapter describes mechanisms underlying the rapid activation of mRNA translation by insulin. A brief overview of translation is provided as a framework for discussing insulin action (see Figs. 1 and 2 for schematic representations).

## mRNA

In eukaryotic cells essentially all mRNAs are monocistronic, are capped at their 5' end with m7GpppN (were N is any nucleotide), and possess a 3' polyA tail of variable length.[4,5] Between the cap and the start codon, which is most often the first downstream AUG, is the 5'

---

*Corresponding Author: John C. Lawrence, Jr.—Department of Pharmacology, University of Virginia Health System, P.O. Box 800735, 1300 Jefferson Park Avenue, Charlottesville, Virginia, 22908-0735 U.S.A. Email: jcl3p@virginia.edu

*Mechanisms of Insulin Action*, edited by Alan R. Saltiel and Jeffrey E. Pessin.
©2007 Landes Bioscience and Springer Science+Business Media.

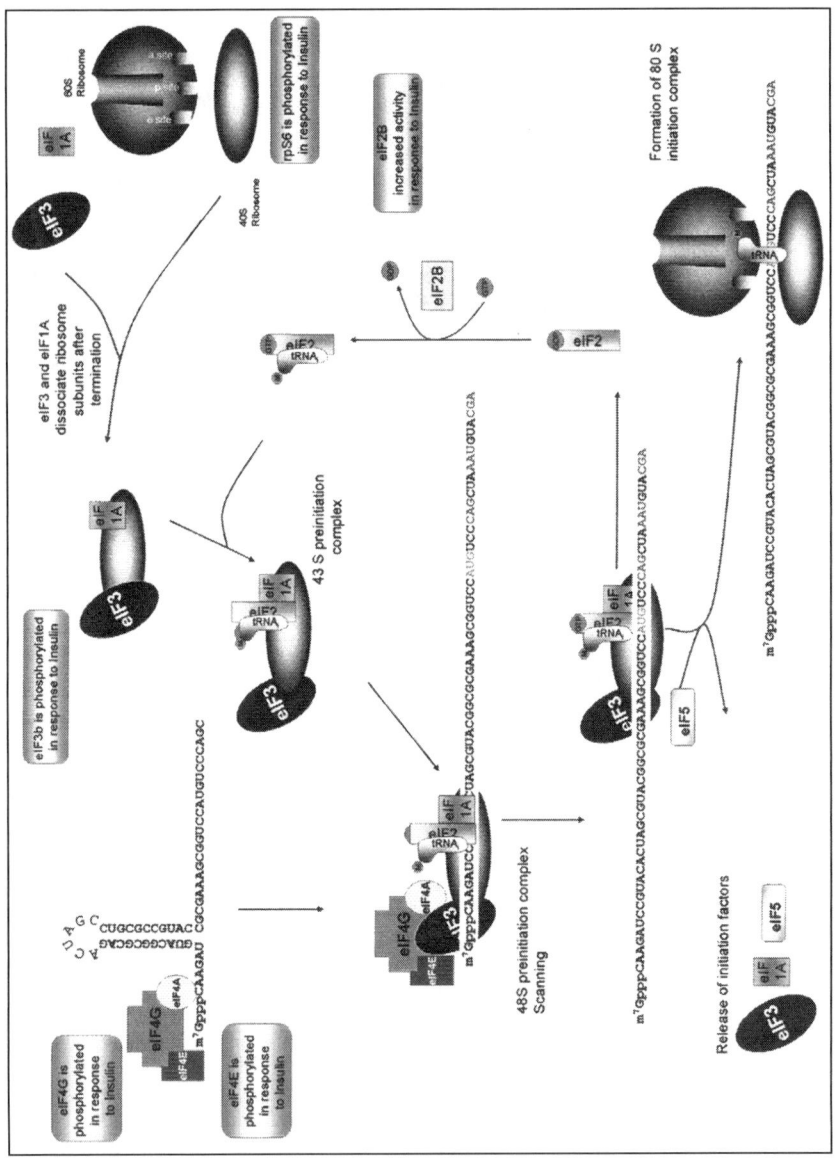

Figure 1. Eukaryotic translation initiation. Steps in initiation that are regulated by insulin are highlighted by red boxes.

# Control of Protein Synthesis by Insulin 73

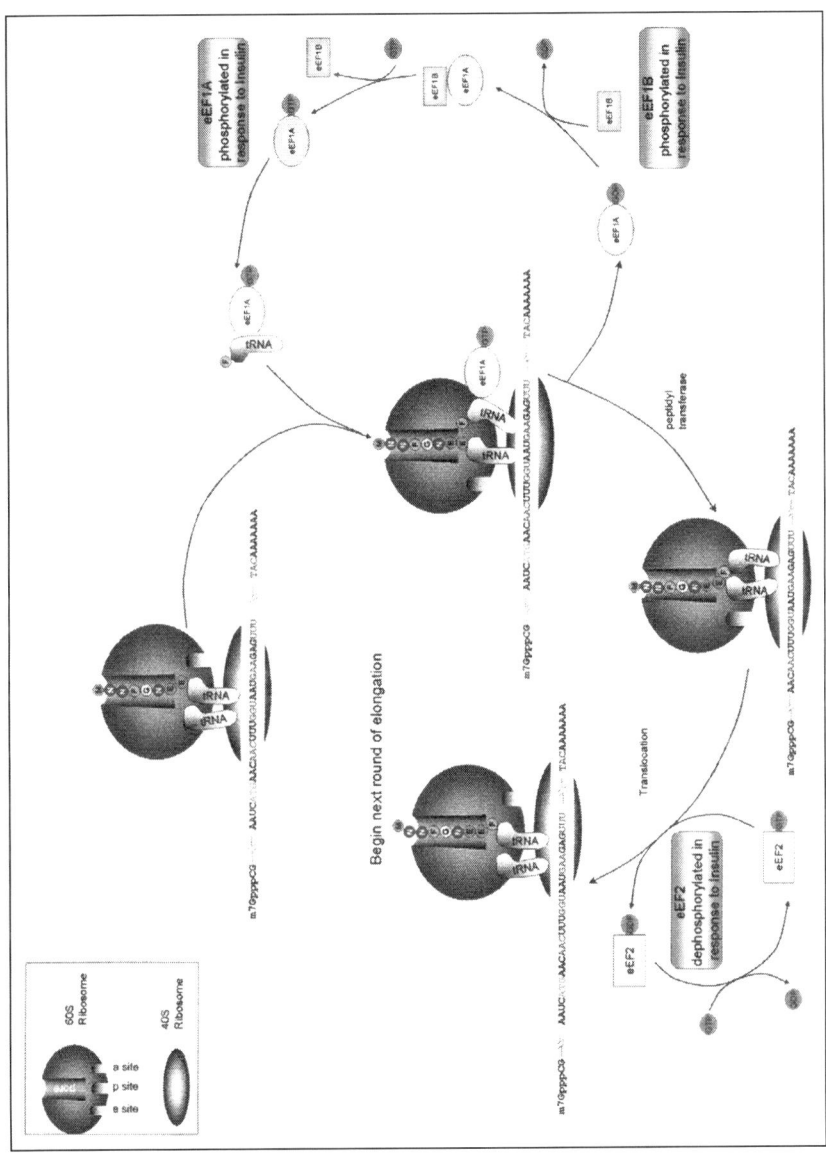

Figure 2. Eukaryotic translation elongation. Steps in elongation that are regulated by insulin are highlighted by red boxes.

untranslated region (UTR). The 5'UTR ranges in size from just a few bases to several hundred bases, and it often contains regions of secondary structure formed by pairing of bases in the RNA.[6] Secondary structure may serve to slow the scanning 40S ribosome to ensure selection of the proper start codon; however, it can be a significant barrier to translation.[7] All things being equal, the efficiency of translation is inversely related to the stability of the structured region; the effect of structure on translation is greatest when the structure is located near the 5' end of the message.[6] Thus, messages with highly structured 5'UTRs are translated poorly and are referred to as "weak" messages. As G/C forms a more stable pair than A/T, 5'UTRs with a high G/C content are more likely to have more stable regions of secondary structure.[6] Messages encoding many receptors, oncogenes, and signal transduction proteins have 5'UTRs with relatively high G/C contents.[8] These messages are expected to compete less well for initiation of translation than "strong" messages, which have unstructured 5'UTRs and are transcribed from genes encoding house keeping proteins. Ornithine decarboxylase (ODC) mRNA has a very highly structured 5'UTR, and it is one of the best characterized examples of a weak message.[7] In 3T3 fibroblasts insulin promoted a several fold increase in the translation of ODC mRNA.[7] The effect was significantly greater than the effect of insulin on global protein synthesis, supporting the view that the translation of "weak" messages is selectively increased by the hormone.

The 5' UTR may also contain sequence motifs that allow selective control of translation. One family of mRNAs crucial for cell and tissue growth encode ribosomal proteins and other proteins involved in protein synthesis. The defining characteristic of these mRNAs is the presence of a tract of 9-12 pyrimidines (TOP) found immediately adjacent to the 5' cap.[9] Translation of TOP messages is preferentially increased by insulin. The mechanism has not been determined. There is evidence that the response involves activation of ribosomal protein S6 kinase 1 (S6K1),[10] although other evidence argues against the involvement of this kinase.[11] While additional research will be needed to resolve this issue, it is clear that by stimulating translation of TOP messages, insulin increases components of the translational machinery, thus increasing the protein synthetic capacity of cells.

Some mRNAs contain internal ribosomal entry sites (IRES), highly structured regions that directly recruit ribosomal subunits and allow translation to initiate without the usual complement of initiation factors required for cap-dependent translation. IRES elements were discovered in picornaviral mRNA, where they function to maintain synthesis of viral proteins following virally-mediated eIF4G destruction, which inhibits host cap-dependent translation.[12] An increasing number of mRNAs that are translated in a cap-independent manner are being discovered.[13] For example, IRES-mediated initiation is involved in the increase in fibroblast growth factor-2 (FGF-2) translation in response to heat shock and oxidative stress, which decrease translation of most cellular proteins.[14] Another example is provided by vascular endothelial growth factor (VEGF) translation, which is increased in myocytes via IRES-dependent translation during hypoxic conditions that inhibit cap-dependent translation.[15] Such hypoxic conditions may exist in tissues in which circulation is compromised as a result of diabetic complications.

## Ribosomes

The functional eukaryotic ribosome consists of two subunits, 40S and 60S, both of which are complex macromolecular structures assembled from rRNA and protein components.[4,5] The small subunit functions independently of the large subunit in the initial stages of translation, involving scanning and selection of the start codon. Protein is synthesized after the two subunits have joined to create the 80S ribosome. Peptide bonds are formed in a ribozyme-mediated reaction catalyzed by the rRNA of the large subunit.[16]

Several ribosomal proteins are phosphorylated in response to insulin. S6 (rpS6), which is one of 26 proteins found in the 40S subunit, was one of the first proteins shown to undergo increased phosphorylation in response to insulin.[17] This discovery provided evidence that insulin controlled protein synthesis by phosphorylating components of the translation apparatus. The phosphorylation of rpS6 is also seen when growth arrested cells reenter the cell cycle.[18]

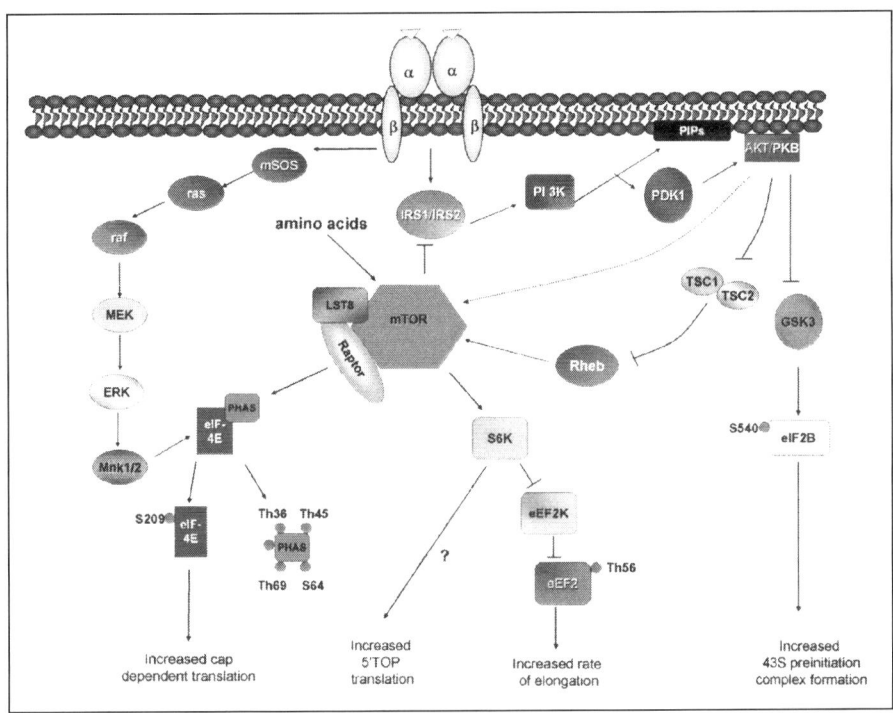

Figure 3. Signal transduction pathways involved in protein translation regulated by insulin signaling.

rpS6 is essential for ribosomal assembly, as deletion of the rpS6 gene results in the inability to produce functional 40s subunits.[19] rpS6 is located between the binding sites for the tRNA and the mRNA.[20,21] Phosphorylation of rpS6 is accompanied by an increase in total cellular protein synthesis, although the actual role of rpS6 phosphorylation remains uncertain.

Many kinases have been described that are able to phosphorylate rpS6.[22,23] In the case of insulin action, S6K1 appears to be the major kinase involved (Fig. 3). Insulin treatment results in a dramatic activation of S6K1 in a variety of cells and tissue. Both the phosphorylation of rpS6 and the activation of S6K1 are inhibited by rapamycin,[24] a potent and selective inhibitor of the mammalian target of rapamycin, mTOR. mTOR a member of a family of Ser/Thr protein kinases having catalytic domains homologous to that in phosphatidyl inositol 3-kinase (PI 3-kinase).[25] mTOR has emerged as a central element in the control of mRNA translation, and it is described later in this chapter.

## Initiation

To initiate translation the large and small ribosomal subunits must dissociate (Fig. 1). This process is facilitated by eIF1A and eIF3,[26] which bind to the small subunit and prevent its interaction with the large subunit. Met-tRNA$_i$ is delivered to the small ribosomal subunit in a ternary complex with eIF2 and GTP.[4,5] eIF3 and eIF1A strengthen the interaction between the ternary complex and 40S ribosome, thus stabilizing the 43S preinitiation complex, which contains eIF1A, eIF2, GTP, Met-tRNA$_i$, eIF3 and small ribosomal subunit. In a parallel and apparently independent process, eIF4F binds to the m$^7$Gppp cap at the 5' end of the mRNA.[4,5] eIF4F is a complex of three subunits—eIF4G, eIF4A, and eIF4E. eIF4A is an mRNA helicase, eIF4E is the cap-binding subunit, and eIF4G is a scaffolding protein that binds eIF4E and eIF4A. eIF4G also binds eIF3, which links eIF4F to the 43S preinitiation complex. The complex of eIF4F, mRNA and the components of the 43S preinitiation complex is known as the 48S initiation complex.

After the small ribosomal subunit is positioned at the 5' end of the mRNA in the 48S initiation complex, it then moves along, or scans, the 5-UTR to select the start codon.[27] This is usually the first AUG encountered, although scanning continues if the AUG is not found in the appropriate sequence context, often referred to as the Kozak consensus motif.[28] When the start codon is found, GTP on eIF2 is hydrolyzed to GDP, the initiation factors dissociate, the large ribosomal subunit binds, and Met-tRNA$_i$ becomes fixed in the p site. At this point the initiation phase is complete.

Initiation is generally the rate limiting phase in translation.[29] Not surprisingly, insulin and many other hormones that control protein synthesis exert their effects by either increasing or decreasing the rate of translation initiation. The effects of insulin involve changes in the activity or availability of initiation factors. Some of the initiation factors and effects of insulin are described below.

## *eIF2*

Before eIF2 can bind Met-tRNA$_i$ to form the ternary complex for initiation, GDP must be replaced with GTP. Nucleotide exchange on eIF2 is a potentially rate-limiting step in translation, and the reaction is controlled by several mechanisms.

GDP binds eIF2 with 400-fold higher affinity than GTP.[30] Thus, decreases in the GTP to GDP ratio, as occur with reductions in ATP, would be expected to inhibit translation by reducing the GTP for GDP exchange on eIF2. Presumably this mechanism functions to inhibit protein synthesis, thus conserving ATP, when the cellular energy supply is inadequate.

Phosphorylation of Ser51 in the α subunit of eIF2 is perhaps the best characterized mechanism for inhibiting nucleotide exchange on eIF2. Phosphorylation of this site markedly increases the affinity of eIF2 for the nucleotide exchange protein, eIF2B.[31] As the amount of eIF2 vastly exceeds that of eIF2B, phosphorylation of Ser51 sequesters the exchange factor and prevents it from acting on other eIF2 complexes.

The kinases that phosphorylate Ser51 are members of a family having highly homologous catalytic domains, but divergent regulatory domains that confer sensitivity to different stimuli.[32] The heme-controlled repressor (HCR) of translation was the first of these enzymes discovered.[33] HCR is inhibited by heme. Thus, as heme levels are depleted, the kinase is released from inhibition, resulting in phosphorylation of Ser51 and inhibition of translation. In this way HCR allows translation of globin to proceed only when the supply of heme is adequate for hemoglobin synthesis. Protein kinase R (PKR) is activated by double-stranded RNA, and is an important component of the host defense to viral infections.[34] Activation of the kinase by viral RNA turns off cellular translation, thereby limiting the synthesis of the proteins needed for viral replication.

In *Saccharomyces cerevisiae* phosphorylation of Ser51 is controlled by the kinase, GCN2.[35] This enzyme is interesting in the respect that its regulatory domain is highly homologous to His-tRNA synthetase. GCN2 is activated by increases in uncharged tRNA, which increases as the amino acid supply decreases. GCN2 function is also increased by dephosphorylation of Ser577 in the kinase.[36] Rapamycin-treatment of yeast was recently shown to reduce the phosphorylation of GCN2 and to increase the phosphorylation of eIF2α.[37] These exciting findings, which link GCN2 and the Tor signaling pathway in yeast, provide a reason to investigate the possible control of mGCN2 by mTOR.

PERK is the most recent member of the Ser51 kinase family to be discovered.[38] PERK spans the endoplasmic reticulum membrane, with its kinase domain facing out and its regulatory domain on the luminal side. The regulatory domain senses unfolded protein in the ER and triggers activation of the kinase, thereby attenuating translation and preventing accumulation of improperly folded proteins in the ER.[39] PERK is highly expressed in pancreatic β cells. Inactivating mutations in PERK in humans lead to Wolcott-Rallison syndrome.[40] The most dramatic phenotype in this syndrome is pancreatic β cell death with infancy-onset diabetes.

## eIF2B

eIF2B is composed of five subunits.[41] The β subunit binds GTP and participates in the nucleotide exchange function of the factor. The ε subunit is phosphorylated and appears to serve a regulatory role.[42] Insulin exerts both long term and acute effects on eIF2B activity. Inducing diabetes in animals decreases the amount of eIF2B in skeletal muscle, an effect that contributes to the reduction of translation initiation in diabetic muscle.[43] Thus, insulin is required to maintain levels of the eIF2B protein. Insulin also promotes a rapid increase in the activity of eIF2B. This response has been observed both in cultured cells[44] and in skeletal muscle in vivo.[42] Activation of eIF2B is associated with dephosphorylation of Ser540 in the ε subunit.[45] This site can be phosphorylated in vitro and in vivo by glycogen synthase kinase 3 (GSK-3) (Fig. 3), a kinase that is inactivated by insulin.[46] Since GSK-3 also phosphorylates and inactivates glycogen synthase,[47] the inhibition of GSK-3 potentially contributes to the stimulatory effects of the hormone on both glycogen synthesis and mRNA translation. In Chinese hamster ovary (CHO) cells selective inhibitors of the kinase reduced the phosphorylation of Ser540, but the inhibitors were without effect on eIF2B activity.[48] These results support the conclusion that the site is phosphorylated by GSK3 in cells, but provide reason to question whether inhibition of GSK3 is sufficient to explain the stimulatory effect of insulin on eIF2B activity.

There is evidence that insulin may control eIF2B activity through a mechanism involving Nck-1, an adaptor protein that is phosphorylated by the insulin receptor.[49] Nck-1 directly binds the β subunit of eIF2B, an interaction dependent on the first and third SH3 domains in Nck-1.[50] Insulin increases the amount of Nck-1 associated with ribosomes, and over-expressing Nck-1 in human embryonic kidney (HEK) 293 cells increased both cap-dependent and -independent translation, which would be consistent with a stimulatory effect on eIF2B activity.[50]

## eIF3

eIF3, which is the largest initiation factor, has several important roles in translation.[51] By binding to the 40S ribosomal subunit, eIF3 prevents association of the large and small ribosomal subunits, thus enhancing dissociation of the subunits, which must occur before translation initiation can proceed. eIF3 enhances binding of the ternary complex of eIF2, GTP, and Met-tRNA$_i$. eIF3 also provides a critical link between the small ribosomal subunit and the factors that are associated with eIF4G. Several of the eleven eIF3 subunits, which range in Mr from 35,000 to 170,000, contain RNA binding domains. Insulin has been shown to increase the phosphorylation of the eIF3b subunit,[52] although no functional effect of phosphorylating this subunit has been described.

Intriguingly, eIF3 subunits or homologous proteins have other important roles in the cell. For example, significant homology exists between some of the subunits of eIF3 and components of the COP9 signalsome.[53] This signaling complex was first detected in plants, where it is involved in adaptation to light. More recently, signalsome components have been found in yeast and mammalian cells, where the functions are less well understood.[54] Intriguingly, all of the subunits of eIF3 have counterparts in the lid of the proteosome.[55]

## eIF4A

eIF4A is a member of the DEAD box family of helicases, whose members participate in many aspects of RNA processing, including transcription, ribosome synthesis, translation, and degradation.[56] ATP binding and hydrolysis are believed to cause conformational changes in eIF4A aiding in its function as a helicase. Melting of secondary structure in the 5'-UTR of mRNA allows the binding of the 40S ribosome to the RNA and facilitates scanning.[57] While eIF4A alone is active as a helicase, its efficiency is markedly increased when it is part of eIF-4F.[58,59] Although insulin is not believed to directly control eIF4A activity, it may indirectly increase activity of the helicase by increasing availability of eIF4E, which results in an increase in eIF4F.[29] Another factor, eIF4B, is required for the helicase activity of eIF4A. eIF4B has been reported to undergo increased phosphorylation in response to insulin treatment;[7] however, no effect of insulin on the function of the factor has been demonstrated.

## eIF4E

The mRNA cap-binding protein, eIF4E, is needed to position the initiation machinery at the 5'end of the message, ensuring that scanning begins upstream of the start codon. It is the least abundant of the known initiation factors (estimated levels between 0.02-0.2 copies per ribosome),[60,61] and its availability appears to limit translation, at least in certain cell types. Binding of eIF4E to the cap can be the deciding factor in whether a particular mRNA will be translated. Overexpression of eIF4E increases preferentially the translation of mRNAs having structured 5'UTRs.[62,63] Presumably, this is due to enhanced formation of eIF4F, which facilitates melting of secondary structure in the 5'UTR by bringing eIF4A to this region. eIF4E is phosphorylated in response to insulin and many other hormones and growth factors that increase protein synthesis.[29] Because of the potential implications with respect to the control of cap-dependent translation, the phosphorylation of eIF4E has been intensively investigated.

Ser209 is the major phosphorylation site in eIF4E.[64-66] The kinase directly responsible for phosphorylating this site is Mnk1, a member of the Rsk family of protein kinases.[67,68] Mnk1 binds directly to the scaffold, eIF4G, and is thus positioned to phosphorylate of eIF4E.[67,69] Like other members of the Rsk family, Mnk1 is phosphorylated and activated by MAP kinases. The link to the MAP kinase pathway accounts for the wide variety of stimuli that increase phosphorylation of eIF4E.

Considering the extensive investigation, it is surprising that the role of eIF4E phosphorylation is still unclear. In earlier studies phosphorylated eIF4E purified from cells exhibited increased affinity for the $m^7$GpppN cap,[70] and the eIF4E found in the 48S complex was mostly in the phosphorylated form,[71] supporting the view that phosphorylation activates eIF4E. In more recent studies, direct phosphorylation of Ser209 in purified eIF4E by using Mnk1 was found to decrease the affinity of eIF4E for the cap.[72,73] The explanation for the discrepancy between this and the earlier finding of increased affinity is uncertain. When overexpressed in cells, eIF4E having a Ser209 to Ala mutation was equally effective as wild type eIF4E in initiating translation.[74] On the other hand, mutation of the corresponding site in *Drosophila melanogaster* eIF4E slowed development and decreased size of the flies,[75] suggesting that eIF4E phosphorylation is needed for normal growth and development. More research will be needed to define the physiological function of eIF4E phosphorylation.

## eIF4E-Binding Proteins

In contrast to the uncertainty regarding the role of eIF4E phosphorylation, control of the availability of eIF4E by a family of eIF4E-binding proteins has been established as an important mechanism. Three members of this family, known as PHAS-I, PHAS-II, and PHAS-III (also known as 4E-BP1, 4E-BP2, and 4E-BP3) are expressed in mammalian cells.[76,77] PHAS-I (Phosphorylated Heat- and Acid-stable protein stimulated by Insulin) was first detected in $^{32}$P-labeled adipocytes as one of several proteins that underwent increased phosphorylation in response to insulin.[78,79]

Nonphosphorylated PHAS-I binds tightly to eIF4E, and inhibits cap-dependent translation.[80] PHAS-I does not inhibit binding of eIF4E to capped mRNA.[80,81] Indeed, it actually appears to increase the affinity of eIF4E for the cap site.[82] Translation is repressed because PHAS-I inhibits eIF4E binding to eIF4G.[83,84] The inhibition is likely to be competitive, as the eIF4E binding site in PHAS-I and eIF4G are homologous. The other two PHAS proteins also bind eIF4E, and prevent binding to eIF4G.[76,85]

When phosphorylated in response to insulin, PHAS-I dissociates from eIF4E,[80,81] allowing eIF4E to engage eIF4G, a necessary step in the recruitment of eIF4G and other initiation factors to the 5' end of the mRNA. PHAS-II is phosphorylated in response to insulin,[86] which promotes dissociation of the PHAS-II/eIF4E complex.[85] In contrast, insulin has little, if any, effect on the amount of PHAS-III bound to eIF4E,[85,87] and effects of insulin on the phosphorylation of PHAS-III have not been reported.

In adipocytes insulin stimulates the phosphorylation of PHAS-I in four sites, all of which conform to a (Ser/Thr)Pro motif.[88,89] These sites correspond to Thr36, Thr45, Ser64, and

Thr69 in rat or mouse PHAS-I, and to Thr37, Thr46, Ser65, and Thr70 in the human protein, which contains an additional amino acid residue in the $NH_2$ terminal region. The four sites are conserved in the other two PHAS isoforms, as well as in PHAS protein in other species.[77] Phosphorylation of PHAS-I in cells occurs in an ordered fashion, with that of Thr36 and Thr45 occurring first, followed by Thr69, and finally Ser64.[90,91]

Phosphorylation of Ser111 in PHAS-I has also been described.[92] Ser111, which is followed by Gln, can be phosphorylated in vitro by casein kinase II,[93] as well as by ATM, ATR, and hSMG-1.[94] Like mTOR, the latter three enzymes belong to a family of protein kinases having catalytic domains more homologous to that in the lipid kinase PI 3-kinase than that in the ACG family of Ser/Thr protein kinases. ATM has been reported to be activated by insulin and to phosphorylate Ser111.[95] This finding was exciting when considered in the context of the previous proposal that phosphorylation of Ser111 serves as a priming event that must occur before the other sites can be phosphorylated.[92] However, recent results indicate that mutation of this site to Ala has little, if any, effect on the phosphorylation of the other sites in PHAS-I over-expressed in HEK293 cells.[85,96] Thus, a role of Ser111 as a priming site seems very unlikely. Phosphorylation of Ser100 has also been reported;[96] however, the relative level of phosphorylation was low and it did not change in response to insulin.

The phosphorylation events most important for the dissociation of PHAS-I and eIF4E have not been conclusively identified. In vitro, phosphorylation of Ser64 has the most dramatic effect on decreasing the affinity of PHAS-I for eIF4E.[88,90] However, mutating Ser64 to Ala had no discernable effect on the amount of PHAS-I bound to eIF4E in HEK293 cells, in either the absence or presence of insulin.[89,90] Phosphorylation of Thr45 also significantly decreases affinity of PHAS-I for eIF4E in vitro.[97] Presumably, phosphorylation of Thr45 or some combination of the three ThrPro sites leads to dissociation of eIF4E.

PHAS-I phosphorylation is controlled by the PKB signaling pathway.[98-100] Thus, inhibitors of PI 3-kinase, which blunt activation of PKB, reduce the phosphorylation of PHAS-I.[98] Moreover, overexpression of active PKB increases phosphorylation of PHAS-I, and phosphorylation is inhibited by overexpression of kinase-dead PKB.[98,99] The phosphorylation of PHAS-I is also attenuated by rapamycin, a finding that provided the first evidence that the effect of insulin was mediated by mTOR.[101]

## eIF4G

The scaffolding protein, eIF4G, binds eIF4E, eIF4A, eIF3, Mnk1, and polyA binding protein (Pab1).[102,103] It has been known for more than ten years that insulin stimulates the phosphorylation of eIF4G.[52] Three phosphorylation sites (Ser1108, Ser1148, and Ser1192) have been identified in eIF4G expressed in HEK293 cells;[104] however, the effect of phosphorylation of these sites remains unclear. Insulin increases the amount of eIF-4G bound to eIF-4E,[83,84] but this response is believed to be due to an increase in availability of eIF-4E resulting from phosphorylation of PHAS-I rather than a direct effect on eIF4G, itself. It will be important to determine whether the association of other proteins with eIF4G is controlled by insulin. eIF4G increases eIF4E's affinity for the cap structure,[105,106] similar to the effect of PHAS-I mentioned above. Therefore, taking these points into consideration, eIF4E is more likely to dissociate from the cap when not bound to either eIF4G or PHAS-I.

The control of eIF4G levels in cells is interesting in several respects. The eIF4G message has a TOP motif,[9] suggesting that its translation should be enhanced in response to insulin-stimulated activation of S6K1. The 5' end of the mRNA also contains an IRES element, which allows eIF4G to be translated in a cap-independent manner.[107] eIF4G is cleaved by certain viral proteases.[108] Cleavage by polio virus produces a fragment that is unable to bind eIF4E, but which retains binding of other initiation factors. The fragment participates in the synthesis of viral proteins, which occur by cap-independent mechanisms, but it is unable to mediate cap-dependent translation.[109,110] eIF4G protein is particularly sensitive to proteolysis by cellular proteases, most likely because of the PEST region in the N-terminal end of the protein.[111,112] Following cellular stress, proteolytic cleavage of eIF4G is evident within minutes, possibly due to activation of calpains.[111]

## eIF5

eIF5 functions to facilitate binding of the 60s ribosomal subunit with the 40S ribosome by stimulating the intrinsic GTPase activity of eIF2.[113] Hydrolysis of GTP decreases the affinity of eIF2 for both Met-tRNA$_i$ and the 40 S ribosome, resulting in the release of eIF2 from the ribosome.[114] The remaining initiation factors (eIF4F complex, eIF1A and eIF3) are released after eIF2, although the chain of events leading to their dissociation is not well defined. eIF5 may be phosphorylated by casein kinase II, an enzyme that has been reported to be activated by insulin.[115] However, whether eIF5 is phosphorylated in response to insulin treatment is not known.

## Elongation

Elongation is a cycle of three major steps.[4,5,116] In the first, aminoacyl-tRNAs (aa-tRNA) are delivered to the a site of the ribosome as ternary complexes with eEF1A and GTP (Fig. 2). The appropriate aa-tRNA is selected by anticodon-codon pairing between the tRNA and the mRNA. In the second step, a peptide bond is formed between the peptidyl-tRNA (or Met-tRNA$_i$) bound in the p site and the aa-tRNA. In the third step, the peptidyl-tRNA translocates to the p site, and the next codon in line is positioned in the a site. Translocation is catalyzed by eEF2 in a reaction driven by GTP hydrolysis (Fig. 2). The cycle repeats until a stop codon is reached, at which point the elongation phase is over.

In eukaryotic cells elongation occurs at a rate of between 3 to 8 amino acids per second.[117,118] Under conditions in which initiation is not rate limiting, insulin may increase protein synthesis by stimulating elongation. Stimulation of elongation provides a means for stimulation of total protein synthesis, whereas effects on initiation can be relatively selective for certain classes of mRNA, depending on the mechanisms involved.

## eEF1A

eEF1A is homologous to members of the Ras family of small GTP-binding proteins. Depending on the cell type, the factor accounts for 1-3% of the soluble protein a cell,[119] an abundance indicative of the important function of eEF1A of delivering all aa-tRNAs (except Met-tRNA$_i$) to the ribosome. The translation of eEF1A mRNA is controlled in parallel to that of ribosomal proteins due to the presence of a TOP motif in the mRNA encoding the factor.

The accuracy of translation is absolutely dependent on correct pairings between the anticodons of the aa-tRNAs and the codons of the mRNA. An aa-tRNA becomes fixed in the a site following the hydrolysis of GTP and release of eEF1A.[116] Increasing the GTPase activity of eIF1A decreases the time allowed for dissociation of mismatched aa-tRNAs. Consequently, the rate of hydrolysis by eEF1A is inversely related to the fidelity of translation. Insulin has been reported to increase the phosphorylation of eEF1A,[120] although the effect of phosphorylation was not specifically addressed. Increasing the GTPase activity of eEF1A could increase the elongation rate, but the incidence of mistakes in the translated product would be expected to increase. For this reason, it would be surprising if the eEF1A-mediated step of elongation were markedly accelerated by insulin.

## eEF1B

Exchange of GTP for GDP on eEF1A is essential for recycling eEF1A, a reaction catalyzed by eEF1B. eEF1B is composed of three subunits ($\alpha\beta\gamma$).[116] The $\alpha$ and $\beta$ subunits are homologous and have GTP exchange activity; the $\gamma$ subunit may link the other two subunits. Both the $\alpha$ and $\beta$ subunits were reported to be phosphorylated in response to insulin in 3T3L1 cells.[120] A partially purified preparation containing both eEF1A and eEF1B from insulin-treated cells supported higher rates of elongation in a rabbit reticulocyte lysate system than such a preparation from control cells.[120] Whether increasing the rate of exchange of GTP for GDP on eEF1B in cells would facilitate elongation is still unclear.

## eEF2

Conformational changes in eEF2 produced by GTP hydrolysis drive translocation.[116] Unlike eEF1A, eEF2 does not appear to require a guanine nucleotide exchange factor for its activity or recycling. eEF2 contains a unique amino acid, dipthamide, which is generated by modification of His715.[121] Dipthamide is ADP ribosylated by diphtheria toxin, which inhibits eEF2 activity. eEF2 activity is also inhibited by phosphorylation in Thr56, which is located within the GTP binding domain.[122,123] Thr56 is phosphorylated by eEF2 kinase, previously known as calmodulin-kinase III.[124,125] As its former name implies, the activity of this enzyme is increased in response to elevations in cytosolic calcium, contributing to the inhibition of translation under these conditions. eEF2 kinase, itself, is controlled by phosphorylation, which decreases kinase activity.[116]

Insulin treatment of CHO-IR cells, which over-express the human insulin receptor, produced a rapid phosphorylation and inactivation of eEF2 kinase.[126] This effect of insulin was inhibited by treating the cells with rapamycin.[125,126] A prime candidate for phosphorylating eEF2 kinase is S6K1, which has been shown to phosphorylate Ser540 in eEF2 kinase and to inhibit eIF2 kinase activity.[127]

## Termination

Elongation continues until one of three stop codons (UGA, UAA, UAG) is encountered.[128] At this point eRF1-GTP binds to the codon in place of aa-tRNA-GTP-eEF1A. Following hydrolysis of GTP by eRF1, peptidyltRNA is hydrolyzed, releasing the newly synthesized protein from the ribosome. Active eRF1 is regenerated by eRF3, which catalyzes the exchange of GTP for GDP on eRF1. There is no known link between insulin and termination.

## The mTOR Signaling Pathway

As mentioned in the previous sections, several steps in initiation and elongation are controlled by the mTOR signaling pathway. mTOR is the 2549 amino acid mammalian counterpart of the Tor1p and Tor2p proteins in *Saccharomyces cerevisiae*, that are involved in nutrient sensing and cell growth.[129] In view of its central role of insulin in controlling nutrient stores, it is not surprising that insulin also utilizes the mTOR pathway to control translation.[130]

### Structure of mTOR

mTOR has several structural features that are shared with other members of the PI 3-kinase-related family of protein kinases.[25,131] As in other family members, the kinase domain in mTOR is located in the COOH terminal region of the protein. The extreme COOH terminal region forms the FATC domain, which is highly conserved in all members of the PI 3-kinase-related family of protein kinases. The FAT domain, which is another conserved region, is located on the $NH_2$ terminal side of the FKBP12-rapamycin binding domain (FRB) in mTOR. The sequence containing the FRB, which is located just upstream of the catalytic domain in mTOR, is not found in other members of the family, except in SMG-1 where there is some homology.[132] mTOR is the only kinase that binds rapamycin-FKBP12 with high affinity. A region located on the $NH_2$ terminal side of the FAT domain and comprising approximately half of the mTOR protein contains 20 HEAT motifs, that presumably serve as docking sites for either regulators or targets of mTOR. Aside from the FRB and catalytic domains, the functions of the domains in mTOR have not been established.

### mTOR-Interacting Proteins

Recent studies have identified two proteins that associate with mTOR.[133-138] The first of these, raptor, is large protein possessing a unique $NH_2$ terminal region followed by three HEAT motifs and seven WD-40 repeats.[134,138] The second, mLST8 (also known as GβL[136]), consists almost entirely of seven WD-40 repeats and is homologous in sequence to the beta subunits of heterotrimeric G proteins. Binding of mLST8 appears to increase the kinase activity of mTOR.[136]

The effect of raptor is less clear, and there is evidence for both stimulation and inhibition of mTOR. Additional research will be needed to resolve this discrepancy.[136,138]

It is striking that the two mTOR interacting proteins, raptor and mLST8, contain a total of fourteen WD40 repeats, which are frequently sites for protein-protein interactions. Thus, it is likely that many proteins that interact with the mTOR complex remain to be discovered.

## Phosphorylation of PHAS-I and S6K1 by mTOR

The best characterized substrates for mTOR are PHAS-I and S6K1. Overexpressing mTOR in cells results in the phosphorylation of both PHAS-I and S6K1.[139,140] In vitro mTOR phosphorylates the four Ser/ThrPro sites in PHAS-I that are phosphorylated in response to insulin.[88] The preferred site for mTOR in S6K1 is Thr389,[141] which is flanked by hydrophobic residues. Thus, the nature of the sites phosphorylated by mTOR in PHAS-I and S6K1 are distinctly different. Based on studies of the ACG family of protein kinases, it would be very unusual for a protein kinase to phosphorylate sites as different as the sites in PHAS-I and S6K1. However, it may not be reasonable to expect mTOR to obey the rules for ACG kinases, given mTOR's homology to PI 3 kinase, which is able to phosphorylate both lipid and protein. At any rate, the differences between the sites in PHAS-I and S6K1 indicate that primary amino acid sequence is not the sole determinant for recognition by mTOR.

There is evidence that raptor is involved in substrate recognition by mTOR. Raptor binds directly to both PHAS-I and S6K1.[135,136] The high-affinity binding of these proteins to raptor is dependent on the TOS (Tor signaling) motif, which is formed by the last five amino acids (PheGluMetAspIle) in PHAS-I and a sequence near the $NH_2$ terminus of S6K1 (PheAspIleAspLeu,).[133,137,142] A Phe to Ala point mutation in the TOS motif is sufficient to disrupt binding of either protein to raptor in vitro.[133,135,136] Another motif, referred to as the RAIP motif for the amino acids involved (Arg,Ala,Ile,Pro), located in the $NH_2$ terminal region of PHAS-I.[143] Removing the RAIP motif either by $\Delta 16$ $NH_2$ terminal truncation or by mutating IlePro to AlaAla, markedly decreases phosphorylation of PHAS-I in cells.[143] These mutations also abolish high-affinity binding to raptor and reduce phosphorylation by mTOR.[136] The findings that mutations in the TOS and/or RAIP motifs both inhibit binding of PHAS-I and S6K to raptor and decrease the phosphorylation of the two proteins in cells, support the view that mTOR directly phosphorylates these proteins in vivo.

## Control of mTOR Activity

Studies in adipocytes and skeletal muscle indicate that mTOR activity and phosphorylation are controlled by insulin. Incubating 3T3-L1 adipocytes increased mTOR activity measured in immune complex kinases assays with recombinant PHAS-I as substrate.[144,145] Insulin has been shown to promote the phosphorylation of Ser2448[146-148] in both fat cells and skeletal muscle.

Interestingly, there is evidence that mTOR is controlled by the PI 3-kinase pathway, which is believed to mediate most, if not all, of the important metabolic actions of insulin. Inhibition of PI 3-kinase abolished both the activation of mTOR by insulin and the insulin-stimulated phosphorylation of Ser2448.[145,148] PKB, which is activated by products of the PI 3-kinase reaction, has been implicated in the phosphorylation of this site. PKB directly phosphorylates Ser2448 in vitro.[148] Moreover, this site is phosphorylated in response to PKB activation in cells, and overexpression of a dominant-negative PKB blocks phosphorylation of the site.[145,148]

Ser2448 forms a part of the epitope for the antibody, mTAb1,[149] whose binding to mTOR is abolished when the site is phosphorylated.[145] mTAb1 binding markedly increases the kinase activity of mTOR,[139] and a mutant mTOR lacking the 20 amino acids that form the mTAb1 epitope, exhibits increased kinase activity.[148,150] Thus, it seems logical to suspect that phosphorylation of Ser2448 leads to activation of mTOR. On the other hand, the effects on S6K1 activity and PHAS-I phosphorylation of overexpressing wild type and mTOR with a Ser2448 to Ala mutation were indistinguishable.[148] While arguing against an important role of Ser2448 phosphorylation, the results do not exclude the possibility that there are redundant mechanisms of activation. Also, the recent findings that mTOR functions in a complex containing mLST8 and raptor,[133,135-137] which were not overexpressed in the previous study in which Ser2448 was mutated, complicates the interpretation.

While the question of whether Ser2448 phosphorylation is important in the control of mTOR function remains open, recent studies in several laboratories have definitively placed PKB upstream of mTOR and have identified other key components of the signaling pathway controlling mTOR function. PKB phosphorylates TSC2 (also known as tuberin), a protein that functions in a heterodimeric complex with TSC1 (also known as hamartin) to suppress mTOR signaling.[151-156] Mutations in the genes encoding TSC1 and TSC2 lead to tuberous sclerosis.[157] This genetic disorder is transmitted in an autosomal-dominant manner and causes serious defects in a variety of organ systems, due to the development of hamartomas. These typically benign tumors often contain very large cells, consistent with the expected effect of an over-active mTOR pathway. Experimentally over-expressing TSC1-TSC2 decreases the phosphorylation of the two mTOR targets, PHAS-I and S6K.[151,155] Likewise, reducing TSC1-TSC2 increases the phosphorylation of S6K-1.[153] Phosphorylation of TSC2 has been reported to promote dissociation of the TSC1-TSC2 complex,[155,156] although there is conflicting evidence on this point.[152,154] Phosphorylation in response to PKB has also been suggested to change the subcellular distribution of TSC1-TSC2,[156] and to accelerate the degradation of the proteins.[154,155] All of these effects represent potential mechanisms through which PKB activation could attenuate TSC1-TSC2 function and derepress mTOR.

TSC1 contains a potential transmembrane spanning domain and TSC2 has a GTPase activating protein (GAP) domain, suggesting that the complex controls a small GTP-binding protein. The search for the target of TSC1-TSC2 was recently rewarded. TSC2 functions as a GAP for Rheb,[158,159] a ras homologue enriched in brain.[160] Epistatic studies in *Drosophila melanogaster* place Rheb downstream of PKB and TSC1-TSC2, and upstream of mTOR.[161,162] When overexpressed in mammalian cells, Rheb increases the phosphorylation of the mTOR targets, PHAS-I and S6K1.[158] Insulin increases GTP-Rheb, suggesting that it is the GTP-bound form of Rheb that increases mTOR function. The current model is that insulin-activation of PKB leads to TSC2 phosphorylation, which decreases Rheb GAP activity (Fig. 3), thereby increasing the active GTP-bound form of Rheb. How Rheb actually increases mTOR activity is still a mystery. This critical piece of the puzzle will need to be found before insulin action on mTOR is fully understood.

In mammalian cells mTOR also functions in a nutrient sensing pathway, allowing translation to proceed when the supply of amino acids is adequate. Although the mechanism is not clear, branched-chain amino acids, in particular leucine, activate mTOR, thereby increasing the phosphorylation of both PHAS-I and S6K1.[163,164] This pathway is clearly distinct from the one utilized by insulin, as it is not blocked by low doses of wortmanin, which abolish the stimulatory effects of insulin on PHAS-I and S6K1.[165,166]

## Concluding Remarks

mRNA translation is an extremely complex process mediated by a myriad of translation factors. Insulin stimulates both the initiation and elongation phases of translation. Several eIFs and eEFs have been shown to be controlled by insulin, although the mechanisms involved in the control of translation are far from been fully defined. The effect of insulin-stimulated phosphorylation of a number of factors remains to be elucidated, and new targets of insulin are sure to be discovered. The mTOR signaling pathway, which functions in an important role in a nutrient sensing pathway conserved from yeast to man, also has a key role in the control of translation by insulin. How insulin controls the activity of mTOR is an outstanding question.

## References

1. Kimball SR, Farrell PA, Jefferson LS. Role of insulin in translational control of protein synthesis in skeletal muscle by amino acids or exercise. J Appl Physiol 2002; 93(3):1168-80.
2. Proud CG, Denton RM. Molecular mechanisms for the control of translation by insulin. Bio Chem J 1997; 328(Pt 2):329-41.

3. O'Brien RM, Granner DK. Gene Regulation. In: Leroith D, Taylor SI, Olefsky JM, eds. Diabetes Mellitus: A Fundamental and Clinical Text. Philadelphia: Lippincott Williams and Wilkins, 2000:291-312.
4. Hershey JW. Translational control in mammalian cells. Annu Rev Biochem 1991; 60:717-755.
5. Merrick WC. Mechanism and regulation of eukaryotic protein synthesis. Microbiol Rev 1992; 56(2):291-315.
6. van der Velden AW, Thomas AA. The role of the 5' untranslated region of an mRNA in translation regulation during development. Int J Biochem Cell Biol 1999; 31(1):87-106.
7. Manzella JM, Rychlik W, Rhoads RE et al. Insulin induction of ornithine decarboxylase. Importance of mRNA secondary structure and phosphorylation of eucaryotic initiation factors eIF-4B and eIF-4E. J Biol Chem 1991; 266(4):2383-9.
8. Kozak M. Influences of mRNA secondary structure on initiation by eukaryotic ribosomes. Proc Natl Acad Sci USA 1986; 83(9):2850-2854.
9. Meyuhas O. Synthesis of the translational apparatus is regulated at the translational level. Eur J Biochem 2000; 267(21):6321-6330.
10. Loreni F, Thomas G, Amaldi F. Transcription inhibitors stimulate translation of 5' TOP mRNAs through activation of S6 kinase and the mTOR/FRAP signalling pathway. Eur J Bio Chem 2000; 267(22):6594-6601.
11. Tang H, Hornstein E, Stolovich M et al. Amino acid-induced translation of TOP mRNAs is fully dependent on phosphatidylinositol 3-kinase-mediated signaling, is partially inhibited by rapamycin, and is independent of S6K1 and rpS6 phosphorylation. Mol Cell Biol 2001; 21(24):8671-8683.
12. Sonenberg N. Picornavirus RNA translation continues to surprise. Trends Genet 1991; 7(4):105-106.
13. Vagner S, Galy B, Pyronnet S. Irresistible IRES. Attracting the translation machinery to internal ribosome entry sites. EMBO Rep 2001; 2(10):893-898.
14. Creancier L, Morello D, Mercier P et al. Fibroblast growth factor 2 internal ribosome entry site (IRES) activity ex vivo and in transgenic mice reveals a stringent tissue-specific regulation. J Cell Biol 2000; 150(1):275-281.
15. Huez I, Creancier L, Audigier S et al. Two independent internal ribosome entry sites are involved in translation initiation of vascular endothelial growth factor mRNA. Mol Cell Biol 1998; 18(11):6178-6190.
16. Nissen P, Hansen J, Ban N et al. The structural basis of ribosome activity in peptide bond synthesis. Science 2000; 289(5481):920-930.
17. Smith CJ, Rubin CS, Rosen OM. Insulin-treated 3T3-L1 adipocytes and cell-free extracts derived from them incorporate 32P into ribosomal protein S6. Proc Natl Acad Sci USA 1980; 77(5):2641-2645.
18. Thomas G, Siegmann M, Gordon J. Multiple phosphorylation of ribosomal protein S6 during transition of quiescent 3T3 cells into early G1, and cellular compartmentalization of the phosphate donor. Proc Natl Acad Sci USA 1979; 76(8):3952-3956.
19. Volarevic S, Stewart MJ, Ledermann B et al. Proliferation, but not growth, blocked by conditional deletion of 40S ribosomal protein S6. Science 2000; 288(5473):2045-2047.
20. Nygard O, Nika H. Identification by RNA-protein cross-linking of ribosomal proteins located at the interface between the small and the large subunits of mammalian ribosomes. EMBO J 1982; 1(3):357-362.
21. Nygard O, Nilsson L. Translational dynamics. Interactions between the translational factors, tRNA and ribosomes during eukaryotic protein synthesis. Eur J Biochem 1990; 191(1):1-17.
22. Thomas G. The S6 kinase signaling pathway in the control of development and growth. Biol Res 2002; 35(2):305-313.
23. Avruch J, Belham C, Weng Q et al. The p70 S6 kinase integrates nutrient and growth signals to control translational capacity. Prog Mol Subcell Biol 2001; 26:115-154.
24. Cheatham L, Monfar M, Chou MM et al. Structural and functional analysis of pp70S6k. Proc Natl Acad Sci USA 1995; 92(25):11696-11700.
25. Abraham RT. Identification of TOR signaling complexes: More TORC for the cell growth engine. Cell 2002; 111(1):9-12.
26. Majumdar R, Bandyopadhyay A, Maitra U. Mammalian translation initiation factor eIF1 functions with eIF1A and eIF3 in the formation of a stable 40 S preinitiation complex. J Biol Chem 2003; 278(8):6580-6587.
27. Kozak M. Pushing the limits of the scanning mechanism for initiation of translation. Gene 2002; 299(1-2):1-34.
28. Kozak M. The scanning model for translation: An update. J Cell Biol 1989; 108(2):229-241.
29. Gingras AC, Raught B, Sonenberg N. eIF4 initiation factors: Effectors of mRNA recruitment to ribosomes and regulators of translation. Annu Rev Biochem 1999; 68:913-963.

30. Panniers R, Rowlands AG, Henshaw EC. The effect of Mg2+ and guanine nucleotide exchange factor on the binding of guanine nucleotides to eukaryotic initiation factor 2. J Biol Chem 1988; 263(12):5519-5525.
31. Webb BL, Proud CG. Eukaryotic initiation factor 2B (eIF2B). Int J Biochem Cell Biol 1997; 29(10):1127-31.
32. Dever TE. Translation initiation: Adept at adapting. Trends Biochem Sci 1999; 24(10):398-403.
33. Tahara SM, Traugh JA, Sharp SB et al. Effect of hemin on site-specific phosphorylation of eukaryotic initiation factor 2. Proc Natl Acad Sci USA 1978; 75(2):789-793.
34. Clemens MJ, Elia A. The double-stranded RNA-dependent protein kinase PKR: Structure and function. J Interferon Cytokine Res 1997; 17(9):503-524.
35. Hinnebusch AG. The eIF-2 alpha kinases: Regulators of protein synthesis in starvation and stress. Semin Cell Biol 1994; 5(6):417-426.
36. Cherkasova VA, Hinnebusch AG. Translational control by TOR and TAP42 through dephosphorylation of eIF2alpha kinase GCN2. Genes Dev 2003; 17(7):859-872.
37. Kubota H, Obata T, Ota K et al. Rapamycin-induced translational derepression of GCN4 mRNA involves a novel mechanism for activation of the eIF2alpha kinase GCN2. J Biol Chem 2003.
38. Harding HP, Zhang Y, Ron D. Protein translation and folding are coupled by an endoplasmic-reticulum-resident kinase. Nature 1999; 397(6716):271-274.
39. Harding HP, Ron D. Endoplasmic reticulum stress and the development of diabetes: A review. Diabetes 2002; 51(Suppl 3):S455-S461.
40. Zhang P, McGrath B, Li S et al. The PERK eukaryotic initiation factor 2 alpha kinase is required for the development of the skeletal system, postnatal growth, and the function and viability of the pancreas. Mol Cell Biol 2002; 22(11):3864-3874.
41. Proud CG. Regulation of eukaryotic initiation factor eIF2B. Prog Mol Subcell Biol 2001; 26:95-114.
42. Kimball SR, Horetsky RL, Jagus R et al. Expression and purification of the alpha-subunit of eukaryotic initiation factor eIF2: Use as a kinase substrate. Protein Expr Purif 1998; 12(3):415-419.
43. Karinch AM, Kimball SR, Vary TC et al. Regulation of eukaryotic initiation factor-2B activity in muscle of diabetic rats. Am J Physiol 1993; 264(1 Pt 1):E101-E108.
44. Welsh GI, Stokes CM, Wang X et al. Activation of translation initiation factor eIF2B by insulin requires phosphatidyl inositol 3-kinase. FEBS Lett 1997; 410(2-3):418-22.
45. Welsh GI, Miller CM, Loughlin AJ et al. Regulation of eukaryotic initiation factor eIF2B: Glycogen synthase kinase-3 phosphorylates a conserved serine which undergoes dephosphorylation in response to insulin. FEBS Lett 1998; 421(2):125-30.
46. Welsh GI, Proud CG. Glycogen synthase kinase-3 is rapidly inactivated in response to insulin and phosphorylates eukaryotic initiation factor eIF-2B. Biochem J 1993; 294(Pt 3):625-9.
47. Cohen P, Frame S. The renaissance of GSK3. Nat Rev Mol Cell Biol 2001; 2(10):769-776.
48. Wang X, Janmaat M, Beugnet A et al. Evidence that the dephosphorylation of Ser(535) in the epsilon-subunit of eukaryotic initiation factor (eIF) 2B is insufficient for the activation of eIF2B by insulin. Biochem J 2002; 367(Pt 2):475-481.
49. Lee CH, Li W, Nishimura R et al. Nck associates with the SH2 domain-docking protein IRS-1 in insulin-stimulated cells. Proc Natl Acad Sci USA 1993; 90(24):11713-11717.
50. Kebache S, Zuo D, Chevet E et al. Modulation of protein translation by Nck-1. Proc Natl Acad Sci USA 2002; 99(8):5406-5411.
51. Hershey JW, Asano K, Naranda T et al. Conservation and diversity in the structure of translation initiation factor EIF3 from humans and yeast. Biochimie 1996; 78(11-12):903-907.
52. Morley SJ, Traugh JA. Differential stimulation of phosphorylation of initiation factors eIF-4F, eIF-4B, eIF-3, and ribosomal protein S6 by insulin and phorbol esters. J Biol Chem 1990; 265(18):10611-6.
53. Wei N, Tsuge T, Serino G et al. The COP9 complex is conserved between plants and mammals and is related to the 26S proteasome regulatory complex. Curr Biol 1998; 8(16):919-922.
54. Hoareau AK, Bochard V, Rety S et al. Association of the mammalian proto-oncoprotein Int-6 with the three protein complexes eIF3, COP9 signalosome and 26S proteasome. FEBS Lett 2002; 527(1-3):15-21.
55. Morris-Desbois C, Rety S, Ferro M et al. The human protein HSPC021 interacts with Int-6 and is associated with eukaryotic translation initiation factor 3. J Biol Chem 2001; 276(49):45988-45995.
56. Rogers Jr GW, Komar AA, Merrick WC. eIF4A: The godfather of the DEAD box helicases. Prog Nucleic Acid Res Mol Biol 2002; 72:307-331.
57. Lorsch JR, Herschlag D. The DEAD box protein eIF4A. 2. A cycle of nucleotide and RNA-dependent conformational changes. Biochemistry 1998; 37(8):2194-2206.
58. Shimogori T, Suzuki T, Kashiwagi K et al. Enhancement of helicase activity and increase of eIF-4E phosphorylation in ornithine decarboxylase-overproducing cells. Biochem Biophys Res Commun 1996; 222(3):748-752.

59. Pause A, Methot N, Svitkin Y et al. Dominant negative mutants of mammalian translation initiation factor eIF-4A define a critical role for eIF-4F in cap-dependent and cap-independent initiation of translation. EMBO J 1994; 13(5):1205-1215.
60. Hiremath LS, Webb NR, Rhoads RE. Immunological detection of the messenger RNA cap-binding protein. J Biol Chem 1985; 260(13):7843-7849.
61. Duncan R, Hershey JW. Identification and quantitation of levels of protein synthesis initiation factors in crude HeLa cell lysates by two-dimensional polyacrylamide gel electrophoresis. J Biol Chem 1983; 258(11):7228-7235.
62. De Benedetti A, Rhoads RE. Overexpression of eukaryotic protein synthesis initiation factor 4E in HeLa cells results in aberrant growth and morphology. Proc Natl Acad Sci USA 1990; 87(21):8212-8216.
63. Pelletier J, Sonenberg N. Insertion mutagenesis to increase secondary structure within the 5' noncoding region of a eukaryotic mRNA reduces translational efficiency. Cell 1985; 40(3):515-526.
64. Shibata S, Morino S, Tomoo K et al. Effect of mRNA cap structure on eIF-4E phosphorylation and cap binding analyses using Ser209-mutated eIF-4Es. Biochem Biophys Res Commun 1998; 247(2):213-216.
65. Whalen SG, Gingras AC, Amankwa L et al. Phosphorylation of eIF-4E on serine 209 by protein kinase C is inhibited by the translational repressors, 4E-binding proteins. J Biol Chem 1996; 271(20):11831-11837.
66. Makkinje A, Xiong H, Li M et al. Phosphorylation of eukaryotic protein synthesis initiation factor 4E by insulin-stimulated protamine kinase. J Biol Chem 1995; 270(24):14824-14828.
67. Waskiewicz AJ, Flynn A, Proud CG et al. Mitogen-activated protein kinases activate the serine/threonine kinases Mnk1 and Mnk2. EMBO J 1997; 16(8):1909-20.
68. Knauf U, Tschopp C, Gram H. Negative regulation of protein translation by mitogen-activated protein kinase-interacting kinases 1 and 2. Mol Cell Biol 2001; 21(16):5500-5511.
69. Pyronnet S, Imataka H, Gingras AC et al. Human eukaryotic translation initiation factor 4G (eIF4G) recruits mnk1 to phosphorylate eIF4E. EMBO J 1999; 18(1):270-279.
70. Minich WB, Balasta ML, Goss DJ et al. Chromatographic resolution of in vivo phosphorylated and nonphosphorylated eukaryotic translation initiation factor eIF-4E: Increased cap affinity of the phosphorylated form. Proc Natl Acad Sci USA 1994; 91(16):7668-7672.
71. Rau M, Ohlmann T, Morley SJ et al. A reevaluation of the cap-binding protein, eIF4E, as a rate-limiting factor for initiation of translation in reticulocyte lysate. J Biol Chem 1996; 271(15):8983-90.
72. Zuberek J, Wyslouch-Cieszynska A, Niedzwiecka A et al. Phosphorylation of eIF4E attenuates its interaction with mRNA 5' cap analogs by electrostatic repulsion: Intein-mediated protein ligation strategy to obtain phosphorylated protein. RNA 2003; 9(1):52-61.
73. Scheper GC, van Kollenburg B, Hu J et al. Phosphorylation of eukaryotic initiation factor 4E markedly reduces its affinity for capped mRNA. J Biol Chem 2002; 277(5):3303-3309.
74. McKendrick L, Morley SJ, Pain VM et al. Phosphorylation of eukaryotic initiation factor 4E (eIF4E) at Ser209 is not required for protein synthesis in vitro and in vivo. Eur J Biochem 2001; 268(20):5375-5385.
75. Lachance PE, Miron M, Raught B et al. Phosphorylation of eukaryotic translation initiation factor 4E is critical for growth. Mol Cell Biol 2002; 22(6):1656-1663.
76. Lawrence Jr JC, Brunn GJ. Insulin signaling and the control of PHAS-I phosphorylation. Prog Mol Subcell Biol 2001; 26:1-31.
77. Rousseau D, Gingras AC, Pause A et al. The eIF4E-binding proteins 1 and 2 are negative regulators of cell growth. Oncogene 1996; 13(11):2415-2420.
78. Hu C, Pang S, Kong X et al. Molecular cloning and tissue distribution of PHAS-I, an intracellular target for insulin and growth factors. Proc Natl Acad Sci USA 1994; 91(9):3730-3734.
79. Belsham GJ, Denton RM. The effect of insulin and adrenaline on the phosphorylation of a 22 000-molecular weight protein within isolated fat cells; possible identification as the inhibitor-1 of the 'general phosphatase'. Biochem Soc Trans 1980; 8(3):382-383.
80. Pause A, Belsham GJ, Gingras AC et al. Insulin-dependent stimulation of protein synthesis by phosphorylation of a regulator of 5'-cap function. Nature 1994; 371(6500):762-7.
81. Lin TA, Kong X, Haystead TA et al. PHAS-I as a link between mitogen-activated protein kinase and translation initiation. Science 1994; 266(5185):653-6.
82. Youtani T, Tomoo K, Ishida T et al. Regulation of human eIF4E by 4E-BP1: Binding analysis using surface plasmon resonance. IUBMB Life 2000; 49(1):27-31.
83. Mader S, Lee H, Pause A et al. The translation initiation factor eIF-4E binds to a common motif shared by the translation factor eIF-4 gamma and the translational repressors 4E-binding proteins. Mol Cell Biol 1995; 15(9):4990-7.
84. Haghighat A, Mader S, Pause A et al. Repression of cap-dependent translation by 4E-binding protein 1: Competition with p220 for binding to eukaryotic initiation factor-4E. EMBO J 1995; 14(22):5701-9.

85. Fergusson G, Mothe-Satney I, Lawrence Jr JC. Serine phosphorylation sites in PHAS-I are dispensable for insulin stimulated dissociation from eIF4E. (In press).
86. Lin TA, Lawrence Jr JC. Control of the translational regulators PHAS-I and PHAS-II by insulin and cAMP in 3T3-L1 adipocytes. J Biol Chem 1996; 271(47):30199-204.
87. Kleijn M, Scheper GC, Wilson ML et al. Localisation and regulation of the eIF4E-binding protein 4E-BP3. FEBS Lett 2002; 532(3):319-323.
88. Mothe-Satney I, Brunn GJ, McMahon LP et al. Mammalian target of rapamycin-dependent phosphorylation of PHAS-I in four (S/T)P sites detected by phospho-specific antibodies. J Biol Chem 2000; 275(43):33836-43.
89. Fadden P, Haystead TA, Lawrence Jr JC. Identification of phosphorylation sites in the translational regulator, PHAS-I, that are controlled by insulin and rapamycin in rat adipocytes. J Biol Chem 1997; 272(15):10240-7.
90. Mothe-Satney I, Yang D, Fadden P et al. Multiple mechanisms control phosphorylation of PHAS-I in five (S/T)P sites that govern translational repression. Mol Cell Biol 2000; 20(10):3558-67.
91. Gingras AC, Gygi SP, Raught B et al. Regulation of 4E-BP1 phosphorylation: A novel two-step mechanism. Genes Dev 1999; 13(11):1422-1437.
92. Heesom KJ, Avison MB, Diggle TA et al. Insulin-stimulated kinase from rat fat cells that phosphorylates initiation factor 4E-binding protein 1 on the rapamycin-insensitive site (serine-111). Biochem J 1998; 336(Pt 1):39-48.
93. Fadden P, Haystead TA, Lawrence Jr JC. Phosphorylation of the translational regulator, PHAS-I, by protein kinase CK2. FEBS Lett 1998; 435(1):105-109.
94. Abraham RT. Cell cycle checkpoint signaling through the ATM and ATR kinases. Genes Dev 2001; 15(17):2177-2196.
95. Yang DQ, Kastan MB. Participation of ATM in insulin signalling through phosphorylation of eIF-4E-binding protein 1. Nat Cell Biol 2000; 2(12):893-8.
96. Wang X, Li W, Parra JL et al. The C terminus of initiation factor 4E-binding protein 1 contains multiple regulatory features that influence its function and phosphorylation. Mol Cell Biol 2003; 23(5):1546-57.
97. Yang D, Brunn GJ, Lawrence Jr JC. Mutational analysis of sites in the translational regulator, PHAS-I, that are selectively phosphorylated by mTOR. FEBS Lett 1999; 453(3):387-390.
98. Gingras AC, Kennedy SG, O'Leary MA et al. 4E-BP1, a repressor of mRNA translation, is phosphorylated and inactivated by the Akt(PKB) signaling pathway. Genes Dev 1998; 12(4):502-13.
99. Takata M, Ogawa W, Kitamura T et al. Requirement for Akt (protein kinase B) in insulin-induced activation of glycogen synthase and phosphorylation of 4E-BP1 (PHAS-1). J Biol Chem 1999; 274(29):20611-8.
100. Kohn AD, Barthel A, Kovacina KS et al. Construction and characterization of a conditionally active version of the serine/threonine kinase Akt. J Biol Chem 1998; 273(19):11937-43.
101. Lin TA, Kong X, Saltiel AR et al. Control of PHAS-I by insulin in 3T3-L1 adipocytes. Synthesis, degradation, and phosphorylation by a rapamycin-sensitive and mitogen-activated protein kinase-independent pathway. J Biol Chem 1995; 270(31):18531-8.
102. Keiper BD, Gan W, Rhoads RE. Protein synthesis initiation factor 4G. Int J Biochem Cell Biol 1999; 31(1):37-41.
103. Hentze MW. eIF4G: A multipurpose ribosome adapter? Science 1997; 275(5299):500-501.
104. Raught B, Gingras AC, Gygi SP et al. Serum-stimulated, rapamycin-sensitive phosphorylation sites in the eukaryotic translation initiation factor 4GI. EMBO J 2000; 19(3):434-444.
105. Niedzwiecka A, Marcotrigiano J, Stepinski J et al. Biophysical studies of eIF4E cap-binding protein: Recognition of mRNA 5' cap structure and synthetic fragments of eIF4G and 4E-BP1 proteins. J Mol Biol 2002; 319(3):615-635.
106. von der HT, Ball PD, McCarthy JE. Stabilization of eukaryotic initiation factor 4E binding to the mRNA 5'-Cap by domains of eIF4G. J Biol Chem 2000; 275(39):30551-30555.
107. Gan W, LaCelle M, Rhoads RE. Functional characterization of the internal ribosome entry site of eIF4G mRNA. J Biol Chem 1998; 273(9):5006-5012.
108. Prevot D, Darlix JL, Ohlmann T. Conducting the initiation of protein synthesis: The role of eIF4G. Biol Cell 2003; 95(3-4):141-156.
109. Goldstaub D, Gradi A, Bercovitch Z et al. Poliovirus 2A protease induces apoptotic cell death. Mol Cell Biol 2000; 20(4):1271-1277.
110. Haghighat A, Svitkin Y, Novoa I et al. The eIF4G-eIF4E complex is the target for direct cleavage by the rhinovirus 2A proteinase. J Virol 1996; 70(12):8444-8450.
111. Neumar RW, DeGracia DJ, Konkoly LL et al. Calpain mediates eukaryotic initiation factor 4G degradation during global brain ischemia. J Cereb Blood Flow Metab 1998; 18(8):876-881.
112. Ventoso I, MacMillan SE, Hershey JW et al. Poliovirus 2A proteinase cleaves directly the eIF-4G subunit of eIF-4F complex. FEBS Lett 1998; 435(1):79-83.

113. Das S, Maitra U. Functional significance and mechanism of eIF5-promoted GTP hydrolysis in eukaryotic translation initiation. Prog Nucleic Acid Res Mol Biol 2001; 70:207-231.
114. Asano K, Phan L, Valasek L et al. A multifactor complex of eIF1, eIF2, eIF3, eIF5, and tRNA(i)Met promotes initiation complex assembly and couples GTP hydrolysis to AUG recognition. Cold Spring Harb Symp Quant Biol 2001; 66:403-415.
115. Maiti T, Bandyopadhyay A, Maitra U. Casein kinase II phosphorylates translation initiation factor 5 (eIF5) in Saccharomyces cerevisiae. Yeast 2003; 20(2):97-108.
116. Browne GJ, Proud CG. Regulation of peptide-chain elongation in mammalian cells. Eur J Biochem 2002; 269(22):5360-8.
117. Lodish HF, Jacobsen M. Regulation of hemoglobin synthesis. Equal rates of translation and termination of - and -globin chains. J Biol Chem 1972; 247(11):3622-3629.
118. Palmiter RD. Differential rates of initiation of conalbumin and ovalbumin messenger ribonucleic acid in reticulocyte lysates. J Biol Chem 1974; 249(21):6779-6787.
119. Slobin LI. The role of eucaryotic factor Tu in protein synthesis. The measurement of the elongation factor Tu content of rabbit reticulocytes and other mammalian cells by a sensitive radioimmunoassay. Eur J Biochem 1980; 110(2):555-563.
120. Chang YW, Traugh JA. Insulin stimulation of phosphorylation of elongation factor 1 (eEF-1) enhances elongation activity. Eur J Biochem 1998; 251(1-2):201-7.
121. Prentice GA, Merrill AR. An enzyme-linked immunosorbent assay for the association of the catalytic domain of diphthamide-specific ribosyltransferases to eukaryotic elongation factor-2. Anal Biochem 1999; 272(2):216-223.
122. Redpath NT, Price NT, Severinov KV et al. Regulation of elongation factor-2 by multisite phosphorylation. Eur J Biochem 1993; 213(2):689-699.
123. Price NT, Redpath NT, Severinov KV et al. Identification of the phosphorylation sites in elongation factor-2 from rabbit reticulocytes. FEBS Lett 1991; 282(2):253-258.
124. Ryazanov AG. Elongation factor-2 kinase and its newly discovered relatives. FEBS Lett 2002; 514(1):26-29.
125. Redpath NT, Price NT, Proud CG. Cloning and expression of cDNA encoding protein synthesis elongation factor-2 kinase. J Biol Chem 1996; 271(29):17547-17554.
126. Redpath NT, Foulstone EJ, Proud CG. Regulation of translation elongation factor-2 by insulin via a rapamycin-sensitive signalling pathway. EMBO J 1996; 15(9):2291-7.
127. Wang X, Paulin FE, Campbell LE et al. Eukaryotic initiation factor 2B: Identification of multiple phosphorylation sites in the epsilon-subunit and their functions in vivo. EMBO J 2001; 20(16):4349-4359.
128. Kisselev L, Ehrenberg M, Frolova L. Termination of translation: Interplay of mRNA, rRNAs and release factors? EMBO J 2003; 22(2):175-182.
129. Zheng XF, Schreiber SL. Target of rapamycin proteins and their kinase activities are required for meiosis. Proc Natl Acad Sci USA 1997; 94(7):3070-3075.
130. Lawrence Jr JC. mTOR-dependent control of skeletal muscle protein synthesis. Int J Sport Nutr Exerc Metab 2001; 11(Suppl):177-85.
131. Gingras AC, Raught B, Sonenberg N. Regulation of translation initiation by FRAP/mTOR. Genes Dev 2001; 15(7):807-26.
132. Denning G, Jamieson L, Maquat LE et al. Cloning of a novel phosphatidylinositol kinase-related kinase: Characterization of the human SMG-1 RNA surveillance protein. J Biol Chem 2001; 276(25):22709-22714.
133. Choi KM, McMahon LP, Lawrence Jr JC. Two motifs in the translational repressor PHAS-I required for efficient phosphorylation by mammalian target of rapamycin and for recognition by raptor. J Biol Chem 2003; 278(22):19667-19673.
134. Kim DH, Sarbassov DD, Ali SM et al. mTOR interacts with raptor to form a nutrient-sensitive complex that signals to the cell growth machinery. Cell 2002; 110(2):163-175.
135. Nojima H, Tokunaga C, Eguchi S et al. The mammalian target of rapamycin (mTOR) partner, raptor, binds the mTOR substrates p70 S6 kinase and 4E-BP1 through their TOR signaling (TOS) motif. J Biol Chem 2003; 278(18):15461-4.
136. Kim DH, Sarbassov DD, Ali SM et al. GbetaL, a positive regulator of the rapamycin-sensitive pathway required for the nutrient-sensitive interaction between raptor and mTOR. Mol Cell 2003; 11(4):895-904.
137. Schalm SS, Fingar DC, Sabatini DM et al. TOS motif-mediated raptor binding regulates 4E-BP1 multisite phosphorylation and function. Curr Biol 2003; 13(10):797-806.
138. Hara K, Maruki Y, Long X et al. Raptor, a binding partner of target of rapamycin (TOR), mediates TOR action. Cell 2002; 110(2):177-89.
139. Brunn GJ, Fadden P, Haystead TA et al. The mammalian target of rapamycin phosphorylates sites having a (Ser/Thr)-Pro motif and is activated by antibodies to a region near its COOH terminus. J Biol Chem 1997; 272(51):32547-32550.

140. Brown EJ, Beal PA, Keith CT et al. Control of p70 s6 kinase by kinase activity of FRAP in vivo. Nature 1995; 377(6548):441-446.
141. Burnett PE, Barrow RK, Cohen NA et al. RAFT1 phosphorylation of the translational regulators p70 S6 kinase and 4E-BP1. Proc Natl Acad Sci USA 1998; 95(4):1432-1437.
142. Schalm SS, Blenis J. Identification of a conserved motif required for mTOR signaling. Curr Biol 2002; 12(8):632-639.
143. Tee AR, Proud CG. Caspase cleavage of initiation factor 4E-binding protein 1 yields a dominant inhibitor of cap-dependent translation and reveals a novel regulatory motif. Mol Cell Biol 2002; 22(6):1674-83.
144. Scott PH, Lawrence Jr JC. Attenuation of mammalian target of rapamycin activity by increased cAMP in 3T3-L1 adipocytes. J Biol Chem 1998; 273(51):34496-34501.
145. Scott PH, Brunn GJ, Kohn AD et al. Evidence of insulin-stimulated phosphorylation and activation of the mammalian target of rapamycin mediated by a protein kinase B signaling pathway. Proc Natl Acad Sci USA 1998; 95(13):7772-7.
146. Reynolds TH, Bodine SC, Lawrence Jr JC. Control of Ser2448 phosphorylation in the mammalian target of rapamycin by insulin and skeletal muscle load. J Biol Chem 2002; 277(20):17657-17662.
147. Navé BT, Ouwens M, Withers DJ et al. Mammalian target of rapamycin is a direct target for protein kinase B: Identification of a convergence point for opposing effects of insulin and amino-acid deficiency on protein translation. Biochem J 1999; 344(Pt 2):427-431.
148. Sekulic A, Hudson CC, Homme JL et al. A direct linkage between the phosphoinositide 3-kinase-AKT signaling pathway and the mammalian target of rapamycin in mitogen-stimulated and transformed cells. Cancer Res 2000; 60(13):3504-3513.
149. Brunn GJ, Hudson CC, Sekulic A et al. Phosphorylation of the translational repressor PHAS-I by the mammalian target of rapamycin. Science 1997; 277(5322):99-101.
150. McMahon LP, Choi KM, Lin TA et al. The rapamycin-binding domain governs substrate selectivity by the mammalian target of rapamycin. Mol Cell Biol 2002; 22(21):7428-38.
151. Tee AR, Fingar DC, Manning BD et al. Tuberous sclerosis complex-1 and -2 gene products function together to inhibit mammalian target of rapamycin (mTOR)-mediated downstream signaling. Proc Natl Acad Sci USA 2002; 99(21):13571-13576.
152. Manning BD, Tee AR, Logsdon MN et al. Identification of the tuberous sclerosis complex-2 tumor suppressor gene product tuberin as a target of the phosphoinositide 3-kinase/akt pathway. Mol Cell 2002; 10(1):151-162.
153. Gao X, Zhang Y, Arrazola P et al. Tsc tumour suppressor proteins antagonize amino-acid-TOR signalling. Nat Cell Biol 2002; 4(9):699-704.
154. Dan HC, Sun M, Yang L et al. Phosphatidylinositol 3-kinase/Akt pathway regulates tuberous sclerosis tumor suppressor complex by phosphorylation of tuberin. J Biol Chem 2002; 277(38):35364-35370.
155. Inoki K, Li Y, Zhu T et al. TSC2 is phosphorylated and inhibited by Akt and suppresses mTOR signalling. Nat Cell Biol 2002; 4(9):648-657.
156. Potter CJ, Pedraza LG, Xu T. Akt regulates growth by directly phosphorylating Tsc2. Nat Cell Biol 2002; 4(9):658-665.
157. Cheadle JP, Reeve MP, Sampson JR et al. Molecular genetic advances in tuberous sclerosis. Hum Genet 2000; 107(2):97-114.
158. Garami A, Zwartkruis FJ, Nobukuni T et al. Insulin activation of Rheb, a mediator of mTOR/S6K/4E-BP signaling, is inhibited by TSC1 and 2. Mol Cell 2003; 11(6):1457-1466.
159. Zhang Y, Gao X, Saucedo LJ et al. Rheb is a direct target of the tuberous sclerosis tumour suppressor proteins. Nat Cell Biol 2003; 5(6):578-581.
160. Yamagata K, Sanders LK, Kaufmann WE et al. Rheb, a growth factor- and synaptic activity-regulated gene, encodes a novel Ras-related protein. J Biol Chem 1994; 269(23):16333-16339.
161. Stocker H, Radimerski T, Schindelholz B et al. Rheb is an essential regulator of S6K in controlling cell growth in Drosophila. Nat Cell Biol 2003; 5(6):559-565.
162. Saucedo LJ, Gao X, Chiarelli DA et al. Rheb promotes cell growth as a component of the insulin/TOR signalling network. Nat Cell Biol 2003; 5(6):566-571.
163. Fox HL, Pham PT, Kimball SR et al. Amino acid effects on translational repressor 4E-BP1 are mediated primarily by L-leucine in isolated adipocytes. Am J Physiol 1998; 275(5 Pt 1):1232-8.
164. Xu G, Kwon G, Marshall CA et al. Branched-chain amino acids are essential in the regulation of PHAS-I and p70 S6 kinase by pancreatic beta-cells. A possible role in protein translation and mitogenic signaling. J Biol Chem 1998; 273(43):28178-84.
165. Anthony JC, Lang CH, Crozier SJ et al. Contribution of insulin to the translational control of protein synthesis in skeletal muscle by leucine. Am J Physiol Endocrinol Metab 2002; 282(5):1092-101.
166. Beugnet A, Tee AR, Taylor PM et al. Regulation of targets of mTOR (mammalian target of rapamycin) signalling by intracellular amino acid availability. Biochem J 2003; 372(Pt 2):555-66.

# Chapter 5

# Hepatic Regulation of Fuel Metabolism

Catherine Clark and Christopher B. Newgard*

## Introduction

It has been recognized for more than a century that the liver plays an important role in maintaining metabolic fuel homeostasis. The purpose of this chapter is to summarize mechanisms by which circulating glucose and lipid concentrations are controlled by hepatic metabolic activities.

Glucose metabolism in the liver is under the tight control of the pancreatic hormones insulin and glucagon. A high insulin/glucagon ratio in the fed state favors pathways of glucose storage and disposal—glycogen synthesis, glycolysis, and the pentose monophosphate shunt—while a low insulin/glucagon ration in the fasted state favors the pathways of glucose production-glycogenolysis and gluconeogenesis.

This tightly regulated control of hepatic glucose metabolism is disrupted in both major forms of diabetes, leading to inappropriate increases in glucose production by the liver. The quest to better understand the molecular and biochemical mechanisms of hepatic dysfunction in diabetes has resulted in the recent discoveries of new transcription factors, regulatory proteins, and allosteric factors, as well as new ideas about the role of spatial organization and compartmentalization in control of metabolic activity in the liver. Moreover, recent years have seen the emergence of new pathways of inter-organ communication in control of fuel homeostasis, with the liver as a central player. This chapter will therefore attempt to integrate these new findings with the bedrock of prior knowledge so as to provide a comprehensive summary of current knowledge of the role of liver in control of intermediary metabolism.

## Glucose Transport

The liver contributes to control of glucose homeostasis by net glucose uptake and storage in fed or anabolic conditions, and net glucose production in fasted or catabolic conditions. These functions are critically dependent upon efficient transport of glucose into and out of hepatocytes in response to changes in physiological conditions. Glucose transport into mammalian tissues is largely achieved via the activities of a family of facilitated glucose transporter proteins, which have varying tissue distribution and kinetic properties.[1,2] The major facilitated glucose transporter isoform expressed in liver and pancreatic islets is GLUT-2, the second member of the family identified.[3,4] GLUT-2 has a higher Km (lower affinity) for glucose than other members of the facilitated glucose transporter family.[4] This property allows GLUT-2 to alter its activity in response to changes in circulating glucose concentrations over the physiological range (4-8 mM), thereby making this transporter isoform highly suitable for its role in glucose homeostasis. All of the facilitated glucose transporters are capable of bidirectional glucose

---

*Corresponding Author: Christopher B. Newgard—Sarah W. Stedman Nutrition and Metabolism Center, Duke Independence Park Facility, 4321 Medical Park Drive, Durham, North Carolina, 27704 U.S.A. Email: Newga002@mc.duke.edu

*Mechanisms of Insulin Action*, edited by Alan R. Saltiel and Jeffrey E. Pessin.
©2007 Landes Bioscience and Springer Science+Business Media.

transport across cellular membranes, with the directionality of activity determined by the relative glucose concentrations on the outside versus the inside of the cell.

Based on its high levels of expression relative to other glucose transporter isoforms in liver, it was assumed until recently that GLUT-2-mediated transport was the major, if not the only mechanism for import and export of glucose in liver cells. This idea was supported by the finding that homozygous knock-out of the GLUT-2 gene in transgenic mice caused hyperglycemia and hypoinsulinemia, accompanied by elevated levels of free fatty acids and ketones.[5] Surprisingly, GLUT-2 knock out mice were also found to exhibit a normal increase in hepatic glucose production in response to injection of glucagon, and isolated hepatocytes from these animals produced glucose at normal rates.[6] This suggested the existence of a GLUT-2-independent pathway of glucose transport. Further investigation of this concept demonstrated that glucose transport in liver cells was inhibited by exposure to reduced temperatures or progesterone. This implied the possible involvement of a membrane traffic-based mechanism for glucose transport in liver cells that complements the facilitated transport mechanism. However, certain metabolic activities are clearly deficient in mouse hepatocytes that lack GLUT-2. These cells exhibit defective activation of glycogenolysis in response to a lowering of blood glucose, and a failure to lower glucose-6-phosphate levels appropriately during fasting.[7] Elevated glucose-6-phosphate levels could in turn result in sustained allosteric activation of glycogen synthase, as well as possible increases in gluokinase activity via modulation of its interaction with the glucokinase regulatory protein (see below). It was suggested that the increase in glucose-6-phosphate was caused by impairment of glucose export during fasting, resulting in rephosphorylation of a cytosolic glucose pool.[8] At the time of writing of this chapter, the potential role of the GLUT-2-independent pathway in increasing hepatic glucose production in diabetes has not been studied, nor is its physiological role fully understood; these topics are clearly deserving of further investigation.

## Glycolysis

The conversion of glucose, an aldose with six carbons, to pyruvate, a three-carbon carboxylic acid, occurs via the series of 10 enzymatic reactions of the glycolytic pathway. A balanced equation for glycolysis is as follows:

Glucose + 2 P$i$ + 2 NAD$^+$ + 2 ADP→←

2 Pyruvate + 2 ATP + 2 NADH + 2 H$^+$ + 2 H$_2$0

Seven of the ten steps in this pathway are catalyzed by enzymes with equilibrium constants that allow the forward (glycolytic) or the reverse (gluconeogenic) direction of the reaction to proceed depending on physiologic changes in the relative concentrations of substrates and products. The three other enzymatic steps of the pathway are catalyzed by glucokinase [GK] (glucose phosphorylation step), phosphofructokinase [PFK] (conversion of fructose-6-phosphate to fructose-1,6-bisphosphate), and pyruvate kinase (conversion of phosphoenolpyruvate to pyruvate), and are considered to be essentially irreversible because of the large release of free energy associated with these reactions. Distinct enzymes have evolved in the liver and in the few other tissues that are capable of gluconeogenesis (e.g., kidney) that can circumvent these otherwise irreversible steps of glycolysis.

Modulation of the concentrations and activities of GK, PFK, and pyruvate kinase are the primary mechanisms for control of glycolytic rate. Until recently it was thought that PFK held most of the control strength (the relative contribution of a single enzyme to the overall flux through a metabolic pathway) for glycolysis in liver cells. Consistent with this idea, PFK is regulated by a complex array of allosteric effectors. However, new studies have uncovered another level of control for the glycolytic pathway, involving GK, its newly discovered binding protein, the glucokinase regulatory protein (GKRP), and regulated movement of GK into and out of the nucleus. This finding has led to a more holistic model of regulation of glycolysis, as summarized for each of the key steps below.

## Glucokinase

The predominant glucose-phosphorylating enzyme in the liver is glucokinase (GK), also known as hexokinase IV. Glucokinase is expressed only in the liver, the islets of Langerhans, and certain specialized neuroendorcrine cells.[8-10] Glucokinase has a sigmoidal substrate dependency (low affinity, high Km) and unlike other members of the hexokinase family (I and II), it is not allosterically inhibited by the product of its reaction, glucose-6-phosphate.[8,11] The glucose concentration of blood in normal individuals ranges from approximately 4 mM in the fasted state to a maximum of 8-9 mM in the fed state, a range compatible with the Kms of GLUT-2 (17 mM) and glucokinase (6 mM).[8,11,12] Thus the kinetics of GLUT-2 and GK allow glucose metabolism in liver cells to be regulated in response to changes in the external glucose concentration.

Glucokinase is regulated chronically by transcriptional control and acutely by the glucokinase regulatory protein (GKRP).[13,14] Levels of GK mRNA are tightly controlled by insulin such that they fall dramatically in fasting or insulinopenic diabetes, whereas they are increased by 20-30-fold by feeding or insulin injection.[15-17] Changes in glucokinase activity due to activation of transcription and new protein synthesis occur with a time frame of approximately 30 to 60 minutes. Although the fed-to-fasted transition occurs gradually, metabolic changes occur rapidly in liver in the transition from the fasted to the fed states, suggesting that other regulatory mechanisms are operative under these conditions.

The discovery of GKRP led to understanding of a new acute regulatory mechanism for modulation of hepatic glucose utilization. GKRP was first discovered as an activity in cellular extracts that could inhibit GK in a fashion antagonized by fructose-1-phosphate.[13] GKRP was subsequently found to be localized to the nucleus, and contributions from several laboratories have led to the following model of its regulatory effects. In the fasted state, GKRP binds GK in the cytosol. Via its nuclear import signal, GKRP then translocates GK to the nucleus. Binding of GKRP inhibits GK activity, and this effect is complemented by sequestration of GK in the nucleus away from the cytosolic pool of glucose.[18-20] In the transition from the fasted-to-fed state or following exposure of isolated hepatocytes to high concentrations of glucose, fructose, or sorbitol, dissociation of GK from GKRP is stimulated, leading to translocation of GK to the cytosol and phosphorylation of glucose to glucose-6-phosphate. This mechanism allows for acute suppression of glucose phosphorylation under fasting conditions and rapid stimulation of glucose phosphorylation after meal ingestion. The post-prandial increase in glucose phosphorylation achieved by GK translocation from the nucleus to the cytoplasm is complemented in later stages by insulin-mediated stimulation of GK gene expression.

In light of the inhibitory activity of GKRP, one might expect that genetic ablation of the gene in transgenic mice would lead to increased GK activity. Instead, both heterozygous and homozygous knock out mice had decreased rather than increased GK activity in liver and normal blood glucose levels in the fasted state.[21,22] Both groups of mice displayed an abnormal excursion of blood glucose concentrations during a glucose tolerance test. These findings appear to be explained by a protective function of GKRP. By sequestering GK in the nucleus when it is not needed metabolically in the fasted state, GKRP protects the GK protein from degradation in the cytosol, thus maintaining a large stable pool of protein that can be rapidly recruited in response to influx of metabolic fuels. This is the first of several examples that will be highlighted in this chapter of important roles of spatial organization and compartmentalization in regulation of hepatic carbohydrate metabolism. The overall mechanism of control of GK activity by GKRP is represented schematically in Figure 1.

A major indication of the critical role of GK in control of hepatic glucose balance and glucose homeostasis came with the discovery that one of the forms of maturity onset diabetes of the young, MODY-2, is caused my mutations in the GK gene. MODY-2 patients are usually heterozygous, with one normal and one mutated GK allele. Patients with MODY-2 have a defect in glucose-stimulated insulin secretion, due to the important role of GK in regulation of glucose-stimulated insulin secretion in islet β-cells.[12,23-25] In addition, MODY-2 patients have impaired hepatic glucose disposal, leading to an increase in net hepatic glucose production.[26]

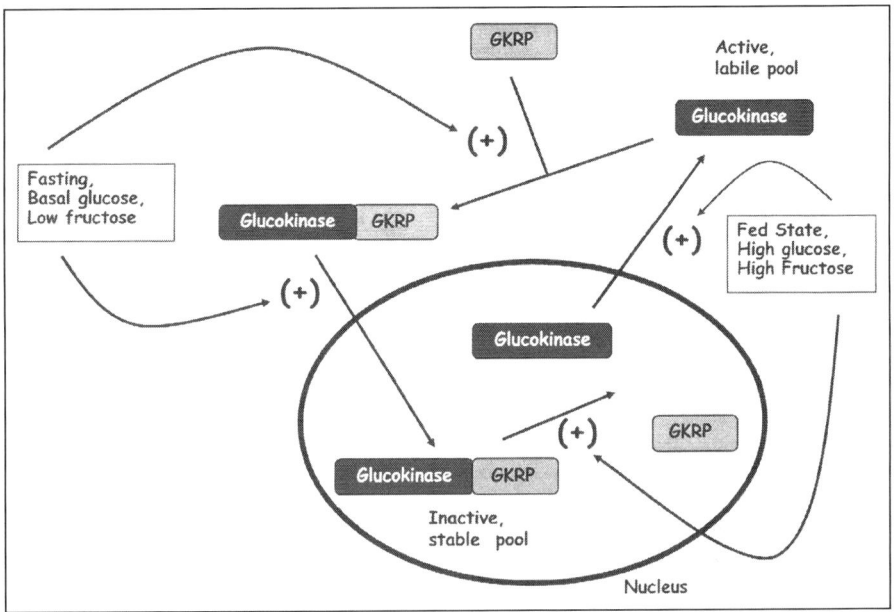

Figure 1. Mechanism of control of hepatic glucokinase activity by the glucokinase regulatory protein (GKRP) (Used with permission from ref. 114).

Patients with GK mutations also store significantly less glycogen in liver than do normal subjects.[27]

Mutations in GK associated with MODY-2 cover a wide range of sites within the protein. The effects of individual mutations have been evaluated by expression of the mutant forms of the enzyme in bacteria and kinetic analyses of the purified, expressed proteins.[28,29] Large decreases in enzymatic activity due to changes in $k_{cat}$, $S_{0.5}$ for glucose, or alterations in Km for ATP are seen in most of the mutant enzymes. One mutation, E300K, causes less effective expression of the mutant GK in heterologous systems, and the expressed protein is less stable at elevated temperatures.[29] Finally, certain mutations can actually lead to a lower $S_{0.5}$ (enhanced affinity) of GK for glucose, such that the mutant GK comes to resemble the low Km hexokinases. Patients with these mutations manifest a form of persistent hyperinsulinism and hypoglycemia due to lowering of the threshold for glucose-stimulated insulin secretion.[30]

Two groups independently produced transgenic mice heterozygous for disruption of a segment of the GK gene that affects gene expression in both the liver and islets of Langherhans.[31,32] This led to a reduction of GK activity of 30-40% in both tissues, mild hyperglycemia, and significant attenuation of glucose-stimulated insulin secretion. Interestingly, knocking out the GK gene completely leads to animals that die at an embryonic stage or shortly after birth.[31-33]

The studies of GK-deficient states described above imply that an increase in liver GK activity may be an effective means of lowering blood glucose in diabetes. In vitro studies of hepatic glucose phosphorylation using recombinant adenoviruses encoding GK or hexokinase I demonstrated that GK overexpression caused marked increases in glycolytic rate and glycogen deposition in hepatocytes, while hexokinase I had minimal effects on these pathways.[34,35] This difference in efficacy is explained by the fact that the common product of the GK and hexokinase I reactions, glucose-6-phosphate, strongly inhibits hexokinase I but not GK.[35]

GK overexpression in liver has also been studied in vivo in normal and diabetic animals. Normal rats infused with a recombinant adenovirus containing the GK cDNA at rates designed to cause moderate (3-fold) or large (7-fold) increases in hepatic GK enzymatic activity, resulted in a significant lowering of blood glucose levels only in the latter group, accompanied by a 67% decline in circulating insulin levels, and a large increase in liver glycogen stores.[36] Studies of transgenic mice containing multiple copies of the GK gene resulted in similar findings.[37] This result appears promising at first glance; however, animals with the 7-fold increase in hepatic GK activity also exhibited a 190% increase in circulating triglycerides and a 310% increase in circulating free fatty acids.[36] This change in lipid profile raises concerns about manipulation of hepatic GK activity as part of a viable therapy for diabetes. However, near-normalization of blood glucose levels was achieved in streptozotocin-diabetic mice in which GK was expressed as a transgene under control of the PEPCK promoter. In this model of type 1 diabetes, the high levels of fatty acids, triglycerides, and ketone bodies were lowered in the transgenic mice, albeit not fully normalized.[38] A potential explanation for the improved lipid profile in the setting of the type 1 diabetes model, as opposed to the findings in normal rats, is that the glucose-lowering effect of glucokinase could result in a greater dependence on lipid oxidation for energy production throughout the body. It is also important to note that the streptozotocin-treated mice were completely insulin deficient, making it difficult to extrapolate these findings to patients with type 1 diabetes or type 2 diabetes, who will have either residual endogenous or exogenously administered insulin available in the blood.

A small molecule activator of GK has been developed and tested in animal models.[39] The drug improves glucose homeostasis in ob/ob mice with type2 diabetes, apparently by activating GK in both the liver and pancreatic islet β-cells.[39] The increase in circulating lipids reported in response to GK overexpression in liver of normal rats would not be anticipated with use of the small molecule activator, as the increase in insulin release should suppress lipolysis and increase lipid storage. In contrast, in animals in which hepatic GK is activated without coincident stimulation of β-cell GK, the fall in blood glucose engendered by stimulation of hepatic glucose uptake leads to a decrease in circulating insulin levels.[36]

The utility of GK as a target for controlling hyperglycemia in diabetes remains an open question. There is no debate about the fact that an increase in GK activity in liver causes a fall in circulating glucose levels in normal rodents, and in models of type 1 and type 2 diabetes. However, the increase in hepatic enzyme activity may require careful titration, as higher levels of GK overexpression could result in perturbation of lipid homeostasis. When using a combined liver/islet GK activator, concerns about circulating hyperlipidemia may be offset by the constitutive increase in insulin secretion. However, new concerns that might arise include: (1) The threat of hypoglycemia brought on by simultaneous stimulation of glucose clearance and insulin secretion. Indeed there was some evidence of hypoglycemia in animals that received the small molecule GK activator;[39] (2) The potential deleterious effect of constitutive increases in insulin release and consequent stimulation of lipid storage, possibly leading to accumulation of stored lipids in liver and muscle. Further investigation of these issues will be required.

## *Phosphofructokinase*

The highly regulated phosphofructokinase (PFK) enzyme catalyzes the conversion of fructose-6-phosphate to fructose-1,6-bisphosphate. ATP serves as the phosphate donor, and PFK exhibits sigmoidal kinetics relative to its substrate fructose-6-phosphate, typical of enzymes that are regulated by allosteric ligands.

PFK is regulated by three major mechanisms in liver cells: (1) It is sensitive to the energy charge of the cell; an increase in the ATP/ADP + AMP ratio inhibits PFK activity; (2) It is inhibited by citrate, the product of the first committed step of the TCA cycle (which draws its precursors from glycolysis); (3) It is activated by fructose-2,6-bisphosphate (F-2,6P$_2$), perhaps the key allosteric regulator of PFK activity in liver due to its tight regulation by the insulin:glucagon ratio.

Fructose-2,6-bisphosphate was discovered as a highly potent activator of PFK activity in the early 1980s.[40-42] At low micromolar concentrations, F-2,6-P$_2$ activates PFK by converting the enzyme from sigmoidal to hyperbolic kinetics with respect to fructose-6-phosphate. This results in an increase of affinity of the enzyme for its substrate (lowering of Km). F-2,6-P$_2$ also alleviates inhibition of PFK by ATP.

A single protein with two distinct catalytic sites catalyzes both the synthesis and degradation of F-2,6P$_2$; the enzyme is 6-phosphofructose-2-kinase/fructose-2,6 bisphosphatase, or more simply, the bifunctional enzyme. Similar to the facilitated glucose transporters and hexokinases, a family of bifunctional enzyme genes are found in humans and other mammalian species, encoding enzymes with distinct patterns of tissue distribution and regulatory properties. The bifunctional enzyme isoform that is preferentially expressed in liver is regulated by phosphorylation of a serine residue near its N-terminus (serine 32). This phosphorylation is catalyzed by cAMP-dependent protein kinase (PKA).[43] Phosphophorylation of the bifunctional enzyme at this site leads to inhibition of its kinase activity and activation of its phosphatase activity, resulting in inhibition of glycolysis via a decrease in the concentration of F2,6-P$_2$.

The discovery of the bifunctional enzyme/F2,6-P$_2$ regulatory mechanism made a huge contribution to the understanding of hormone-mediated regulation of glycolysis in the liver via glucagon and insulin. Glucagon levels are increased in the fasted state, leading to binding of glucagon to its receptor. The glucagon receptor belongs to a family of G-protein coupled receptors, which exert their biologic effect through the cAMP second messenger system. The increase in cAMP levels in response to glucagon binding increases the activity of PKA, which in turn phosphorylates the bifunctional enzyme to favor degradation of F2,6-P$_2$. This regulation is reversed in the fasted-to-fed transition, where following a mixed meal, glucose and insulin levels rise while glucagon levels decrease. Dephosphorylation of the bifunctional enzyme is stimulated under these conditions, leading to activation of its 2-kinase activity and a rapid increase in F-2,6-P$_2$ levels.

Dephosphorylation of the bifunctional enzyme appears to be stimulated by a product of the pentose monophosphate shunt pathway, xylulose-5-phosphate (Xu5P).[44] The rise in plasma glucose concentration in the fasted to fed transition leads to increased glucose uptake by GLUT-2 and GK-mediated glucose phosphorylation to glucose-6-phosphate, which enters the pentose shunt pathway, resulting in increased levels of Xu5P. This leads to increased activity of a Xu5P-activated protein phosphatase, which in turn dephosphorylates the bifunctional enzyme. In addition to this regulation, postprandial increases in insulin levels stimulate the degradation of cAMP via activation of phosphodiesterases, leading to a decrease in PKA activity and the maintenance of the bifunctional enzyme in a nonphosphorylated state. These regulatory events are summarized schematically in Figure 2.

Similar to GK, the bifunctional enzyme has been investigated as a potential target for treatment of hyperglycemia in diabetes. A doubly-mutated form of the bifunctional enzyme (Ser32Ala, His258Ala) has diminished fructose-2,6-bisphosphatase activity and its 2-kinase activity cannot be inhibited by PKA-mediated phosphorylation of Ser32. As a result, the mutated enzyme has a very high kinase/phosphatase ratio, favoring F-2,6-P$_2$ synthesis.[45-47]

Overexpression of the wild-type or doubly-mutated form of the bifunctional enzyme increases F-2,6-P$_2$ levels in isolated cells[45] and in the liver of whole animals,[46,47] with larger increases seen when the mutant form is expressed. The metabolic impact of bifunctional enzyme overexpression in normal mice is similar to that of overexpressed glucokinase in normal rats—blood glucose levels are reduced, but with significant increases in circulating FFA and TG levels.[46] The doubly mutated form of bifunctional enzyme effectively lowers blood glucose levels in mice with streptozotocin-induced type 1 diabetes, and similar to the effect of GK overexpression, partially normalizes circulating FFA and TG levels in these animals.[46] In the KK/HIJ and KK. Cg-A$^y$/J mouse models of type 2 diabetes, adenovirus-mediated expression of the doubly mutated form of the bifunctional enzyme resulted in partial normalization of blood glucose levels, lowering of circulating insulin levels, and a modest reduction in circulating FFA and TG levels.[47] Again, as proposed for animals with GK overexpression, the fall in circulating glucose may have stimulated lipid oxidation for energy production. A better understanding of changes in lipid metabolism and the potential for hypoglycemia is required to fully understand the potential of the bifunctional enzyme as a target of diabetes therapies.

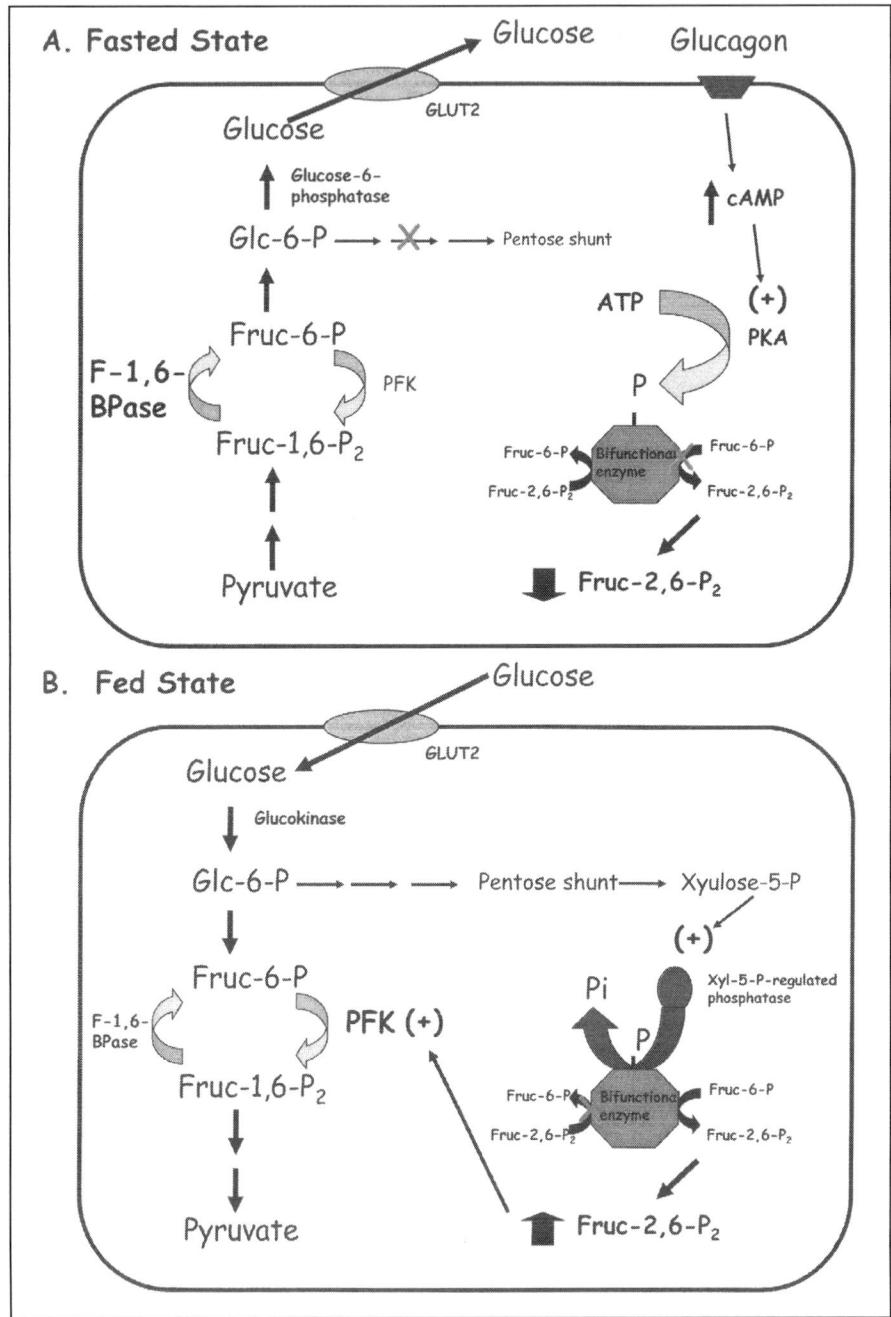

Figure 2. Mechanisms involved in regulation of bifunctional enzyme activity, fructose-2,6-bisphosphate levels and phosphofructokinase (PFK) activity in liver in the fasted (panel A) and fed (panel B) states. See text for details. Abbreviations: F-1,6-BPase: fructose-1,6-bisphosphatase; Xyl-5-P: xyulose-5-phosphate). Fruc-2,6-P2: fructose-2,6-bisphosphatase; Fruc-6-P: fructose-6-phosphate; PKA: protein kinase A/cyclic AMP-dependent protein kinase). (Used with permission from ref. 114.)

## Pyruvate Kinase

Pyruvate kinase (PK) catalyses the conversion of phosphenolpyruvate to pyruvate, and represents the last of the irreversible steps of glycolysis in the liver. It is subject to both transcriptional and allosteric regulation.

Transcription of the gene encoding the liver form of pyruvate kinase (L-PK) is decreased in the fasted or insulinopenic diabetic states, and increased in the fed state or by insulin injection. This pattern of response to changes in physiologic conditions mimics those of GK, PFK, and the bifunctional enzyme, providing for concerted transcriptional induction of the glycolytic pathway in liver during the fasted to fed transition. Providing further coordination is the recent finding that transcription of L-PK is regulated by the same Xu5P-regulated protein phosphatase that dephosphorylates the bifunctional enzyme under anabolic conditions. The Xu5P-regulated protein phosphatase dephosphorylates and activates the carbohydrate responsive element binding protein (ChREBP), a recently discovered transcription factor that stimulates transcription of L-PK, as well as the lipogenic genes acetyl-CoA carboxylase and fatty acid synthase.[48,49] Thus, increases in circulating glucose levels play a direct role in stimulation of L-PK expression via activation of the Xu5P-regulated phosphatase and dephosphorylation of ChREBP.

The effect of glucose to stimulate transcription of L-PK is complemented by allosteric regulation of the enzyme at multiple levels. Fructose 1,6-bisphosphate, the product of the PFK reaction, activates L-PK, and represents a "feed-forward" allosteric activating mechanism through which an early step in the glycolytic pathway activates a later step. In addition, L-PK is similar to PFK in that it is also sensitive to the energy charge of liver, such that a high ATP/ADP + AMP ratio decreases its activity. Like the bifunctional enzyme, L-PK is inhibited by PKA-mediated serine phosphorylation.[50] Phosphorylation of L-PK lowers its affinity (increases Km) for its substrate PEP and renders it more sensitive to energy charge inhibition. Finally, in the fasted to fed transition, the rise in insulin levels plays an indirect role in L-PK activation by antagonizing the actions of glucagon to stimulate PKA-mediated phosphorylation of the enzyme.

## Gluconeogenesis

Gluconeogenesis is the pathway by which glucose is synthesized in mammalian tissues such as liver and kidney. Gluconeogenic substrates include alanine, pyruvate, lactate, and glycerol. To enter the gluconeogenic pathway, alanine and lactate are converted to pyruvate via transamination and lactate dehydrogenase, respectively. Glycerol is generated from lipolysis in adipocytes in the fasted state, circulating to the liver and entering the gluconeogenic pathway after conversion to dihydroxyacetone via the enzymes glycerol kinase and glycerol-3-phosphate dehydrogenase.

Thirteen enzymatic steps convert pyruvate to glucose in the pathway of gluconeogenesis; seven of these are reversible and shared with glycolysis while six are unique to gluconeogenesis and evolved to circumvent the irrevsible steps of glycolysis.

The first four steps, in which pyruvate is converted to phosphenolpyruvate (PEP), represent a second example of spatial organization of hepatic metabolic regulatory enzymes. Carboxylation of pyruvate to oxaloacetate by the mitochondrial enzyme pyruvate carboxylase (PC) is the first step. This is followed by conversion of oxaloacetate to malate by the mitochondrial isoform of malate dehydrogenase. Malate is then transported out of the mitochondria and into the cytosol where it is converted back to oxaloacetate by cytosolic malate dehydrogenase. Finally, oxaloacetate is converted to PEP in a decarboxylation step catalyzed by phosphoenolpyruvate carboxykinase (PEPCK). One mole of ATP is consumed in the PC reaction and one mole of GTP in the PEPCK reaction for every mole of pyruvate converted to PEP. The oxidation of fatty acids, which is promoted in the fasted state, supplies most of the energy for these two reactions, and also generates metabolic products that serve as important regulators of these early steps of the gluconeogenic pathway (see below).

The next six enzymatic steps in which PEP is converted to fructose-1,6-bisphosphate involve reversible enzymes shared with glycolysis. Conversion of fructose-1,6-bisphosphate to fructose-6-phosphate, however, requires a unique enzyme, fructose-1,6-bisphosphatase, because the energy-consuming PFK reaction of glycolysis is not reversible. The next step of conversion of fructose-6-phosphate to glucose-6-phosphate is catalyzed by the shared enzyme hexose phosphate isomerase. Finally, the terminal step of gluconeogenesis is catalyzed by the glucose-6-phosphatase (G6Pase) enzyme complex, which constitutes a third example of spatial organization and compartmentalization in hepatic fuel metabolism. The catalytic subunit of the G6Pase complex, a phosphohydrolase enzyme, is sequestered in the enodplasmic reticulum (ER). As many as three other proteins are also involved—T1, a G6P translocase that delivers G6P to the catalytic subunit, and T2 and T3, putative ER glucose and inorganic phosphate translocases that move the reaction products back to the cytosol.[51,52] Among these translocases, only T1 has been definitely characterized and cloned; glucose and inorganic phosphate may exit the ER via activities of the T1 protein or other as yet undefined mechanisms. In sum, the reactions of gluconeogenesis occur in three cellular locations: the mitochondria, the cytosol, and the ER.

The sequestration of the catalytic subunit of the G6Pase enzyme complex in the ER is a fascinating example of how spatial organization and compartmentalization contributes to the function and regulation of the pathway in which it resides. The catalytic unit itself is a nonspecific phosphohydrolase that can remove phosphate groups from several different phosphorylated sugars. The T1 translocase component of the enzyme complex provides specificity by transporting only G6P into the ER, while excluding closely related sugar phosphates such as fructose-6-phosphate or mannose-6-phosphate.[53] In this way uncontrolled hydrolysis of phosphorylated sugars is prevented and energy is spared.

### *Regulation of Pyruvate Carboxylase and Pyruvate Dehydrogenase*

Gluconeogenesis is most active in fasting and insulinopenic conditions. As the fed to fasted transition occurs gradually, increases in gluconeogenic flux also occur gradually, in concert with coordinated up-regulation of PEPCK, G6Pase, and fructose-1,6-bisphosphate gene expression. This mechanism is complemented by an important interplay between fatty acid oxidation and reciprocal regulation of mitochondrial PC and pyruvate dehyrogenase (PDH) activities.

Regulation of the PC/PDH branchpoint in mitochondrial metabolism represents an important example of the dynamic links between control of lipid and carbohydrate metabolism in liver. In the fasted state, a high glucagon:insulin ratio activates hormone-sensitive lipase in adipose tissue. When activated, hormone-sensitive lipase catalyzes the hydrolysis of triglycerides to free fatty acids (FFAs) and glycerol, leading to increased concentrations of FFA and glycerol in the blood. These metabolites are transported to the liver, where they enter oxidative and gluconeogenic pathways, respectively. Fatty acid oxidation is favored in liver in the fasted state in part because the decrease in glycolytic flux leads to a reduction in malonyl CoA levels, an important allosteric inhibitor of carnitine palmitoyltransferase I (CPT-I), the enzyme that gates the entry of long-chain fatty acyl-CoAs into the mitochondria for oxidation (54). Thus, the falling malonyl CoA levels relieve inhibition of CPT-I, which when combined with the increase in substrate supply, increases the rate of fatty acid oxidation.[54] The products of fatty acid oxidation, acetyl CoA, NADH, and ATP act as allosteric inhibitors of PDH, which catalyzes the conversion of pyruvate to acetyl CoA for oxidation in the TCA cycle.[55,56] Inhibition of PDH is also accomplished by its phosphorylation by members of the family of PDH kinases. PDH kinase activity is stimulated by ATP, acetyl CoA and NADH, which serves as a second link between fatty acid oxidation and suppression of PDH.[57,58] Importantly, acetyl CoA is also a potent activator of PC.[59,60] As summarized in Figure 3, these events divert pyruvate away from oxidation in the TCA cycle towards the first steps of gluconeogenesis. In addition to generating the key allosteric intermediates for achieving this switch, fatty acid oxidation produces ATP that can be utilized in the energy consuming PC and PEPCK reactions.

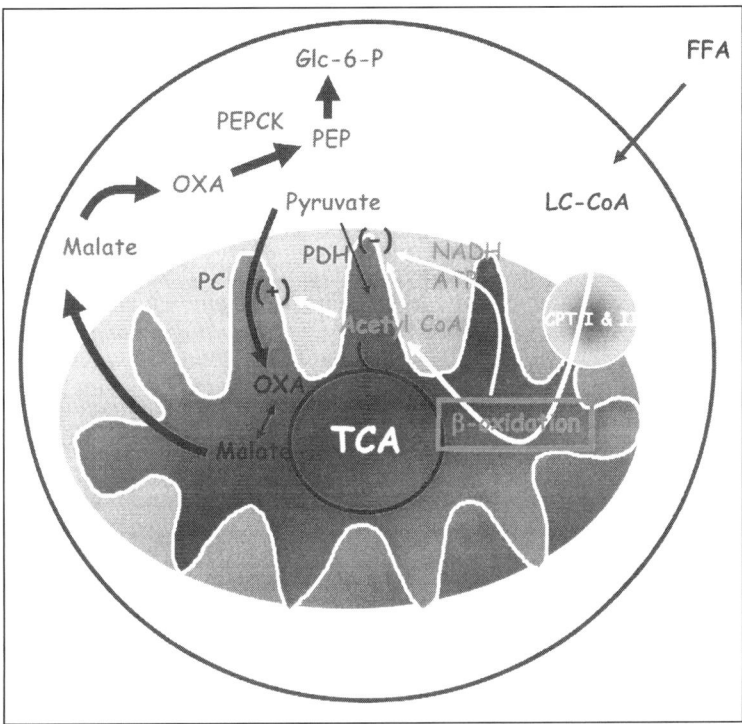

Figure 3. Mechanisms by which fatty acid oxidation suppresses glucose oxidation and initiates gluconeogenesis in liver in the fasted state. See text for details. Abbreviations: FFA: free fatty acids; PDH: pyruvate dehydrogenase; PC: pyruvate carboxylase; PEPCK: phosphoenolpyruvate carboxykinase; OXA: oxaloacetate; PEP: phosphoenolpyruvate. (Used with permission from ref. 114.)

## *Other Allosteric Regulation of Gluconeogenesis — A New Look*

Two other key gluconoegenic enzymes, PEPCK and the G6Pase complex, are not known to be allosterically regulated, but the activity of a third, fructose-1,6-bisphosphatase, had long been thought to be allosterically regulated by F-2,6-$P_2$, the product of the bifunctional enzyme and activator of PFK. However, a recent study provides evidence that this may not be true in vivo.[61] Using NMR to evaluate metabolism of [u-$^{13}$C]glucose and $D_2O$, relative contributions of gastric emptying, glycogenolysis, and gluconeogenesis to plasma glucose were estimated in 24-h fasted rats that received an oral bolus of glucose. The authors found that even though the glucose load caused F-2,6-$P_2$ levels to rise more than 10-fold, the fractional contribution made by gluconeogenesis was unchanged over the 150-min study period, consistent with the idea that F-2,6-$P_2$ had little acute effect on fructose-1,6-bisphosphatase activity in the in vivo setting. These experiments also support the notion that allosteric regulation plays a small role overall in suppressing gluconeogenesis in the transition from the fasted to fed state.

## *Regulation of PEPCK, Fructose-1,6-Bisphosphatase, and Glucose-6-Phosphatase*

Because the transition from fed to fasted state occurs gradually, it is logical to expect that much of the regulation of gluconeogenesis will occur at the transcriptional level. Accordingly, transcription of the PEPCK, fructose-1,6-bisphosphatase, and glucose-6-phosphatase catalytic subunit genes increase in a gradual and coordinated fashion as the glucagon/insulin ratio rises.

Glucocorticoids and cAMP increase PEPCK gene transcription, resulting in increased levels of mRNA, with mRNA stabilization supplementing this effect. cAMP-activated PKA phosphorylates the cAMP response element binding protein (CREBP), which binds to the CRE in the PEPCK gene promoter.[62,63] In this way CREBP initiates the assembly of a larger protein complex, which causes transcription to be initiated from the PEPCK gene promoter.[64] Glucocorticoids use a similar mechanism to bind to the GRE response element in the PEPCK gene promoter.[65,66] This action of glucocorticoid is antagonized by insulin.[65,67]

In concert with decreased expression of key glycolytic enzymes, the gluconeogenic enzymes PEPCK, fructose-1,6 bisphosphatase and glucose-6-phosphatase are increased several fold in response to fasting or insulinopenic diabetes. Conversely, in the transition from the fasted to fed conditions, glycolytic enzymes are induced, whereas the key gluconeogenic enzymes decrease in a coordinated fashion. Transcription of the gene encoding the catalytic subunit of G6Pase is increased in response to glucagon-induced increases in cAMP, and is also stimulated by hyperglycemic and hyperlipidemic conditions associated with type 2 diabetes, in both in vitro and whole animal studies.[68-72] Insulin inhibition of G6Pase transcription dominates the stimulatory effects of cAMP and glucose.[70,73] Expression of the G6Pase catalytic subunit is increased in liver of Zucker diabetic fatty (ZDF) rats relative to lean control animals, suggesting that the normal dominant effect of insulin to suppress G6Pase expression is weakened in these insulin resistant animals.[74]

Recent studies have added understanding to this coordinate regulation of gluconeogenic gene expression via the discovery of an important role of peroxisome proliferator-activated receptor-γ coactivator-1α(PGC-1α). Originally discovered as a regulator of mitochondrial biogenesis and thermogenesis,[75] PGC-1α levels were shown to rise dramatically in liver in fasting and in diabetes, conditions under which gluconeogenic enzymes are coordinately induced.[76] Subsequent studies demonstrated that PGC-1α overexpression activates hepatic expression of the PEPCK, fructose-1,6-bisphosphatase, and G6Pase catalytic subunit genes.[76] These effects appear to be mediated by interaction of PGC-1α with hepatocyte nuclear factor 4-α and the forkhead transcription factor FOXO1.[76,77] These findings have led to a new model of coordinated regulation of gluconeogenic genes, in which glucagon-mediated increases in cAMP levels in liver increase PGC-1 expression, stimulating expression of the gluconeogenic genes. These effects may be antagonized by insulin in the fed state via Akt-1-mediated phosphorylation of FOXO1, resulting in suppression of gluconeogenesis.[77] Possible interactions of PGC-1 and other regulators of gluconeogenic gene expression such as CREB and the glucocorticoid receptor remain to be defined.

Genetic engineering studies have been conducted to evaluate the regulatory role of hepatic PEPCK in control of liver and whole animal gluconeogenesis, with some surprising observations. On the one hand, overexpression studies in transgenic mice provide support for a prominent role of the enzyme in metabolic control, as mice with increased hepatic PEPCK activity exhibit an increase in fasting blood glucose and a clear impairment in glucose disposal in response to a glucose challenge.[78] On the other hand, tissue-specific knock out of PEPCK in liver had no significant change on glucose homeostasis, but did result in fat accumulation in the liver.[79] Further studies of isolated perfused livers from PEPCK knock out and wild type mice revealed that gluconeogenesis from lactate or pyruvate was almost completely abrogated, whereas gluconeogenesis from glycerol was increased.[80] The continued flux of glycerol to glucose coupled with compensatory increases in extra-hepatic (kidney) gluconeogenesis may explain how PEPCK knock-out mice are protected from catastrophic hypoglycemia in the absence of a key gluconeogenic enzyme in liver.

Recent studies have also provided insight into the metabolic effects of altered expression of the key components of the G6Pase complex, the catalytic subunit and the T1 translocase. Overexpression of the G6Pase catalytic subunit in rat hepatocytes caused substantial decreases in glycolytic flux and glycogen deposition, and a parallel increase in gluconeogenesis,[81] whereas overexpression of the T1 translocase caused a smaller but still significant increase in glucose-6-phosphate hydrolysis.[82] Expression of T1 caused a similar inhibition of glycogen formation as observed with the catalytic subunit, but a much smaller impairment of glycolytic flux.[82] The more pronounced effect of T1 on glycogen deposition may be explained by the observation that T1 enhances the rate of glucose-1-phosphate hydrolysis, with no effect on fructose-6-phosphate hydrolysis; glucose-1-phosphate is an intermediate in glycogen synthesis but not glycolysis.[82] Thus, overexpression of either the T1 translocase or the G6Pase catalytic subunit alter activity of the G6Pase enzyme complex, but the catalytic subunit has the more pronounced effect. Consistent with this idea, adenovirus-mediated overexpression of the catalytic subunit in liver of normal rats caused several of the abnormalities associated with early-stage type 2 diabetes, including glucose intolerance, hyperinsulinemia, a marked decrease in hepatic glycogen content, and increased peripheral (muscle) triglyceride stores.[74] These observations clearly highlight the importance of tight control of the balance between glucose phosphorylation and G6P hydrolysis in regulation of hepatic glucose handling and homeostasis.

## Glycogen Metabolism

During the fasted-to-fed transition, the liver is able to convert glucose and other precursors to glycogen for storage, whereas in the fed-to-fasted transition, glycogen is degraded to produce free glucose. Glycogen metabolism is regulated by a complex mix of hormonal and metabolic signals.

### *Regulation of Glycogen Phosphorylase and Glycogen Synthase by Covalent Modification and Allosteric Regulators*

In the catabolic state (e.g., fasting or stress), glucagon or β-adrenergic agonists bind to their receptors in liver cells to initiate a "glycogenolytic cascade."[83] Activation of adenylate cyclase by the GαS heterotrimeric GTP-binding protein leads to an accumulation of cAMP and activation of cAMP-dependent protein kinase (PKA). PKA then mediates the phosphorylation and activation of phosphorylase kinase. Phosphorylation of glycogen phosphorylase on serine 14 by phosphorylase kinase converts this enzyme from its inactive (phosphorylase b) to its active (phosphorlyase a) form.[84] Phosphorylase a catalyzes the cleavage of α-1,4-glycosidic bonds at the nonreducing termini of the glycogen particle, yielding glucose 1-phosphate, which is then converted to glucose 6-phosphate (G6P) by phosphoglucomutase. Hydrolysis of G6P to form free glucose for delivery to the circulation occurs via the G6Pase complex. Activation of glycogen phosphorylase in the fasted stated is accompanied by reciprocal regulation of glycogen synthase. Thus, fasting leads to the phosphorylation of glycogen synthase at a cluster of serine residues, resulting in inactivation of the enzyme.[85] Phosphorylation of glycogen synthase is primarily catalyzed by PKA and glycogen synthase kinase 3 (GSK-3).

Glycogen synthesis predominates in the fasted-to-fed transition. The formation of glycogen in the liver can occur via a direct pathway, in which G6P is produced through the phosphorylation of glucose obtained from the circulation, or an indirect pathway, in which G6P is formed from gluconeogenic precursors (86). Once G6P is formed through either pathway, it is converted to glucose 1-phosphate by phosphoglucomutase and to UDP-glucose by UDP-glucose phosphorylase. UDP-glucose serves as the substrate for glycogen synthase, which catalyzes the addition of glucose residues to the glycogen particle via the formation of α-1,4-glycosidic bonds. A branching enzyme then links blocks of glucose residues through α-1,6-glycosidic linkages to form the highly branched structure of mature glycogen.[86]

Regulation of glycogen synthesis occurs at many levels. The expression of GLUT-2 in the liver allows efficient equilibration of extracellular and intracellular glucose concentrations, and the rate of conversion of glucose to G6P by GK also increases in proportion to changes in glucose over the physiological range. Glucose is an allosteric inhibitor of glycogen phosphorylase, whereas G6P is an activator of glycogen synthase.[84,87] Glycogen phosphorylase and glycogen synthase are dephosphorylated via a common protein phosphatase, protein phosphatase 1 (PP-1), and the two enzymes become better substrates for PP-1 in the presence of glucose and G6P, respectively.[84,87] Importantly, dephosphorylation of glycogen phosphorylase reduces the activity of the enzyme, whereas conversely, dephosphorylation of glycogen synthase increases its

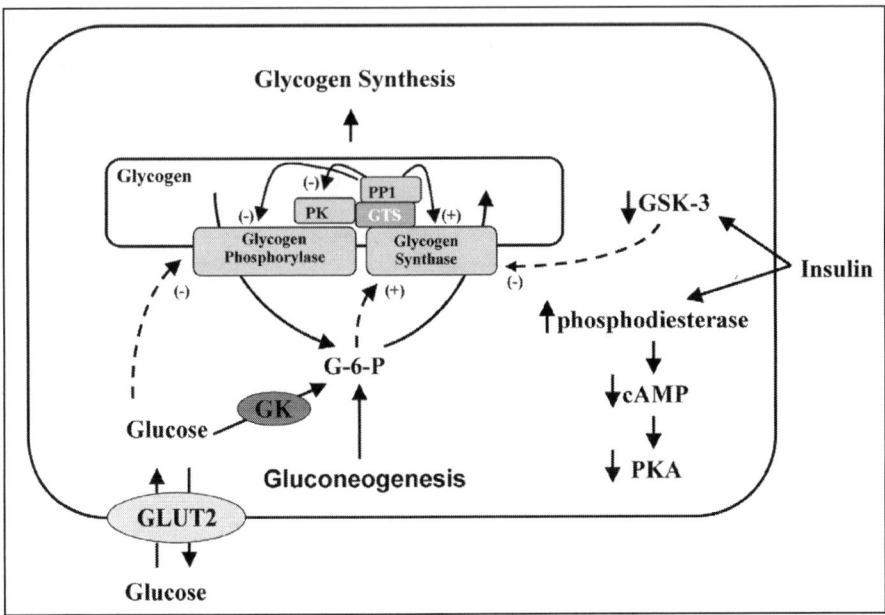

Figure 4. Mechanisms involved activation of glycogenesis in the fed state. See text for details. Abbreviations: GSK-3: glycogen synthase kinase-3; PKA: protein kinase A/cAMP-dependent protein kinase; PK: phosphorylase kinase; GTS: glycogen targeting subunit); PP1: protein phosphatase-1; GK: glucokinase.

activity. The rise in insulin associated with the fed state stimulates phosphorylation of GSK-3 via a branch of the insulin signaling pathway that includes PI-3 kinase and Akt-1 (protein kinase B).[88] Phosphorylation of GSK-3 reduces its kinase activity, thereby resulting in reduced phosphorylation of glycogen synthase and an increase in its enzymatic activity. A summary of regulatory events pertaining to glycogen synthesis that are activated in the fed state is provided in Figure 4 including the organizing role played by glycogen targeting subunits of protein phosphatase 1 (see below).

### Glycogen Targeting Subunits of Protein Phosphatase 1

Protein phosphatase 1 is an enzyme with a wide array of substrates found in diverse regions of mammalian cells. The delivery of PP-1 to specific substrates is mediated by a large family of PP-1-binding proteins. Sorting of PP-1 via these binding proteins allows the enzyme to participate in an array of critical processes, including cell division, vesicle fusion, ion channel function, and glycogen metabolism. Among the family of PP-1 binding proteins are scaffolding proteins that organize glycogen metabolism, known collectively as glycogen-targeting subunits of PP-1 (hereafter referred to generally as glycogen-targeting subunits). Glycogen-targeting subunits bring PP-1 and the enzymes of glycogen metabolism together in a multi-protein complex that is also associated with the glycogen particle, allowing for efficient regulation of glycogen metabolism.[89] This constitutes yet another example of spatial organization of important enzymes and proteins that regulate fuel metabolism in liver.

There are several different isoforms of glycogen-targeting subunits.[89] The first glycogen-targeting subunit to be cloned was $G_M/R_{Gl}$, which is preferentially expressed in striated skeletal muscle.[90,91] Liver preferentially expresses a 35-kDa protein known as $G_L$.[92] Protein targeting to glycogen (PTG),[93] also known as PPP1R5,[94] and a fourth form, PPP1R6,[95] are similar in size to $G_L$ but differ from $G_L$ and $G_M/R_{Gl}$ in that they are expressed in a broad array of tissues. All of the glycogen-targeting subunits are able to bind to glycogen and PP-1, and appear to exhibit differential capacities for binding of glycogen synthase, glycogen phosphorylase, and phosphorylase kinase.[89]

Glycogen-targeting subunits exhibit sequence homology around the PP-1 binding motif and putative-glycogen-binding regions. $G_M/R_{Gl}$ differs from the other members of the family in that it contains a large COOH-terminal domain. This unique protein segment contains a hydrophobic region that has been proposed to mediate binding of $G_M/R_{GL}$ to the sarcoplasmic-reticulum in muscle.[90] In addition, $G_M/R_{Gl}$ contains two potential phosphorylation sites. Site 1 has the sequence RRGS (serine 46 in the human $G_M/R_{Gl}$ sequence) and is not found in other glycogen targeting subunit isoforms, whereas site 2 contains a consensus PKA-mediated phosphorylation site in the middle of the PP-1 binding motif (serine 65 in the human $G_M/R_{Gl}$ sequence).

Evidence that glycogen targeting subunits can exert a regulatory effect on glycogen metabolism came to light when overexpression of PTG was shown to stimulate glycogen synthesis in 3T3 L1 cells or rat hepatocytes, in concert with stimulation of G6P-independent glycogen synthase activity.[93,96] Moreover, overexpression of PTG caused dramatic enhancement of glycogen synthesis even when carbohydrates and insulin were removed from hepatocyte cultures (leaving amino acids as the main carbon source), suggesting that PTG can function to stimulate the indirect (gluconeogenic) pathway of glycogen synthesis. PTG overexpression also impaired the normal glycogenolytic action of forskolin and glucagon.[96]

Subsequent studies compared the metabolic effects of PTG overexpression to those of other glycogen targeting subunit isoforms. To this end, recombinant adenoviruses encoding $G_M/R_{Gl}$, $G_L$, and a truncated version of $G_M/R_{Gl}$ that lacked its C-terminal 700 amino acids, termed $G_M\Delta C$, were expressed in hepatocytes. These studies demonstrated a rank order of glycogenic potency of the targeting subunits as follows: $G_L > PTG > G_M\Delta C > G_M/R_{Gl}$.[97,98] Interestingly, hepatocytes expressing $G_M\Delta C$, but not any of the native targeting subunits, retained the ability to degrade glycogen in response to decreased media glucose concentrations, stimulation with forskolin, or a combination of both.[98]

Overexpression of either GK or various glycogen-targeting subunits stimulates glycogen deposition in isolated hepatocytes. However, cells with overexpressed GK synthesized glycogen in a highly glucose concentration-dependent manner, and also produced larger quantities of lactate than cells with overexpressed targeting subunits. Moreover, co-overexpression of GK and a single glycogen-targeting subunit had additive effects on glycogen synthesis, while co-overexpression of two different glycogen-targeting subunits did not. These results suggest that GK and glycogen-targeting subunits activate distinct pathways of glycogen synthesis. This was confirmed through NMR-based analysis of glycogen biosynthesis, which demonstrated that cells overexpressing glycogen-targeting subunits synthesize a larger proportion of their glycogen from the level of TCA cycle intermediates than cells with overexpressed GK.[98]

Adenovirus-mediated overexpression of PTG in liver of normal rats resulted in improvement of glucose clearance during an oral glucose tolerance test. However, unlike GK, which caused increases in circulating lipids when overexpressed in normal rats, PTG improved whole-body glucose tolerance without perturbing lipid homeostasis. As was the case in hepatocytes, PTG overexpression in liver of normal rats led to increased glycogen synthesis via both direct and indirect pathways, but also impaired the glycogenolytic capacity of the rats in response to fasting. Thus, PTG expression improved glucose tolerance, but caused a form of glycogen storage disease.[99]

The improvement of glucose tolerance without perturbation of lipid homeostasis in rats with overexpressed PTG stimulated further in vivo studies with recombinant adenoviruses encoding $G_L$, $G_M\Delta C$, and $G_M/R_{Gl}$. Results of an oral glucose tolerance test (OGTT) revealed that only $G_M\Delta C$ was capable of lowering blood glucose levels in insulin-resistant, glucose intolerant, high-fat fed rats, despite large and roughly equal increases in glycogen deposition in rats with hepatic overexpression of $G_L$ or $G_M\Delta C$. The relative lack of impact of $G_L$ on blood glucose levels was subsequently explained by the finding that overexpression of $G_L$ markedly increased liver glycogen levels in both fed and fasted animals when compared to control rats. In contrast, rats overexpressing $G_M\Delta C$ had normal glycogen levels in the fed state and also were able to degrade glycogen in response to a fast to near-normal levels. Thus, $G_L$ has a level of glycogenic potency that causes glycogen overstorage in the fasted state, thereby preventing significant synthesis of glycogen during OGTT. In contrast, animals overexpressing $G_M\Delta C$ are able to synthesize additional glycogen during the OGTT, thereby enhancing glucose clearance in response to this challenge.[98]

The utility of $G_M\Delta C$ to reverse hyperglycemia in streptozotocin(STZ)-induced diabetic rats has also been investigated.[100] $G_M\Delta C$ was able to normalize blood glucose levels and glycogen levels even though hepatic GK expression was decreased in the insulinopenic animals. This enhanced glycogen synthesis in the face of low GK expression presumably occurred via activation of the indirect (gluconeogenic) pathway of glycogen synthesis, but this remains to be firmly established. In addition, hyperphagia was observed in STZ treated rats, but overexpression of $G_M\Delta C$ in these animals decreased food intake despite dramatic lowering of circulating insulin and leptin levels. The mechanistic link between enhanced glycogen synthesis and feeding behavior in this model remains to be explored. Overall, these studies show that appropriately regulated augmentation of glycogen storage can correct perturbations in glucose homeostasis caused by either impaired insulin action or impaired insulin secretion.

Targets other than GK, PFK, and the glycogen targeting subunits have been considered for lowering of blood glucose in diabetes. For example, several compounds have been described that inhibit glycogen phosphorylase in vitro.[101,102] Chronic administration of one of these inhibitors, compound 14, to ob/ob mice improved glucose tolerance and lowered circulating glucose levels in the fed state.[101] Furthermore, the effects of insulin to activate glycogen synthesis have been mimicked in vitro with GSK-3 inhibitors,[103] and these compounds also improved glucose tolerance in ZDF rats, with a concomitant enhancement of liver glycogen synthesis.[104] However, GSK-3 plays a role in many other cellular processes,[88] possibly imposing limitations for its future use as a diabetes therapy.

## New Developments

Recent years have witnessed the emergence of examples of inter-organ communication in control of fuel homeostasis, and an important role of the liver in such networks. For example, tissue-specific knock out of the GLUT-4 glucose transporter in adipose tissue results in development of insulin resistance in skeletal muscle and liver, which cannot be ascribed to changes in circulating leptin or free fatty acid levels.[105] Conversely, adenovirus-mediated expression of malonyl CoA decarboxylase in liver of high-fat fed, insulin resistant rats results in lowering of liver triglyceride and circulating free fatty acid levels, accompanied by reversal of muscle and whole-animal insulin resistance.[106] The amelioration of muscle insulin resistance in this model is accompanied by reduction in a specific lipid-derived metabolite, β-OH-butyryl carnitine in muscle. Furthermore, knock out of another lipid metabolizing enzyme, steroyl CoA desaturase-1 (SCD-1), in liver protects against diet-induced obesity and development of metabolic dysregulation and diabetes.[107] Malonyl CoA decarboxylase degrades malonyl CoA to acetyl CoA, thereby lowering the concentrations of the proximal precursor of de novo lipogenesis and decreasing the concentrations of a key allosteric inhibitor of carnitine palmitoyltransferase I, leading to increased rates of fatty acid oxidation. SCD-1 catalyzes the conversion of saturated fatty acids to monounsaturated fatty acids. Knock-out of SCD-1 therefore increases the concentrations of saturated and more readily oxidized fatty acids in the liver, diverting them away from esterification and storage pathways. This effect of SCD-1 appears to involve an activation of 5'-AMP kinase.[107] Consistent with a critical role of lipid partitioning in control of hepatic insulin sensitivity and glucose homeostasis, liver-specific overexpression of lipoprotein lipase increases liver triglyceride stores and causes hepatic insulin resistance.[108]

Two other emergent mechanisms for control of hepatic metabolism involving signals generated at a distance are worthy of consideration in closing this chapter. First, studies over a number of years have identified a hepatoportal glucose sensor of unknown molecular design, by which the positive glucose concentration gradient that occurs between the hepatoportal vein and arterial blood following absorption of a carbohydrate-containing meal triggers increased glucose disposal in the liver and muscle.[109,110] The effect on hepatic glucose balance in liver appears to be mediated in part via signals sent through the hepatic branch of the vagus nerve. More recently, signaling through this mechanism has been shown to be absent in GLUT-2 knock-out mice in which normal insulin secretion is maintained by transgenic expression of GLUT-1 in islets.[111] Second, further investigation of autonomic nervous regulation of liver glucose metabolism has revealed the presence of hypothalamic sensors that mediate the effects of increased free fatty acids on hepatic autoregulation.[112] Autoregulation refers to a lipid mediated decrease in hepatic glycogenolysis that compensates for lipid-mediated increases in hepatic gluconeogenesis (see Fig. 3), allowing hepatic glucose production to remain balanced in normal humans and experimental animals. This autoregulatory mechanism is impaired in type 2 diabetes, contributing to an increased rate of glucose production in the disease.[113] Recent work demonstrates that elevations in FFA are detected in the hypothalamus in response to hypothalamic esterification of fatty acids, signaling via hypothalamic ATP-sensitive potassium channels, and propagation of the signal via efferent hepatic vagal branch outflow.[112] The next phase of these important studies will be to define the mechanisms by which changes in nervous system activity alter activities of key enzymes, transporters, and their regulatory proteins in coordinating appropriate physiological responses in normal subjects, and how such mechanisms go awry in patients with metabolic syndrome and type 2 diabetes.

## References

1. Bell GI, Kayano T, Buse JB et al. Molecular biology of mammalian glucose transporters. Diabetes Care 1990; 13:198-208.
2. Thorens B, Charron MJ, Lodish HF. Molecular physiology of glucose transporters. Diabetes Care 1990; 13:209-18.
3. Thorens B, Sarkar HK, Kaback HR et al. Cloning and functional expression in bacteria of a novel glucose transpoter in liver, intestine, kidney, and beta-pancreatic islet cells. Cell 1988; 55:281-90.
4. Johnson JH, Newgard CB, Milburn JL et al. The high Km glucose transporter of islets of langerhans is functionally similar to the low affinity transporter of liver and has an identical primary sequence. J Biol Chem 1990; 265:6548-6551.
5. Guillam MT, Hummler E, Schaerer E et al. Early diabetes and abnormal postnatal pancreatic islet development in mice lacking GLUT-2. Nature Genetics 1997; 17:327-330.
6. Guillam MT, Burcelin R, Thorens B. Normal hepatic glucose production in the absence of GLUT2 reveals an alternative pathway for glucose release from hepatocytes. Proc Natl Acad Sci USA 1998; 95:12317-12321.
7. Burcelin R, del Carmen Munoz M, Guillan MT et al. Liver hyperplasia and paradoxical regulation of glycogen metabolism and glucose-sensitive gene expression in GLUT2-null hepatocytes. Further evidence for the existence of a membrane-based glucose release pathway. J Biol Chem 2000; 275:10930-10936.
8. Matschinsky FM. Regulation of pancreatic beta-cell glucokinase: From basics to therapeutics. Diabetes 2002; 51(Suppl 3):S394-404.
9. Hughes SD, Quaade C, Milburn JL et al. Expression of normal and novel glucokinase mRNAs in anterior pituitary and islet cells. J Biol Chem 1991; 266:4521-4530.
10. Jetton TL, Liang Y, Pettepher CC et al. J Biol Chem 1994; 269:3641-54.
11. Wilson JE. Regulation of mammalian hexokinase activity. In: Beitner R, ed. Regulation of carbohydrate metabolism. Boca Raton: CRC Press, 1984:45-85.
12. Newgard CB. Regulatory role of glucose transport and phosphorylation in pancreatic islet β-cells. Diabetes Reviews 1996; 4:191-205.
13. Malaisse WJ, Malaisse-Lagae F, Davies DR et al. Regulation of glucokinase by a fructose-1-phosphate-sensitive protein in pancreatic islets. Eur J Biochem 1990; 190:539-545.
14. Dethuex M, Vandekerckhove J, Van Schaftingen E. Cloning and sequencing of rat liver cDNAs encoding the regulatory protein of glucokinase. FEBS Lett 1994; 321:111-115.
15. Iynedjian PB, Ucla C, Mach B. Molecular cloning of glucokinase cDNA: Developmental and dietary regulation of glucokinase mRNA in rat liver. J Biol Chem 1987; 262:6032-38.
16. Andreone TL, Printz RL, Pilkis SJ et al. The amino acid sequence of rat liver glucokinase deduced from cloned cDNA. J Biol Chem 1989; 264:363-69.

17. Iynedjian PB, Pilot P-R, Nouspikel T et al. Differential expression and regulation of the glucokinase gene in liver and islets: Implications for control of glucose homeostasis. Proc Natl Acad Sci USA 1989; 86:7838-42.
18. Toyoda Y, Miwa I, Satake S et al. Nuclear location of the regulatory protein of glucokinase in rat liver and translocation of the regulator to the cytoplasm in response to high glucose. Biochem Biophys Res Comm 1995; 215:467-473.
19. Brown KS, Kalinowski SS, Megill JR et al. Glucokinase regulatory protein may interact with glucokinase in the hepatocyte nucleus. Diabetes 1997; 46:179-186.
20. Shiota C, Coffey J, Grimsby J et al. Nuclear import of hepatic glucokinase depends upon glucokinase regulatory protein, whereas export is due to a nuclear export signal sequence in glucokinase. J Biol Chem 1999; 274:37125-37130.
21. Farrelly D, Brown KS, Tieman A et al. Mice mutant for glucokinase regulatory protein exhibit decreased liver glucokinase: A sequestration mechanism in metabolic regulation. Proc Natl Acad Sci USA 1999; 96:14511-14516.
22. Grimsby J, Coffey JW, Dvorozniak MT et al. Characterization of glucokinase regulatory protein-deficient mice. J Biol Chem 2000; 275:7826-7831.
23. Vehlo G, Frougel P, Clement K et al. Primary pancreatic beta-cell secretory defect caused by mutations in glucokinase gene in kindreds of maturity onset diabetes of the young. Lancet 1992; 340:444-48.
24. Byrne MM, Sturis J, Clement K et al. Insulin secretory abnormalities in subjects with hyperglycemia due to glucokinase mutations. J Clin Invest 1994; 93:1122-30.
25. Newgard CB, Matschinsky FM. Substrate Control of Insulin Release. In: Jefferson J, Cherrington A, eds. Handbook of Physiology, Vol II. Oxford Univ. Press, 2001:125-152.
26. Frougel P, Zouali H, Vionnet N et al. Familial hyperglycemia due to mutations in glucokinase. N Engl J Med 1993; 328:697-702.
27. Velho G, Petersen KF, Perseghin G et al. Impaired hepatic glycogen synthesis in glucokinase-deficient (MODY-2) subjects. J Clin Invest 1996; 98:1755-1761.
28. Gidh-Jain M, Takeda J, Xu LZ et al. Glucokinase mutations associated with noninsulin-dependent (type 2) diabetes mellitus have decreased enzymatic activity: Implications for structure/function relationships. Proc Natl Acad Sci USA 1993; 90:1932-36.
29. Liang Y, Kesavan P, Wang L et al. Variable effects of maturity-onset-diabetes of youth (MODY)-associated glucokinase mutations on substrate interactions and stability of the enzyme. Biochem J 1995; 309:167-73.
30. Glaser B, Kesavan P, Heyman M et al. Familial hypoinsulinism caused by an inactivating glucokinase mutation. N Engl J Med 1998; 338:226-240.
31. Grupe A, Hultgren B, Ryan A et al. Transgenic knockouts reveal a critical requirement for pancreatic β-cell glucokinase in maintaining glucose homeostasis. Cell 1995; 83:69-78.
32. Bali D, Svetlanov A, Lee H-W et al. Animal model for maturity-onset diabetes of the young generated by disruption of the mouse glucokinase gene. J Biol Chem 1995; 270:21464-67.
33. Terauchi Y, Sakura H, Yasuda K et al. Pancreatic β-cell-specific targeted disruption of glucokinase gene. J Biol Chem 1995; 270:30253-256.
34. O'Doherty RM, Lehman DL, Seoane J et al. Differential metabolic effects of adenovirus-mediated glucokinase and hexokinase I overexpression in rat primary hepatocytes. J Biol Chem 1996; 271:20524-20530.
35. Seoane J, Gomez-Foix AM, O'Doherty RM et al. Glucose-6-phosphate produced by glucokinase, but not hexokinase is the signal for the activation of hepatic glycogen synthase. J Biol Chem1996; 271:23756-23760.
36. O'Doherty RM, Lehman DL, Telemaque-Potts S et al. Metabolic impact of glucokinase overexpression in liver: Lowering of blood glucose in fed rats is accompanied by hyperlipidemia. Diabetes 1999; 48:2022-20277.
37. Niswender KD, Shiota M, Postic C et al. Effects of increased glucokinase gene copy number on glucose homeostasis and hepatic glucose metabolism. J Biol Chem 1997; 272:22570–22575.
38. Ferre T, Pujol A, Riu E et al. Correction of diabetic alterations by glucokinase. Proc Natl Acad Sci USA 1996; 93:7225-7230.
39. Grimsby J, Sarabu R, Corbett WL et al. Allosteric activators of glucokinase: Potential role in diabetes therapy. Science 2003; 301:370-373.
40. Van Schaftingen E, Hue L, Hers HG. Fructose-2,6-bisphosphate, the probable structure of the glucose and glucagon-sensitive stimulator of phosphofructokinase. Biochem J 1980; 192:897-901.
41. Pilkis SJ, El-Maghrabi MR, Pilkis J et al. Fructose-2,6-bisphosphate. A new activator of phosphofructokinase. J Biol Chem 1981; 256:3171-3174.
42. Uyeda K, Furuya E, Sherry AD. The structure of "activation factor" for phosphofructokinase. J Biol Chem 1981; 256:8679-8684.

43. Murray KJ, El-Maghrabi MR, Kountz PD et al. Amino acid sequence of the phosphorylation site of rat liver 6-phosphofructo-2-kinase/fructose-2,6-bisphosphatase. J Biol Chem 1984; 259:7673-7681.
44. Nishimura M, Uyeda K. Purification and characterization of a novel xyulose 5-phosphate-activated protein phosphatase catalyzing dephosphorylation of fructose-6-phosphate, 2 kinase: Fructose-2,6-bisphosphatase. J Biol Chem 1995; 270:26341-26346.
45. Argaud D, Lange AJ, Becker TC et al. Adenovirus-mediated overexpression of liver 6-phosphofructo-2-kinase/fructose-2,6-bisphosphatase in gluconeogenic rat hepatoma cells. Paradoxical effect on Fru-2,6-P2 levels. J Biol Chem 1995; 270:24229-24236.
46. Wu C, Okar DA, Newgard CB et al. Overexpression of 6-phosphofructo-2-kinase/fructose-2, 6-bisphosphatase in mouse liver lowers blood glucose by suppressing hepatic glucose production. J Clin Invest 2001; 107:91-98.
47. Wu C, Okar DA, Newgard CB et al. Increasing fructose 2,6-bisphosphate overcomes hepatic insulin resistance of type 2 diabetes. Am J Physiol Endocrinol Metab 2002; 282:E38-45.
48. Yamashita H, Takenoshita M, Sakurai M et al. A glucose-responsive transcription factor that regulates carbohydrate metabolism in the liver. Proc Natl Acad Sci USA 2001; 98:9116-9121.
49. Kabashima T, Kawaguchi T, Wadzinski BE et al. Xyulose-5-phosphate mediates glucose-induced lipogenesis by xyulose 5-phosphate-activated protein phosphatase in rat liver. Proc Natl Acad Sci USA 2003; 100:5107-5112.
50. Engstrom D, Ekman P, Humble E et al. Pyruvate kinase. Enzymes 1987; 18:47-75.
51. Nordlie RC. Metabolic regulation by multifunctional glucose-6-phosphatase. Curr Topic Cell Reg 1974; 8:33-117.
52. Arion WJ, Lange AJ, Walls EH et al. Evidence for the participation of independent translocation for phosphate and glucose 6-phosphate in the microsomal glucose-6- phosphatase system. Interactions of the system with orthophosphate, inorganic pyrophosphate, and carbamyl phosphate. J Biol Chem 1980; 255:10396-10406.
53. Lange AJ, Arion WJ, Beaudet AL. Type 1b glycogen storage disease is caused by a defect in the glucose-6-phosphate translocase of the glucose-6-phosphatase system. J Biol Chem 1980; 255:8381-8384.
54. McGarry JD. Banting lecture 2001: Dysregulation of fatty acid metabolism in the etiology of type 2 diabetes. Diabetes 2002; 51:7-18.
55. Tsai CS, Burgett MW, Reed LJ. α-keto acid dehydrogenase complexesXX: A kinetic study of the pyruvate dehydrogenase complex from bovine kidney. J Biol Chem 1973; 248:8348-8352.
56. Denton RM, Randle PJ, Bridges BJ et al. Regulation of mammalian pyruvate dehydrogenase. Mol Cell Biochem 1975; 9:27-53.
57. Sugden MC, Holness MJ. Recent advances in mechanisms regulating glucose oxidation at the level of the pyruvate dehydrogenase complex by PDKs. Am J Physiol 2003; 284:E855-862.
58. Pettit FH, Pelley JW, Reed LJ. Regulation of pyruvate dehydrogenase kinase and phosphatase by acetyl CoA/CoA and NADH/NAD ratios. Biochem Biophys Res Comm 1975; 65:575-582.
59. Ashman LK, Wallace JC, Keech DB. Desensitization of pyruvate carboxylase against acetyl CoA stimulation by chemical modification. Biochem Biophys Res Comm 1975; 51:924-931.
60. Warren GB, Tipton KF. The role of acetyl CoA in the reaction pathway of pig-liver pyruvate carboxylase. Eur J Biochem 1974; 47:549-554.
61. Jin ES, Uyeda K, Kawaguchi T et al. Increased hepatic fructose-2,6-bisphosphate after an oral glucose load does not affect gluconeogenesis. J Biol Chem 2003; 278:28427-28433.
62. Quinn PG, Wong TW, Magnuson MA et al. Identification of the basal and cAMP regulatory elements in the promoter of the phosphoenolpyruvate carboxykinase gene. Mol Cell Biol 1988; 8:3467-75.
63. Park EA, Roesler WJ, Liu J et al. The role of the CCAAT/enhancer-binding protein in the transcriptional regulation of the gene for phosphoenolpyruvate carboxykinase (GTP). Mol Cell Biol 1990; 10:6264-72.
64. Waltner-Law M, Duong D, Daniels MC et al. Elements of the Glucocorticoid and Retinoic Acid Response Units are Involved in cAMP-mediated Expression of the PEPCK Gene. J Biol Chem 2003; 278:in press.
65. Sasaki K, Cripe TP, Koch SR et al. Multihormonal regulation of phosphoenolpyruvate carboxykinase gene transcription: Dominant role of insulin. J Biol Chem 1984; 259:15242-51.
66. Duong DT, Waltner-Law M, Sears R et al. Insulin inhibits hepatocellular glucose production by disrupting the association of CBP and RNA polymerase II with the PEPCK gene promoter. J Biol Chem 2002; 277:32234-242.
67. Sasaki K, Granner DK. Regulation of phosphoenolpyruvate carboxykinase gene transcription by insulin and cAMP: Reciprocal actions on initiation and elongation. Proc Nat Acad Sci USA 1988; 85:2954-2958.

68. Liu Z, Barrett EJ, Dalkin AC et al. Effect of acute diabetes on rat hepatic glucose-6-phosphatase activity and its messenger RNA level. Biochem Biophys Res Comm 1994; 205:680-686.
69. Massillon D, Barzilai N, Chen W et al. Glucose regulates in vivo glucose-6-phosphatase gene expression in the liver of diabetic rats. J Biol Chem 1996; 271:9871-9874.
70. Argaud D, Zhang Q, Pan W et al. Regulation of rat liver glucose-6-phosphatase gene expression in different nutritional and hormonal states: Gene structure and 5'-flanking sequence. Diabetes 1996; 45:1563-71.
71. Mithieux G, Vidal H, Zitoun C et al. Glucose-6-phosphatase mRNA and activity are increased to the same extent in kidney and liver of diabetic rats. Diabetes 1996; 45:891-896.
72. Argaud D, Kirby TL, Newgard CB et al. Glucose stimulation of glucose-6-phosphatase gene expression in primary hepatocytes and Fao hepatoma cells. Requirement for glucokinase expression. J Biol Chem 1997; 272:12854-12861.
73. O'Brien RM, Streeper RS, Ayala JE et al. Insulin-regulated gene expression. Biochem Soc Trans 2001; 29:552-8.
74. Trinh KY, O'Doherty RM, Anderson P et al. Perturbation of fuel homeostasis caused by overexpression of the glucose-6-phosphatase catalytic subunit in liver of normal rats. J Biol Chem 1998; 273:31615-31620.
75. Wu Z, Puigserver P, Andersson U et al. Mechanisms controlling mitochondrial biogenesis and respiration through the thermogenic coactivator PGC-1. Cell 1999; 98:115-24.
76. Yoon JC, Puigserver P, Chen G et al. Control of hepatic gluconeogenesis through the transcriptional activator PGC-1. Nature 2001; 413:131-8.
77. Puigserver P, Rhee J, Donovan J et al. Insulin-regulated hepatic gluconeogenesis through FOXO1-PGC-1alpha interaction. Nature 2003; 423:550-5.
78. Valera A, Pujol A, Pelegrin M et al. Transgenic mice overexpressing phosphoenolpyruvate carboxykinase develop noninsulin-dependent diabetes mellitus. Proc Natl Acad Sci USA 1994; 91:9151-4.
79. She P, Shiota M, Shelton KD et al. Phosphoenolpyruvate carboxykinase is necessary for integration of hepatic energy metabolism. Mol Cell Biol 2000; 20:6508-17.
80. She P, Burgess SC, Shiota M et al. Mechanisms by which liver-specific PEPCK knockout mice preserve euglycemia during starvation. Diabetes 2003; 52:1649-54.
81. Seoane J, Trinh K, O'Doherty RM et al. Metabolic impact of adenovirus-mediated overexpression of the glucose-6-phosphatase catalytic subunit in primary hepatocytes. J Biol Chem 1997; 272:26972-26977.
82. An J, Li Y, van De Werve G et al. Overexpression of the P46 (T1) translocase component of the glucose-6-phosphatase complex in hepatocytes impairs glycogen accumulation via hydrolysis of glucose 1-phosphate. J Biol Chem 2001; 276:10722-10779.
83. Sutherland EW, Robinson GA. The role of cyclic AMP in the control of carbohydrate metabolism. Diabetes 1969; 18:797-819.
84. Newgard CB, Hwang PK, Fletterick RJ. The family of glycogen phosphorylases: Structure and function. CRC critical reviews in biochemistry and molecular biology 1989; 24:69-99.
85. Lawrence Jr JC, Roach PJ. New insights into the role and mechanism of glycogen synthase activation by insulin. Diabetes 1997; 46:541-7.
86. McGarry JD, Kuwajima M, Newgard CB et al. From dietary glucose to liver glycogen—The full circle round. Annual Review of Nutrition 1986; 7:51-73.
87. Ferrer JC, Favre C, Gomis RR et al. Control of glycogen deposition. FEBS Lett 2003; 546:127-32.
88. Frame S, Cohen P. GSK3 takes centre stage more than 20 years after its discovery. Biochem J 2001; 359:1-16.
89. Newgard CB, Brady MJ, O'Doherty RM et al. Organizing glucose disposal: Emerging roles of the glycogen targeting subunits of protein phosphatase-1. Diabetes 2000; 49:1967-1977.
90. Tang PM, Bondor JA, Swiderek KM et al. Molecular cloning and expression of the regulatory (RG1) subunit of the glycogen-associated protein phosphatase. J Biol Chem 1991; 266:15782-15789.
91. Chen YH, Hansen L, Chen MX et al. Sequence of the human glycogen-associated regulatory subunit of type 1 protein phosphatase and analysis of its coding region and mRNA level in muscle from patients with NIDDM. Diabetes 1994; 43:1234-1241.
92. Doherty MJ, Moorhead G, Morrice N et al. Amino acid sequence and expression of the hepatic glycogen-binding (GL)-subunit of protein phosphatase-1. FEBS Lett 1995; 375:294-298.
93. Printen JA, Brady MJ, Saltiel AR. PTG, a protein phosphatase 1-binding protein with a role in glycogen metabolism. Science 1997; 275:1475-1478.
94. Doherty MJ, Young PR, Cohen PT. Amino acid sequence of a novel protein phosphatase 1 binding protein (R5) which is related to the liver-and muscle-specific glycogen binding subunits of protein phosphatase 1. FEBS Lett 1996; 399:339-343.

95. Armstrong CG, Browne GJ, Cohen P et al. PPP1R6, a novel member of the family of glycogen-targetting subunits of protein phosphatase 1. FEBS Lett 1997; 418:210-214.
96. Berman HK, O'Doherty RM, Anderson P et al. Overexpression of protein targeting to glycogen (PTG) in rat hepatocytes causes profound activation of glycogen synthesis independent of normal hormone- and substrate-mediated regulatory mechanisms. J Biol Chem 1998; 273:26421-26425.
97. Gasa R, Jensen P-B, Berman H et al. Differential regulatory and metabolic properties of glycogen targeting subunits (PTG, $G_L$, $G_M$) expressed in hepatocytes. J Biol Chem 2000; 275:26396-26403.
98. Gasa R, Clark C, Yang R et al. Reversal of diet-induced glucose intolerance by hepatic expression of a variant glycogen-targeting subunit of protein phosphatase-1. J Biol Chem 2002; 277:1524-1530.
99. O'Doherty RM, Jensen PB, Anderson P et al. Activation of direct and indirect pathways of glycogen synthesis by hepatic overexpression of protein targeting to glycogen. J Clin Invest 2000; 105:479-488.
100. Yang R, Newgard CB. Hepatic expression of a targeting subunit of protein phosphatase-1 in streptozotocin-diabetic rats reverses hyperglycemia and hyperphagia despite depressed glucokinase expression. J Biol Chem 2003; 278:23418-23425.
101. Martin WH, Hoover DJ, Armento SJ et al. Discovery of a human liver glycogen phosphorylase inhibitor that lowers blood glucose in vivo. Proc Natl Acad Sci USA 1998; 95:1776-1781.
102. Treadway JL, Mendys P, Hoover DJ. Glycogen phosphorylase inhibitors for treatment of type 2 diabetes mellitus. Expert Opin Investig Drugs 2001; 10:439-354.
103. Coghlan MP, Culbert AA, Cross DA et al. Selective small molecule inhibitors of glycogen synthase kinase-3 modulate glycogen metabolism and gene transcription. Chem Biol 2000; 7:793-803.
104. Cline GW, Johnson K, Regittnig W et al. Effects of a novel glycogen synthase kinase-3 inhibitor on insulin-stimulated glucose metabolism in Zucker diabetic fatty (fa/fa) rats. Diabetes 2002; 51:2903-2910.
105. Abel ED, Peroni O, Kim JK et al. Adipose-selective targeting of the GLUT4 gene impairs insulin action in muscle and liver. Nature 2001; 409:729-733.
106. An J, Muoio DM, Shiota M et al. Hepatic expression of malonyl CoA decarboxylase reverses muscle, liver, and whole animal insulin resistance. Nat Med 2004; 10(3):268-274.
107. Ntambi JM, Miyazaki M, Dobrzyn A. Regulation of stearoyl-CoA desaturase expression. Lipids 2004; 29:1061-1065.
108. Kim JK, Fillmore JJ, Chen Y et al. Tissue-specific overexpression of lipoprotein lipase causes tissue-specific insulin resistance. Proc Natl Acad Sci USA 2001; 98:7522-7527.
109. Moore MC, Rossetti L, Pagliassotti MJ et al. Neural and pancreatic influences on net hepatic glucose uptake and glycogen synthesis. Am J Physiol 1996; 271:E215-E222.
110. Burcelin R, Crivelli V, Perrin C et al. GLUT4, AMP kinase, but not the insulin receptor, are required for hepatoportal glucose sensor-stimulated muscle glucose utilization. J Clin Invest 2003; 111:1555-1562.
111. Burcelin R, Dolci W, Thorens B. Glucose sensing by the hepatoportal sensor is GLUT2-dependent. In vivo analysis of GLUT2-null mice. Diabetes 2000; 49:1643-1648.
112. Lam TKT, Pocai A, Gutierrez-Juarez R et al. Hypothalamic sensing of circulating fatty acids is required for glucose homeostasis. Nature Medicine 2005; (online ahead of print).
113. Boden G, Chen X, Capulong E et al. Effects of free fatty acids on gluconeogenesis and autoregulation of glucose production in type 2 diabetes. Diabetes 2001; 50:810-816.
114. Newgard CB. Regulation of glucose metabolism in the liver. In: DeFronzo RA, Ferrannini E, Keen H et al. International Textbook of Diabetes. 3rd ed. John Wiley and Sons, 2004:253-276.

## CHAPTER 6

# Insulin Action Gene Regulation

Calum Sutherland, Richard M. O'Brien and Daryl K. Granner*

## Introduction

Insulin regulates metabolism by altering the concentration of critical proteins or by inducing post-translational modifications of preexisting molecules. The latter represents a well-recognized action of insulin, and it has been extensively studied for many years.[1-3] By contrast, it is only recently that considerable advances have been made in understanding several aspects of insulin-regulated gene expression.

Although insulin could potentially affect any of the multiple steps in the flow of information from gene to protein, it appears that transcription, mRNA stability and translation represent the primary sites of insulin action. This chapter will principally focus on insulin-regulated gene transcription. It is now clear that insulin can have positive and negative effects on the transcription of specific genes within the same cell.[4,5] In addition, the genes regulated by insulin encode proteins involved in a variety of biologic phenomena (Fig. 1). Many, but not all, of these mRNAs direct the synthesis of enzymes that have a well-established metabolic connection to insulin action (Fig. 1). Not unexpectedly, this type of regulation is mostly seen in the primary tissues associated with the metabolic actions of insulin namely liver, muscle and adipose tissue but insulin also regulates gene expression in tissues not commonly associated with metabolic effects.[4,5]

The *cis/trans* model of transcriptional control underpins the current understanding of how insulin regulates gene transcription at a molecular level. Briefly stated, the fidelity and frequency of initiation of transcription of eukaryotic genes is mediated by the interaction of *cis*-acting DNA elements with *trans*-acting factors. The specific sequence of the *cis*-acting element determines which *trans*-acting factor(s) can bind. In addition, the concept of hormone response units, comprising multiple *cis*-acting elements working in concert, adds a further level of complexity to this basic concept. For example, the combination of *cis*-acting elements is likely to dictate the precise protein complex that can interact with the gene promoter, and thus the direction, magnitude and regulation of the hormonal response. Importantly, it has become clear that the presence of a *cis*-acting element within a gene promoter does not always predict the response to a hormone.

Even at the simplest level of the *cis/trans* model, there are several potential mechanisms by which insulin could influence transcriptional initiation, including regulation of *trans*-acting factor subcellular localisation, stability and activity (DNA binding and/or transactivation potential). Insulin achieves this regulation by initiating signal transduction pathways that relay the information to the factors, usually resulting in the post-translational modification of the factor (or a regulator) in a manner that alters its function.

*Corresponding Author: Daryl K. Granner—Department of Molecular Physiology and Biophysics, 707 Light Hall, Vanderbilt University Medical School, Nashville, Tennessee 37232-0615, U.S.A. Email: daryl.granner@mcmail.vanderbilt.edu

*Mechanisms of Insulin Action*, edited by Alan R. Saltiel and Jeffrey E. Pessin.
©2007 Landes Bioscience and Springer Science+Business Media.

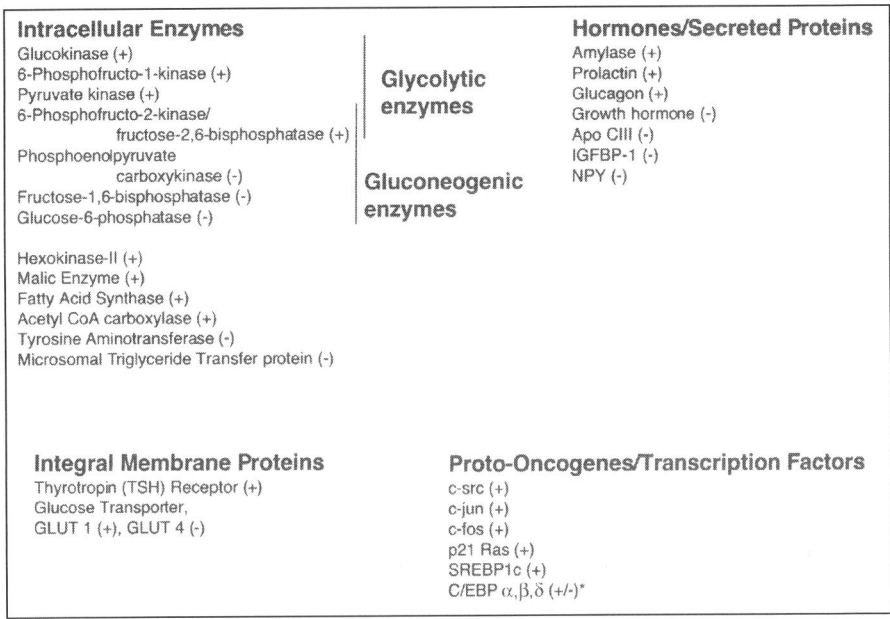

Figure 1. Insulin regulates the transcription of genes involved in a variety of biologic phenomena. (+): insulin stimulates transcription; (-) insulin inhibits transcription. *Both stimulatory and inhibitory effects have been reported for this gene product.

Several *cis*-acting elements that mediate the effect of insulin on gene transcription have been defined. These are referred to as insulin response sequences or elements (IRSs/IREs). To date the general consensus is that there is no consensus IRS/IRE. In this chapter the identification of the best characterized IRSs/IREs and putative *trans*-acting factors with which they are associated is described in detail. Most of these transcriptional regulators were not initially characterized as insulin regulated factors, but have subsequently been implicated in the regulation of specific gene promoters by insulin.

The physiological importance of insulin-regulated gene transcription is apparent from studies on the glycolytic and gluconeogenic pathways, which can be viewed as a series of three opposing substrate cycles.[6] Insulin and glucagon have antagonistic actions on the expression of the genes encoding all of the key regulatory enzymes in these three cycles.[6] The biochemical mechanisms that mediate this antagonism are of considerable interest and it is already apparent that different mechanisms are utilized with different genes. Thus, for example, insulin inhibits the stimulation of PEPCK gene transcription by cAMP (see "Phosphoenolpyruvate Carboxykinase (PEPCK)" section), whereas cAMP inhibits the stimulation of hepatic glucokinase gene transcription by insulin.[7] Most importantly, although not yet clearly defined, defects in gene expression are likely to play a key role in the pathophysiology of Type 2DM; for example, reduced expression of GLUT2 or glucokinase may be involved in the β cell insulin secretory defect;[8,9] increased PEPCK gene expression may lead to an increase in hepatic glucose production (HGP);[10,11] whereas reduced GLUT4/HKII expression may be the cause of reduced peripheral glucose utilization (PGU).[10,12] In addition, altered TNF-α, resistin, adiponectin, and PTP1B gene expression have been implicated in the generation of insulin resistance.[13-17]

# Insulin Signal Transduction and Gene Expression

## Key Signalling Molecules

### The Receptor

The insulin receptor (IR) is a heterotetrameric membrane glycoprotein composed of two α subunits and two β subunits linked by disulphide bonds. Binding of insulin to the α subunit leads to a conformational change that promotes autophosphorylation, and activates the intrinsic tyrosine kinase domain of the β subunit. The receptor autophosphorylation on specific tyrosine residues enhances the ability of the IR to recruit and regulate target proteins.[18,19]

### The Receptor Substrates

Proximal substrates of the receptor include members of the insulin receptor substrate family (IRS). There are four closely related IRS's, namely IRS-1, IRS-2, IRS-3 and IRS-4,[19-21] as well as the functionally related Shc,[22] and p62dok,[23,24] that all act as molecular adapters to relay the insulin signal (see ref. 25 for review). IRS-1 and IRS-2 are widely expressed while IRS-4 is expressed in brain, kidney and thymus. Rodents express IRS-3 predominantly within adipose tissue, however this isoform does not appear to exist in human tissue.

The IRS's are phosphorylated on a number of tyrosine residues by insulin receptor tyrosine kinase. Once phosphorylated, the IRS proteins subsequently interact with various effector proteins that contain src homology (SH) 2 domains. These include Grb-2 (growth receptor bound protein-2), the regulatory subunit of PI3-kinase (p85), the tyrosine kinases fyn and csk and the tyrosine phosphatase SHP2.[20,26,27]

### The Effectors

#### PI3-Kinase

Phosphoinositide 3-kinase (PI3-kinase) is activated by insulin as well as many growth factors and cytokines (for review see refs. 27-30). PI3-kinase catalyses the phosphorylation of the 3' hydroxyl of phosphatidylinositols. Phosphatidylinositols (PtdIns), PtdIns(4)P and PtdIns(4,5)$P_2$ can all serve as substrates for the PI3-kinases in vitro. Upon PI3-kinase stimulation, cellular levels of PtdIns(3,4)$P_2$ and PtdIns(3,4,5)$P_3$ but not PtdIns(3)P, are elevated.[27,31,32] Although PtdIns(3,4)$P_2$ can be produced from PtdIns(4)P via the activation of PI3-kinase, the bulk of this product is generated from the dephosphorylation of PtdIns(3,4,5)$P_3$ by a 5'phosphatase (for review see ref. 33). Thus, PtdIns(3,4,5)$P_3$ is the major phosphatidylinositol that is generated directly upon activation of PI3-kinase.

A role for PI3-kinase has been established in almost all of the major actions of insulin such as the regulation of glycogen synthesis, protein synthesis, lipolysis, glucose uptake, membrane trafficking, cytoskeletal arrangement and apoptosis (see reviews refs. 30, 34). Interestingly, the regulation of transcription of genes involved in metabolism by insulin also appears predominantly dependent on PI3-kinase activation (Table 1 and refs. 35-37).

The products of the PI3-kinase reaction, PtdIns (3,4,5)$P_3$ and PtdIns (3,4)$P_2$, regulate cellular processes through interactions with proteins that contain pleckstrin homology (PH) domains. The best studied PH domain targets for these lipids are members of the AGC subfamily of protein kinases.[38,39] These include PDK1 (3'-phosphoinostide-dependent kinase 1) and PKB (protein kinase B). The activation of other members of this family, such as SGK (serum and glucocorticoid-induced kinase), p70 S6 kinase (S6K), p90rsk and MSK (mitogen and stress activated protein kinase), requires phosphorylation by PDK1,[39] therefore they are indirectly regulated by the binding of PtdIns (3,4,5)$P_3$ to PDK1. However, each of these protein kinases also receive signalling inputs from other insulin regulated pathways. For example, S6K activation requires mTOR activity,[40] while p90rsk activation requires p42/p44 MAPK activity (see below ref. 41). Thus, inhibitors of mTOR (rapamycin) and the p42/p44 MAPK pathway (e.g., PD98059) block activation of S6K and p90rsk, respectively (Fig. 2). Both of these

*Table 1. Regulation of gene transcription by insulin*

| GENE | Insulin Effect | Signalling Pathway | Tissue/Cells | Ref. |
| --- | --- | --- | --- | --- |
| PEPCK | I (90-100%) | PI 3-Kinase | Liver/Hepatoma | 35 |
| Glucose 6-Phosphatase | I (90-100%) | PI 3-Kinase | Liver/Hepatoma | 181 |
| IGF Binding Protein-1 | I (90-100%) | PI 3-Kinase/ mTOR | Liver/Hepatoma | 193 |
| CYP2E1 | I (40-70%) | PI 3-Kinase/ mTOR + c-src | Liver/Hepatoma | 198 |
| Microsomal Triglyceride Transfer Protein | I (50%) | MAP Kinase | Hepatoma | 199 |
| Surfactant Protein-A | I (20-80%) | PI 3-Kinase/ mTOR | Lung | 200 |
| Fatty Acid Synthase | S (3-5-fold) | PI3-Kinase | Adipose + Liver | 201 |
| Apo A1 | S (2-3-fold) | MAP Kinase + PKC | Hepatoma | 122 |
| c-fos | S (10-12-fold) | MAP Kinase | Liver/Hepatoma | 202 |
| Plasminogen Activator Inhibitor-1 | S (2-4-fold) | PI3-Kinase + MAP Kinase | Hepatoma/CHO-IR | 115 |
| SHARP2 | S (3-5-fold) | PI3-Kinase | Liver | 203 |
| pip92 | S (3-6-fold) | MAP Kinase | Hepatoma | 202 |
| Fos Related Antigen-1 | S (3-10-fold) | MAP Kinase | Liver/CHO-IR | 77 |
| Hexokinase-II | S (2-5-fold) | PI3-Kinase/ mTOR | L6 myotubes | 204 |
| p85_ (PI 3-kinase) | S (2-fold) | PI3-Kinase + mTOR | Muscle/C2C12 | 205 |
| Leptin | S (5-fold) | PI3-Kinase + MAP Kinase | Adipocytes | 206 |
| Prolactin | S (10-20-fold) | PI3-Kinase | Adipocytes | 207 |
| Uncoupling Protein-1 | S (15-fold) | PI3-Kinase + MAP Kinase | Adipocytes (Brown) | 208 |
| Glucokinase | S (1.5-4-fold) | PI3-kinase (Class II) | B-Cells | 209 |
| Insulin | S (1.5-3-fold) | PI3-Kinase/ mTOR | B-Cells | 209 |
| Activity-regulated Cytoskeletal-gene (ARC) | S (2-10-fold) | MAP Kinase + src | Neuroblastoma | 210 |
| TNFα | S (2.5-fold) | MAP Kinase | Macrophages | 211 |
| vEGF | S (6-fold) | PI3-kinase | Fibroblasts | 212 |
| LDL-receptor | S (8-20-fold) | PI3-Kinase + MAP Kinase | Ovary | 213 |

I: inhibited, S: stimulated

inhibitors are known to antagonise insulin regulation of a number of genes, implicating these protein kinases in this action of insulin (Table 1).

### *The p42/p44 MAPK Cascade*

The adaptor molecules Shc and Grb2 bind either singly or in combination to the IRS's through their SH2 or PTB domains.[42] Grb2 is complexed to the Ras guanine exchange factor mSOS (son of sevenless). Recruitment of mSOS from the cytosol to the plasma membrane activates Ras (a 21 kDa GTPase). In its active GTP bound form Ras associates with the N-terminal region of the serine/threonine kinase Raf, bringing it to the plasma membrane to become activated.[43,44] Activated Raf forms a stable complex with another protein kinase termed MKK1 (mitogen activated protein kinase kinase1, also known as MEK1). Phosphorylation of MKK1 by Raf increases MKK1 activity.[45-47] MKK1 in turn, phosphorylates and activates p42/p44 mitogen activated protein kinase (MAPK). p42/44 MAPKs are members of the MAPK superfamily that also include the p38 MAPK and c-jun N-terminal protein kinase (JNK) isoforms.[48] These latter MAPKs are activated predominantly by cellular stresses such as osmotic stress, oxidative stress, UV irradiation, heat stress and cytokines.[48,49] In some cells insulin is a very weak activator of p38 MAPK and JNK, but the role of these kinases in insulin action is unclear, as strong activation of either of these molecules can promote insulin resistance.[50,51]

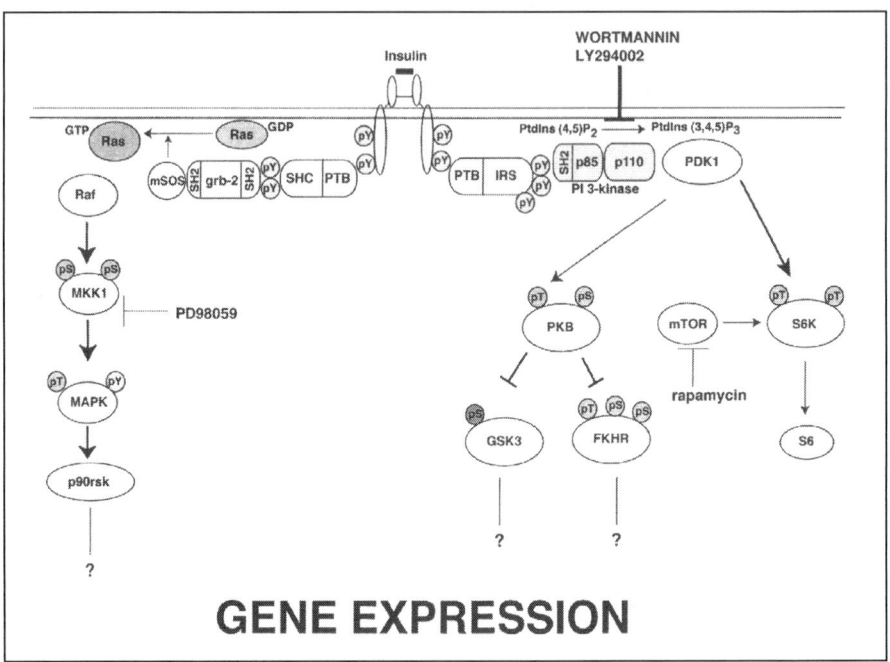

Figure 2. Insulin signaling pathways potential molecular connections between the insulin receptor at the cell surface and gene expression. Commonly used pharmacological inhibitors are given along with their site of action. Arrows indicate activation, I indicates inhibition. Abbreviations; pY phosphotyrosine, pS phosphoserine, pT phosphothreonine.

Once activated, p42/p44 MAPK phosphorylates many downstream substrates that are involved in numerous cellular processes such as proliferation, differentiation, cell survival and gene transcription. In addition, MAPK is involved in the activation of several downstream serine/threonine protein kinases, such as the p90rsk isoforms (RSKs 1-3), MSK1/MSK2 (mitogen and stress activated protein kinases) and MNK1/MNK2 (MAPK interacting kinases).[52] Once activated, p90rsk phosphorylates downstream targets that are involved in gene transcription, cell cycle regulation and cellular metabolism. The p42/p44 MAPK is key in the mitogenic actions of many hormones and growth factors. Indeed, this molecule appears crucial in the regulation of immediate early genes by insulin in a variety of tissues (Table 1).

### *Other Insulin Regulated Signalling Molecules*

There are multiple protein kinase C isoforms, most (if not all) of which require phosphorylation by PDK1 for activity (see ref. 53 for review). Meanwhile, there is increasing evidence that atypical PKC's ($\lambda/\zeta$) have an important role in some aspects of insulin action.[54-58] In addition, many of the actions of insulin can be mimicked when cells are incubated with phorbol esters (which activate classical and novel although not atypical PKCs).

Insulin treatment of cells can lead to the generation of hydrogen peroxide ($H_2O_2$).[59-62] Treatment of cells with this agent can mimic many of the effects of insulin.[60,63,64] The mechanism of $H_2O_2$ generation is not fully understood, nor is the mechanism by which the $H_2O_2$ mediates insulin action, however this molecule is one of many that can generate reactive oxygen species which are known to influence the activity of the transcription factor NFκB (for review see ref. 65). Interestingly, recent work has identified the phosphotyrosine phosphatases PTP1B and TCP1α, as targets for $H_2O_2$ inhibition (for review see ref. 17). PTP1B, and possibly TCP1α? are key modulators of insulin action, as they dephosphorylate and inhibit the insulin receptor and the IRSs.[62,66]

The insulin receptor also regulates a complex of proteins including cbl, CAP, caveolin and flotillin, which together represent one branch of the insulin signaling pathways that mobilize GLUT4 transporters (for review see ref. 67). As yet there is little evidence that these signaling molecules are directly involved in the regulation of gene transcription.

## *Key Insulin/IGF-1-Regulated Transcription Factors*

### Proto-Oncogenes/Immediate Early Genes

Proto-oncogenes and/or immediate early genes (e.g., jun, fos, elk, fra), were among the first genes shown to be regulated by insulin signaling pathways. Transcription factors such as ELK1, Sap1a, heat shock factor (HSF-1) and serum response factor (SRF) are direct substrates of MAPK. Homodimers of the c-jun family or heterodimers of this family with members of either the c-fos or ATF families, comprise the activator protein-1 (AP1) complex.[68] AP-1 was first identified as the mediator of phorbol ester stimulation of the SV40 enhancer.[69] These bZIP transcription factors (both jun-jun and jun-fos dimers) bind to the consensus sequence TGA(G/C)TCA known as the TPA response element (TRE). Interestingly, insulin can regulate the AP-1 motif at two levels. First, insulin stimulates transcription of the genes encoding both c-fos and c-jun.[70] Second, insulin may mediate an effect on AP-1 through an alteration in the phosphorylation state and transactivation potential of c-fos and c-jun. For example, both p90$^{rsk}$ and MAP kinase can phosphorylate c-fos in vitro[71] although it is not known whether this occurs in vivo. Moreover, one report suggests that insulin may augment the phosphorylation and transactivation potential of c-fos via an unidentified kinase, distinct from MAP kinase.[72] Although relatively little is known about the regulation of c-fos phosphorylation, the phosphorylation of c-jun by multiple kinases, at multiple sites, has been studied in detail (see ref. 68 for review). MAP kinase, as well as several other insulin-regulated serine kinases, including casein kinase II and GSK-3, phosphorylate c-jun in vitro,[73] however, the precise role of these three kinases in insulin regulated gene transcription through AP-1 promoter elements has yet to be established (see refs. 68, 74, 75 for review).

Insulin also regulates both the activity and expression of the c-fos-related antigen-1 (Fra-1).[76,77] The regulation of Fra-1 gene expression by insulin is complex but requires the MAP kinase pathway.[77] In addition, the MAP kinase pathway, but not the PI3-kinase pathway, is required for insulin-induced Fra-1 phosphorylation.[77]

### FOXO Family of Transcription Factors

Initial interest in these transcription factors as mediators of insulin action arose when DAF 16 (the *Caenorhabditis elegans* (*C. elegans*) homologue of FOXO transcription factors) was found to be genetically linked to molecules related to the mammalian insulin signaling pathway. These are DAF-2 (DAF standing for dauer arrest phenotype), AGE-1, PDK1 and AKT 1 and 2, the mammalian homologues of the insulin/IGF1 receptor, PI3-kinase, PDK1, and PKB, respectively. Genetic studies in *C. elegans* indicate that DAF-16 lies downstream of these molecules, and that activation of this pathway leads to the inactivation of DAF-16.[78,79] By analogy, insulin is proposed to inhibit FOXO activity, through PI3-kinase, PDK1 and PKB activation. Subsequently, FKHR, AFX and FKHRL1 (related members of the FOXO family) have been shown to be phosphorylated on three sites by PKB in vitro.[80-84] These phosphorylation events are all PI3-kinase- and PDK1- but not mTOR- or MAPK-dependent.[85] FKHR-induced transcription is inhibited by insulin treatment or by the overexpression of a constitutively active Myr-PKB.[81,82,86-88] Inhibition is the result of PKB-mediated phosphorylation of FKHR, and subsequent nuclear exclusion (see reviews refs. 89, 90). Interestingly, FKHR or FKHR-L1 can bind to DNA sequences in vitro, that are required for insulin regulation of genes involved in gluconeogenesis (i.e., PEPCK type IRS's, (Table 2).[91,92] Thus, it is likely that this family of *trans*-acting factors plays a key role in this important action of insulin (see "Coordinated Regulation of PEPCK, G6Pase, IGFBP-1 and TAT Gene Expression?" section for a more detailed discussion).

Table 2. Insulin response sequences

| IRS Type | Sequence | Example of Gene Regulated | Potential Binding Factor | Effect of Insulin |
|---|---|---|---|---|
| Serum Response Element (SRE) | CC(A/T)$_6$GG | c-fos, b-actin | serum response factor | ↑ |
| AP-1 motif | TGA(G/C)TCA | Collagenase-1<br>Malic Enzyme | AP-1 (c-fos, Fra-1/2, c-jun, etc) | ↑ |
| Sterol Response Element | CANNTG (E-box) and TCACNCCAC | Fatty Acid Synthase | SREBP | ↑ |
| TTF-2 motif | C(T/A)(A/G)A(A/G)(C/A)AAACA | Thyroglobulin<br>Thyroperoxidase | TTF-2 | ↑ |
| Sp1 motif | GGGGCGGG | Apo A1<br>PAI-1 | Sp1 | ↑ |
| Type A PEPCK-like motif (TIRE) | TGTTTTG | PEPCK<br>IGFBP-1<br>G6Pase (IRS2) | Foxo1/others | → |
| Type B PEPCK-like motif (TIRE) | TGTTTTT | G6Pase (IRS1) | Foxo1/others | → |
| Type B PEPCK-like motif (TIRE) | TATTTTA | G6Pase (IRS3) | unknown | → |

## SREBP-1c

Mutagenesis studies on the promoters of the cholesterol biosynthetic genes identified the sterol regulatory element-1 (SRE1; 5'-ATCACCCAC-3') as the promoter sequence required for regulation of expression by cholesterol.[93] The Sterol Regulatory Element-Binding Proteins (SREBPs) are a group of proteins that belong to the basic helix-loop-helix leucine zipper (bHLH-Zip) family of transcription factors (for review see refs. 94-96). There are currently three members of the family; namely SREBP-1a, SREBP-1c and SREBP-2. The first two are encoded from the use of alternative first exons of the same gene, while SREBP-2 is expressed from a separate gene.[97] The three SREBPs have a common structure (SREBP-1 is approximately 50% identical to SREBP-2), including an N-terminal transcription factor domain of around 480 residues, a middle region of 80 amino acids containing two hydrophobic transmembrane segments, and a C-terminal regulatory domain consisting of 590 amino acids. SREBPs also differ from other bHLH-Zip family members in that they are synthesised as a precursor attached to the endoplasmic reticulum (ER) or the nuclear envelope in a hairpin fashion. The N and C-terminal ends face into the cytoplasm leaving a 'lumenal loop' of 31 amino acids projecting into the ER. In order to function as a transcription factor, the N-terminal domain must be cleaved off and this can then move to the nucleus as mature SREBP (for review see ref. 96). This cleavage occurs as a two-step proteolytic cascade. Firstly, SREBP cleavage activating protein (SCAP) binds to the C-terminal half of SREBP and transports it to the Golgi apparatus where the two proteases required for the proteolytic cascade are located. The first proteolytic cleavage (Site 1) is catalysed by the protease S1P (see ref. 98 for review). The second cleavage (catalyzed by S2P) occurs N-terminal of site 1, within the first membrane spanning segment, but only following site 1 cleavage. Insulin may stimulate cleavage and hence the activation of SREBP-1c through regulation of the SCAP inhibitor protein termed Insig-2.[99] Insig-2a mRNA is reduced by insulin, as well as by feeding.

SREBP-1 (also known as ADD-1) is responsible for promoting the differentiation of cultured rat adipocytes.[100] Interestingly, the SREBPs also bind to the SRE-related sequence CACGTG, better known as an E-box.[101] A single amino acid residue is responsible for this dual DNA binding specificity. Thus, whereas most bHLH proteins contain a conserved E-K-X-R sequence, SREBP1 has an atypical tyrosine residue at position 320 (resulting in the sequence E-K-X-Y). Cotransfection experiments with combinations of wild type and Y320R SREBP-1c demonstrate that the homodimeric Y320R mutant binds to the core E-box motif but not to the SRE1. Meanwhile, the heterodimeric Y320R/Y320 and wild-type homodimer bind to both the SRE1 and E-box sequences. Hence it is the presence of the atypical tyrosine at 320 which allows the dual DNA binding specificity.[101]

The predominant form of SREBP in liver is SREBP-1c,[96,102] even though it is a weaker activator of transcription than SREBP-1a (which is constitutively expressed at low levels). However, in cultured cell lines the opposite occurs with SREBP-1a the predominant form.[103] During fasting SREBP-1 mRNA and protein levels fall but are restored by refeeding.[104] Concomitant increases in the mRNA levels of lipogenic enzymes (such as fatty acid synthase) are observed during refeeding and this is partially blunted in SREBP-1 null mice.[105] No changes in SREBP-2 are detected with fasting and refeeding. The response of SREBP-1c to feeding is probably mediated by insulin, since insulin stimulation of rat hepatocytes induces SREBP-1c mRNA and protein, with no change of SREBP-1a or SREBP-2.[106] The effect of insulin on hepatic SREBP-1c expression requires PI-3K activity while expression of constitutively active PKB leads to the accumulation of SREBP-1c mRNA.[106] Interestingly, the SREBP-1 gene promoter contains an SRE1 site, therefore allowing possible auto-induction. There is also evidence that insulin can directly influence the transcriptional activity of SREBP-1c[96,107] and that this may involve the MAPK pathway.[107] As well as phosphorylation, SREBP-1 undergoes CBP/p300 mediated acetylation, resulting in its stabilization,[108] although it is unclear if this process is regulated by insulin.

One of the best studied SREBP-1 regulated genes is Fatty acid synthetase (FAS). FAS plays a central role in de novo lipogenesis in mammals. Paulauskis and Sul demonstrated that insulin increases FAS gene transcription in the livers of diabetic mice.[109] The effect is rapid (3.5-fold after 30 minutes) and reaches a maximum 7-fold increase after 2 hours of insulin treatment.[109] cAMP abolishes this stimulation as does cycloheximide, suggesting that on-going protein synthesis is required for this action of insulin. Interestingly, refeeding fasted animals induces FAS promoter activity and promotes SREBP-1c interaction with the FAS promoter.[110] Meanwhile, mutation of the SRE1 site (at -150) within this promoter blunts both SREBP-1c binding and the response to refeeding. In addition, mutation of an E-box motif (at -65) blunts both SREBP-1c binding, and induction of the FAS promoter by refeeding.[110] Therefore both elements are involved in SREBP1c induction of this gene promoter.

Increased levels of SREBP-1c in transgenic mice reduces the expression of the gluconeogenic (insulin repressed) gene PEPCK,[111] while adenoviral-mediated expression of a dominant positive form of SREBP-1c represses PEPCK gene expression in hepatocytes.[111,112] However, the PEPCK gene promoter does not contain a consensus SRE1. Therefore the molecular mechanism by which over expressed active SREBP1c represses PEPCK is not clear, but may involve an interaction with the CREB-binding protein.[111] Conversely, expression of a dominant negative form of SREBP-1c using adenovirus blocks insulin's ability to induce glucokinase and the lipogenic enzymes, and its ability to repress PEPCK.[111]

In summary, insulin regulation of SREBP-1c synthesis and/or cleavage and/or transactivating potential, is likely to underlie the regulation of the fatty acid biosynthetic genes, while subtle, if poorly understood, differences in regulation and function of the SREBP isoforms allow different, if overlapping, patterns of gene regulation by insulin and sterols. As such, SREBP-1c is a key mediator of the regulation of gene transcription by insulin.

### Sp1

Sp1 belongs to a growing family of transcription factors[113,114] that were originally considered basal transcription factors, not involved in hormone-regulated gene expression. However, several genes, including plasminogen activator inhibitor type 1 (PAI-1; 115), and Apo A1[116] are now thought to be stimulated by insulin through Sp1 binding elements (see ref. 117 for review). Conversely, there are a multitude of gene promoters that contain Sp1-binding elements but are not insulin regulated. This demonstrates that the context of the Sp1 element in relation to other promoter elements is of utmost importance in gene regulation by insulin.[118] Indeed, in some cases Sp1 may actually antagonise or oppose the action of insulin through certain IRS sequences.[117]

Sp1 has been linked to the regulation of the glycolytic enzyme glyceraldehyde-3-phosphate dehydrogenase (GAPDH). In their initial studies, Alexander and colleagues demonstrated that insulin causes a 3-fold increase in GAPDH mRNA in the H4IIE rat hepatoma cell line and a 10-fold change in the 3T3 F442A adipocyte cell line.[119,120] Using the transient transfection of GAPDH/CAT fusion genes, they then found that the stimulatory effect of insulin on human GAPDH gene expression is mediated through *cis-acting* sequences located between -488 and +21.[120] Further analysis of the effect of insulin on expression of additional GAPDH/CAT fusion gene constructs suggested that the GAPDH promoter contains two independent insulin response elements, designated IRE-A (located between -488 and -409) and IRE-B (located between -308 and -269). The core sequence of IRE-A was observed to have close homology with sequences in the promoters of a number of other insulin regulated genes.[121] Now, Wong and colleagues have shown that the GAPDH IRE-A element competes with the Apo A1 gene promoter insulin response sequence for binding of Sp1,[116,122] suggesting that Sp1 may also play a role in insulin regulation of GAPDH expression.

The molecular mechanism by which insulin regulates Sp1 is not yet fully elucidated, but is likely to include direct and indirect processes. For example Sp1 binding to the Apo A1 gene promoter (between nucleotides -411 and -404) requires activation of both p42/p44 MAP kinase and protein kinase C.[116,123] Meanwhile, insulin may regulate both Sp1 expression[124] and DNA binding through a complex interplay with other factors such as AP-1 and Egr-1.[125,126]

**Serum Response Factor and Ternary Complex Factor**

c-fos is the cellular homologue of the transforming gene of FBJ murine osteosarcoma virus. Insulin induces c-fos gene transcription in H4IIE cells.[127] A number of studies have implicated the involvement of the MAP kinase pathway in insulin-stimulated c-fos gene transcription.[128-130] For example, overexpression of wild-type $p21^{ras}$ promotes insulin-stimulated c-fos gene expression[128] whereas overexpression of dominant negative mutants of $p21^{ras}$ (and Raf-1) block insulin-stimulated c-fos gene expression.[129,130]

In experiments using c-fos/CAT fusion genes, the c-fos promoter region from -356 to +109 was sufficient to mediate a response to insulin.[131] Mutation of four bases in the c-fos serum response element (SRE), previously shown to abolish the response to serum, also blunted the effect of insulin on this region of the c-fos gene promoter.[131] The c-fos SRE (located between -320 and -299 in the c-fos promoter) transfers the insulin response to a heterologous promoter,[130] conclusively demonstrating the presence of an IRS/IRE. The sequence of the c-fos SRE is compared with other IRSs in Table 2.

Although a number of SRE binding proteins have been identified,[132] the most prominent is the 67 kDa serum response factor ($p67^{SRF}$). Others include NFIL6, Phox 1 and DBF/MAPF1. An Ets domain binding motif (CAGGAT), recognized by $p62^{TCF}$ (ternary complex factor), is located just 5' of the c-fos SRE. In gel retardation assays, using nuclear extracts from insulin treated cells, the formation of an SRE protein complex (designated "band 2") increases.[133,134] Several lines of evidence indicate that band 2 is a ternary complex consisting of the SRE, $p67^{SRF}$ and $p62^{TCF}$[133,134] and that insulin may stimulate its formation through increased phosphorylation of $p62^{TCF}$. MAP kinase phosphorylates Elk-1, a protein highly homologous to $p62^{TCF}$, leading to increased binding[135] and/or transcriptional activity of Elk-1.[136] Taken together, it is possible that a direct connection between an insulin-stimulated protein kinase and an SRE binding protein has been made, and that this potentially explains the mechanism of insulin action on c-fos gene expression. Unfortunately, detailed mutagenesis of the c-fos SRE by Thompson et al[134] demonstrates that the effect of insulin on c-fos gene transcription does not correlate with band 2 formation. However, a second complex (designated "band 3") has now been identified, the formation of which is also increased in extracts from insulin treated cells.[134] The time course for formation of band 3 correlates with the induction of c-fos gene transcription by insulin.[134] $p67^{SRF}$ appears to be the only DNA binding protein in the band 3 complex, but Thompson et al suggest that the effect of insulin is mediated indirectly by other unidentified proteins in the band 3 complex that directly associate with $p67^{SRF}$.[134] Thus, although the MAP kinase pathway is implicated in insulin-stimulated c-fos gene transcription, the precise *trans*-acting factor that mediates this action of insulin (directly or indirectly) remains to be conclusively identified.

**Thyroid Transcription Factor-2**

Insulin like growth factor-1 (IGF-1) stimulates thyroglobulin gene transcription in rat thyroid cell lines,[137] an effect mediated through the -168 to +39 region of the promoter.[138] Mutation of any one of three transcription factor binding sites in this region abolishes IGF-1 regulation.[138] These three elements bind the thyroid transcription factors (TTF)-1, TTF-2 and a ubiquitous factor (UFA), respectively. IGF-1 induces TTF-2 (but not TTF-1 or UFA) binding, and protein synthesis is required for this effect.[138,139] Importantly, this TTF-2 element can confer a stimulatory action of insulin on the expression of a reporter gene when ligated to a heterologous promoter.[140]

TTF-2 is also required for induction of thyroperoxidase expression by insulin. Again, this action of insulin requires two additional transcription factors, namely TTF-1 and nuclear factor-1 (NF-1).[141] Interestingly, NF-1 is also an insulin-induced gene.[141]

As TTF-2 is expressed specifically in thyroid cells, which lack functional insulin receptors,[142] technically this is an IGF-1 response protein. However, it is included in this chapter because the nucleotide sequence of this element has homology with the PEPCK type IRS (Table 2), the IGF-1 and insulin signaling pathways have major overlap, and TTF-2, like FOXO, is a member of the forkhead transcription factor family.[143] It is possible that divergent evolution has provided homologous mechanisms for gene regulation in functionally distinct tissues. Therefore the study of the regulation of TTF-2 by IGF-1 may provide clues to the regulation of gene transcription by insulin in more classical insulin sensitive tissues.

## Key Insulin-Regulated Gene Promoters

### Overview

Insulin is now believed to influence the expression of more than 150 gene products. It is not yet clear how many of these genes are directly regulated at the level of gene transcription but the application of DNA microarray technology is likely to address this in the near future. Obviously it is beyond the scope of this chapter to review the molecular regulation of all of these genes in detail. Therefore, we will focus on the best studied (and possibly best understood) insulin-regulated gene promoters, and present the current understanding of the DNA elements, *trans*-acting factors and signaling pathways that permit their regulation by insulin. In this way we will provide examples of the complexity of insulin regulation of metabolically important gene promoters, the importance of interactions with other hormones, and develop the concept that insulin may regulate every one of its target gene promoters by different mechanisms.

### Phosphoenolpyruvate Carboxykinase (PEPCK)

Phosphoenolpyruvate carboxykinase catalyzes the conversion of oxaloacetate to phosphoenolpyruvate, which is the initial, irreversible step in hepatic and renal gluconeogenesis. The mechanisms that mediate the tissue-specific and hormonally regulated expression of the hepatic cytosolic PEPCK gene have been studied in great detail (for review see refs. 4, 5, 144). The rate of transcription of the PEPCK gene is stimulated by cAMP, retinoic acid, thyroid hormone and glucocorticoids but is inhibited by insulin and glucose. In H4IIE cells the inhibitory effects are dominant, since insulin and glucose prevent cAMP- and glucocorticoid-stimulated PEPCK gene transcription. Insulin primarily inhibits the initiation of PEPCK gene transcription but also reduces the rate of transcript elongation.[145]

Complex hormone response units, composed of multiple *cis*-acting elements, are required to manifest the full response to each of these hormones. For example, in the PEPCK gene a glucocorticoid response unit (GRU) mediates the stimulatory action of glucocorticoids.[146] Previous studies[146-150] showed that this GRU consists of a tandem array (5' to 3') of four accessory factor binding sites (gAF1, from -455 to -431; gAF2 from -420 to -403, gAF3 from -327 to -321; and the cyclic AMP response element [CRE] from -93 to -86) and two glucocorticoid receptor binding sites (GR1 and GR2, from -395 to -349). gAF1, gAF2, gAF3 and the CRE do not function as glucocorticoid response elements themselves. However, when any of these are mutated the glucocorticoid response is reduced by 50-60% and when any combination of two is mutated the promoter is no longer responsive to glucocorticoids; thus GR1 and GR2 are inert by themselves.

Similarly, Hanson and colleagues have shown that four *cis*-acting elements (designated the CRE, P3[I], P3[II] and P4) are required for cAMP-stimulated PEPCK gene transcription in HepG2 cells.[144] More recent evidence implicates gAF3 in the cAMP response as well as the glucocorticoid response in H4IIE cells.[151] Together these elements form what can be termed the cAMP response unit (CRU). The CRE, located between -93 and -86 relative to the transcription start site, contains the consensus cAMP response element sequence T(G/T)ACGTCA found in many, but not all, cAMP regulated genes. The other three elements required for the cAMP response in HepG2 cells form a complex unit located between -285 and -238 in relation to the transcription initiation site. Thus, the PEPCK promoter exemplifies an emerging paradigm: complex hormone response units and not simple hormone response elements are prevalent in eukaryotic promoters.[152]

At least two cis-acting elements mediate the action of insulin on PEPCK gene transcription. One element is located between -437 and -402 (distal) and the other(s) between -271 and +69 (proximal). The distal IRS has been analyzed in detail in the context of a heterologous promoter.[88,153,154] The core sequence (located between -413 and -407) is shown in Table 2. This element, that we call the PEPCK-like IRS, may also mediate the negative effect of insulin on the expression of the insulin-like growth factor binding protein-1 (IGFBP-1), G6Pase and tyrosine aminotransferase (TAT) genes (see later). In all of these gene promoters the IRS coincides with an element required for full induction of transcription by glucocorticoids. In the case of PEPCK this is gAF2. Again, in all three genes hepatic nuclear factor-3 (HNF-3 also known as FoxA2) may be the accessory factor required for full induction of gene transcription by glucocorticoids.[154-156] However, the available data suggests that HNF-3 does not directly mediate the action of insulin.[154] Importantly, in vivo footprinting studies reveal no change in the footprint over the PEPCK IRS following treatment of H4IIE cells with insulin.[157]

The phosphatase inhibitor, okadaic acid, mimics the action of insulin on PEPCK gene transcription[158] suggesting that, as with most of insulin's actions, changes in protein phosphorylation are involved in these effects. Indeed, regulation of PEPCK gene transcription by insulin requires PI3-kinase activity, but not mTOR or MAP kinase activity (Table 1 and refs. 35-37). In addition, strong activation of PKB/Akt[159] or pharmacological inhibition of the PKB-inhibited kinase glycogen synthase kinase-3 (GSK3),[160] represses PEPCK transcription. This suggests the presence of a linear signaling pathway from PI3-kinase, through PKB and GSK3, to a PEPCK IRS-binding protein. In addition, there is data linking the FOXO transcription factor family with insulin regulation of this gene promoter (see "Coordinated Regulation of PEPCK, G6Pase, IGFBP-1 and TAT Gene Expression?" section).

Interestingly, many other cellular manipulations affect the expression of PEPCK. Repressing manipulations include; over expression of active SREBP-1c,[111] active Ras, active Raf,[37] or LIP;[161] oxidative stress,[64] activation of AMP-activated protein kinase;[162] or treatment of cells with phorbol esters,[163] or EGCG, a constituent of green tea.[164] Thus, much more information will be required before the complicated molecular processes involved in the regulation of PEPCK gene transcription are fully appreciated.

## *Insulin-Like Growth Factor Binding Protein-1 (IGFBP-1)*

The insulin-like growth factor binding proteins are a family of six secreted proteins, designated IGFBP-1 to 6, that specifically bind IGF-I and IGF-II but do not bind insulin. The structure of the IGFBP proteins and the differential, tissue-specific and hormonal regulation of expression of the IGFBP genes has been reviewed elsewhere.[165]

The multihormonal regulation of the PEPCK and IGFBP-1 genes is similar in that cAMP, thyroid hormone and glucocorticoids stimulate the hepatic expression of both genes whereas insulin has a dominant inhibitory effect (see ref. 166 for review). The structural organization of the PEPCK and IGFBP-1 promoters also has similarities. Two glucocorticoid response elements (GRE) and an IRS are present in the IGFBP-1 gene promoter.[167-169] In the human promoter the IRS is located between (-120 and -96) and the two GREs are found between (-110 and -84) and (-198 and -173).[167,170] Powell and colleagues defined the human IGFBP-1 IRS by analyzing the ability of insulin to repress basal IGFBP-1 reporter gene expression.[167] A deletion that abolished the negative effect of insulin was identified. The region encompassed by the deletion was then shown to confer an inhibitory effect of insulin on CAT reporter gene expression directed by a heterologous promoter.[167] The core IRS sequence, T(G/A)TTTTG, is the same as that found in the distal IRS of the PEPCK gene (Table 2), but the PEPCK gene promoter has a single copy of this element whereas the IGFBP-1 gene promoter has two copies arranged as an inverted palindrome.

In a remarkably similar fashion to the distal PEPCK IRS, the IGFBP-1 IRS also acts as an accessory factor binding site that is required for the glucocorticoid response. Thus, mutation of the IGFBP-1 gene IRS abolishes the induction of human IGFBP-1 gene transcription by glucocorticoids even though both GREs are intact.[154,170] As described in the preceding section, the accessory factor is thought to be HNF-3.[154,156]

However, although this similarity in promoter structure exists, many differences in the signaling pathways involved in the regulation of PEPCK and IGFBP1 expression have recently been identified (see "Coordinated Regulation of PEPCK, G6Pase, IGFBP-1 and TAT Gene Expression?" section for details). Therefore, it appears that the presence of an inverted palindrome of the IRS, as opposed to the single sequence in the distal IRS of the PEPCK promoter, results in a distinct molecular link between the insulin receptor and the promoter element.

### Glucose 6-Phosphatase Catalytic Subunit (G6Pase)

Glucose 6-phosphatase converts G6P to glucose, the final step in both gluconeogenesis and glycogenolysis. This enzyme activity is expressed mainly in liver and kidney, and is catalyzed by a multi-component integral membrane system within the endoplasmic reticulum.[171-174] Current thinking suggests that the complete system requires a catalytic subunit (G6Pase), along with specific transporters for glucose, G6P, and inorganic phosphate.[171-174] Inactivating mutations in the G6Pase gene result in reduced hepatic glucose production and are responsible for glycogen storage disease type 1a.[175] In contrast, over expression of G6Pase in liver cells results in increased hepatic glucose production.[176-178]

Insulin acutely inhibits G6Pase enzyme activity, at least partly, and this may be PI3-kinase-dependent,[179] but the major form of inhibition of this activity occurs at the level of gene transcription.[180,181] The overall hormonal regulation of G6Pase gene expression is similar to that for PEPCK and IGFBP-1. Glucocorticoids and cAMP stimulate gene transcription, while insulin represses both basal and induced gene expression.[180,182] However, in contrast to PEPCK, glucose induces G6Pase gene transcription, but this is also antagonised by insulin.[183]

Two regions of the mouse G6Pase gene promoter, designated A (from -231 to -199) and B (from -198 to -159), are required for complete repression of basal G6Pase gene transcription by insulin.[180,182] These regions are highly conserved between species, including human.[92] Region A acts as an accessory element that enhances the action of insulin on region B.[182] Therefore Region B is referred to as the G6Pase IRS, however, the two regions together comprise the G6Pase insulin response unit (IRU). HNF-1 is the accessory factor that binds to Region A,[182] while Region B actually contains three PEPCK-like IRS motifs that are arranged in tandem and designated IRS-1 to IRS-3 (or PEPCK-type B, A and C, respectively, (see Table 2).[180,184] IRS-2 has an identical sequence to the PEPCK IRS, however IRS-1 (TGTTTTT) and IRS-3 (TATTTTA), differ by one and two nucleotides, respectively. Interestingly, these differences in sequence alter the protein binding characteristics of each IRS,[184] suggesting that the molecular mechanism of insulin regulation of G6Pase differs from that for the PEPCK IRS (see "Coordinated Regulation of PEPCK, G6Pase, IGFBP-1 and TAT Gene Expression?" section). Similarly, while the PEPCK IRS also functions as a key binding site for an accessory factor for glucocorticoid induction, the G6Pase IRS motifs may be directly colocalized with a glucocorticoid receptor binding site (O'Brien unpublished observations).

### Tyrosine Aminotransferase (TAT)

The tyrosine aminotransferase gene has served as a paradigm for the hormonal regulation and tissue-specific expression of hepatic genes.[185] Schutz and colleagues defined three far-upstream enhancers that mediate this regulation.[186] An enhancer at -11 kbp mediates liver specific TAT gene expression, whereas enhancers at -3.6 kbp and -2.5 kbp mediate the induction of TAT gene transcription by cAMP and glucocorticoids, respectively. Grange and colleagues have characterized an additional glucocorticoid-dependent enhancer at -5.4 kbp.[187] As described above for the PEPCK gene, all three hormone-dependent enhancers in the TAT gene promoter are actually hormone response units in that multiple accessory factor binding sites are required to manifest the full response to cAMP and glucocorticoids.[185,187]

Regulation of TAT gene expression by insulin has been studied in considerable detail, with several different results, depending on the cell line studied (see ref. 188 for review). Meanwhile, Ganss et al[185] proposed that insulin mediates its negative effect on cAMP- and glucocorticoid-stimulated TAT gene transcription through the -3.6 kbp and -2.5 kbp enhancers, respectively. They suggest that insulin acts to disable the TAT CRE in the -3.6 kbp enhancer but this action of insulin may be mediated indirectly through the well-characterized ability of the hormone to stimulate cAMP phosphodiesterase activity.[189]

Schütz and colleagues have shown that a CCAAT box, CACCC box and an HNF-3 binding site found in the vicinity of the -2.5 kbp TAT GRE are all required for full glucocorticoid-stimulated TAT gene transcription.[186] The HNF-3 binding site, not the CCAAT box or CACCC box, is the site of insulin action in the -2.5 kbp enhancer.[185] This HNF-3 binding site contains a TGTTTGT motif similar to the PEPCK/IGFBP-1 core IRS. Although the detailed mutagenesis analyses required to prove that this sequence within the TAT HNF-3 motif is the site of insulin action have not been reported, it is possible that insulin mediates its negative effect on PEPCK, IGFBP-1, G6Pase and TAT gene transcription through the same *trans*-acting factor.

## Coordinated Regulation of PEPCK, G6Pase, IGFBP-1 and TAT Gene Expression?

As discussed above, it is clear that structurally related DNA sequences are important in insulin inhibition (as well as glucocorticoid induction) of PEPCK, G6Pase, and IGFBP-1 (Table 2). Meanwhile, the FOXO proteins FKHR, FKHR-L1 and AFX are a family of *trans*-acting factors whose activity is repressed by insulin (see "FOXO Family of Transcription Factors" section). Taken together with the finding that the FOXO proteins can bind to the G6Pase, PEPCK, TAT and IGFBP-1 IRS's in vitro,[91,92] a strong case can be made that these factors play an important role in inhibition of one or more of these gene promoters by insulin. In agreement with this possibility, PI3-kinase activity is required for insulin inhibition of the FOXO proteins,[85] as well as G6Pase, PEPCK and IGFBP-1 gene expression.[35,160,162,181,190] However, other studies have brought this simple hypothesis into question, or at least suggest that distinct mechanisms are likely to be involved in the regulation of each of these gene promoters by insulin.

For example, Hall et al placed the IGFBP-1 IRS upstream of a thymidine kinase promoter and mutated single nucleotides in each half of the inverted palindromic sequence.[88] The ability of FKHR-L1 to activate these mutant IRS's, and the effect of insulin on these promoters, with or without FKHR-L1 overexpression was assessed. A clear dissociation was observed between FKHR-L1 binding in vitro and the ability of insulin to repress fusion gene expression in cells.[88] However, a perfect correlation was observed when FKHR-L1 was over expressed. This suggests that the endogenous insulin response factor that binds to the IGFBP-1 IRS is not FKHR-L1. In contrast, fasting IGFBP-1 gene expression is elevated in transgenic mice over expressing constitutively active FKHR.[191] Also, heterozygous deletion of FKHR suppresses the elevation of IGFBP-1 gene expression that occurs in fasting mice with a simultaneous heterozygous deletion of the insulin receptor gene.[191] Surprisingly, reduced expression of FKHR alone in mice has no effect on IGFBP-1 gene expression.[191] Nevertheless, this genetic data would support a role for FKHR in the regulation of the IGFBP-1 gene.

Detailed studies on the signaling mechanisms used by insulin to regulate IGFBP-1 gene expression suggest that there must be other *trans*-acting factors involved in this action of insulin. For example, the mTOR inhibitor rapamycin blocks the regulation of IGFBP-1 expression by insulin,[192,193] demonstrating a requirement for mTOR in the pathway from insulin to the

IGFBP-1 promoter (insulin regulation of FOXO activity is unaffected by rapamycin). In addition, insulin requires PI3-kinase activity, but not MAP kinase or mTOR activity, in order to reduce PEPCK[35] and G6Pase gene expression.[181] Therefore, this rapamycin sensitive pathway is not common to all three of these insulin regulated genes. Similarly, although phorbol ester, okadaic acid and $H_2O_2$ treatment of hepatoma cells mimic insulin and repress PEPCK and G6Pase gene transcription, these agents antagonize insulin repression of IGFBP1 gene transcription.[158,163,190,194] The effect of such cellular manipulations on FKHR activity is not yet fully characterized. However, there is clearly distinct molecular wiring between the insulin receptor and the IGFBP-1 gene promoter compared to the pathway between the insulin receptor and the G6Pase or PEPCK gene promoters.

Similarly, the molecular regulation of the PEPCK and G6Pase gene promoters by insulin is likely to have distinct components. For example, stable expression of FKHR in hepatoma cells induces G6Pase expression, without affecting PEPCK gene expression.[195] Meanwhile, expression of constitutively active FKHR in mice also induces G6Pase expression levels without much effect on PEPCK expression.[191] This implicates FKHR in the regulation of G6Pase but not PEPCK gene expression. However, heterozygous deletion of FKHR blunts the increase in both G6Pase and PEPCK gene expression that is associated with haploinsufficiency of the insulin receptor.[191] Whether this reflects an overall effect of loss of FKHR on insulin sensitivity, or a direct function of FKHR in the regulation of these genes is not yet clear.

Therefore, it is quite likely that although it makes metabolic sense to coordinate the molecular regulation of these gene products, distinct signaling mechanisms and different DNA binding protein complexes have evolved to permit appropriate regulation by insulin.

## Conclusions

Now that multiple IRSs/IREs have been characterized, it is apparent that a unique consensus sequence does not exist (Table 2). Several genes whose transcription is inhibited by insulin, namely PEPCK, G6Pase, IGFBP-1 and TAT, appear to share a related core IRS. However, much remains to be learned concerning the *trans*-acting factors that bind to the identified IRSs/IREs, and about the precise mechanism(s) of insulin signaling to these proteins. Interestingly, it appears that even related IRS/IRE sequences are regulated by different insulin signaling pathways, potentially regulating the binding of distinct *trans*-acting factors to the transcriptional initiation complex. The development of DNA microarray technology combined with genetic manipulation of cells, tissues and animals, is leading to an explosion of studies intended to categorize all insulin-regulated gene promoters through their dependence on insulin signaling molecules and specific *trans*-acting factors. Hopefully this will aid in the clarification of the dependency and sufficiency of each IRS/IRE sequence and the importance of interactions with associated gene promoter elements.

In summary, much remains to be learned, but the importance of these studies is emphasized by the fact that many of the mutations that result in maturity onset diabetes of the young occur in transcription factors,[196,197] and that the altered rate of transcription of genes associated with insulin resistance almost certainly contributes to the serious complications of, if not the development of, type 2 diabetes mellitus, dyslipidemia and hypertriglyceridemia.

### *Acknowledgements*

Due to the limitation of the number of references we apologize to those authors whose work has not been cited. We appreciate the many helpful discussions we have had with members of our laboratories, and with other colleagues. In particular we thank Christopher Lipina for his input, and Deborah Brown, who was a great help in the preparation of this chapter. C.D.S is the recipient of the Diabetes UK Senior Fellowship (RD02/002473). The work discussed was supported by DK35107 and DK07061 (to D.K.G), DK56374 and DK61645 (to R.O'B), and also the Vanderbilt Diabetes Research and Training Center (DK20593).

# References

1. Denton RM. Early events in insulin actions. Adv Cyclic Nucleotide Protein Phosphorylation Res 1986; 20:293-341.
2. Cohen P, Campbell DG, Dent P et al. Dissection of the protein kinase cascades involved in insulin and nerve growth factor action. Biochem Soc Trans 1992; 20:671-74.
3. Avruch J. Insulin signal transduction through protein kinase cascades. Mol Cel Biochem 1998; 182:31-48.
4. O'Brien RM, Granner DK. Regulation of gene expression by insulin. Physiol Rev 1996; 76:1109-61.
5. Sutherland C, O'Brien RM, Granner DK. Genetic regulation of glucose metabolism. In: Jefferson LS, Cherrington AD, eds. Handbook of Physiology. Oxford University Press, 2003:707-34.
6. Granner DK, Pilkis S. The genes of hepatic glucose metabolism. J Biol Chem 1990; 265:10173-78.
7. Iynedjian PB, Jotterand D, Nouspikel T et al. Transcriptional induction of glucokinase gene by insulin in cultured liver cells and its repression by the glucagon-cAMP system. J Biol Chem 1989; 264:21824-29.
8. Unger RH. Diabetic hyperglycemia: Link to impaired glucose transport in pancreatic b cells. Science 1991; 251:1200-05.
9. Permutt MA, Chiu KC, Tanizawa Y. Glucokinase and NIDDM. A candidate gene that paid off. Diabetes 41 1992; 11(1367-1372).
10. Granner DK, O'Brien RM. Molecular physiology and genetics of NIDDM. Diabetes Care 1992; 15(3):369-95.
11. Valera A, Pujol A, Pelegrin M et al. Transgenic mice overexpressing PEPCK develop NIDDM. Proc Natl Acad Sci 1994; 91:9151-54.
12. Kahn CR. Insulin action, Diabetogenes and the cause of Type II Diabetes. Diabetes 1994; 43:1066-84.
13. Hotamisligil GS, Shargill NS, Spiegelman BM. Adipose expression of TNF-a: Direct role in obesity-linked insulin resistance. Science 1993; 259:87-91.
14. Maddux BA, Sbraccia P, Kumakura S et al. Membrane glycoprotein PC-1 and insulin resistance in noninsulin-dependent DM. Nature 1995; 373:448-51.
15. Le Roith D, Zick Y. Recent advances in our understanding of insulin action and insulin resistance. Diabetes Care 2001; 24(3):588-97.
16. Steppan CM, Lazar MA. Resistin and obesity-associated insulin resistance. Trends Endocrinol Metab 2002; 13(1):18-23.
17. Tonks NK. PTP1B: From sidelines to the front lines! FEBS Letters 2003; 546(1):140-8.
18. Kahn CR, White MF. The Insulin receptor and the molecular mechanism of insulin action. J Clin Invest 1988; 82(4):1151-56.
19. White MF, Kahn CR. The insulin signalling system. J Biol Chem 1994; 269(1):1-4.
20. White MF. The IRS-signalling system: A network of docking proteins that mediate insulin action. Mol Cell Biochem 1998; 182:3-11.
21. Uchida T, Myers Jr MG, White MF. IRS-4 mediates protein kinase B signaling during insulin stimulation without promoting antiapoptosis. Mol Cell Biol 2000; 20(1):126-38.
22. Pelicci G, Lanfrancone L, Grignani F et al. A novel transforming protein (SHC) with an SH2 domain is implicated in mitogenic signal transduction. Cell 1992; 70(1):93-104.
23. Yamanashi Y, Baltimore D. Identification of the Abl- and rasGAP-associated 62 kDa protein as a docking protein. Dok Cell 1997; 88(2):205-11.
24. Wick MJ, Dong LQ, Hu D et al. Insulin-receptor-mediated p62dok tyrosine phosphorylation at residues 362 and 398 plays distinct roles for binding GAP and Nck and is essential for inhibiting insulin-stimulated activation of Ras and Akt. J Biol Chem 2001; 278(46):42843-50.
25. Pawson T. Protein modules and insulin signaling networks. Nature 1995; 373:573-80.
26. Kaburagi Y, Yamauchi T, Yamamoto-Honda R et al. The mechanism of insulin-induced signal transduction mediated by the insulin receptor substrate family. Endocr J 1999; 46(Suppl):S25-34.
27. Cantley LC. The PI3-kinase pathway. Science 2002; 296:1655-57.
28. Kapeller R, Cantley LC. PI3-kinase. Bioessays 1994; 16(8):565-76.
29. Vanhaesebroeck B, Leevers SJ, Panayotou G et al. Phosphoinositide 3-kinases: A conserved family of signal transducers. Trends Biochem Sci 1997; 22(7):267-72.
30. Shepherd PR, Withers DJ, Siddle K. Phosphoinositide 3-kinase: The key switch mechanism in insulin signalling. Biochem J 1998; 333(3):471-90.
31. Stephens LR, Hughes KT, Irvine RF. Pathway of phosphatidylinositol(3,4,5)-trisphosphate synthesis in activated neutrophils. Nature 1991; 351(6321):33-39.
32. Hawkins PT, Jackson TR, Stephens LR. Platelet-derived growth factor stimulates synthesis of PtdIns(3,4,5)P3 by activating a PtdIns(4,5)P2 3-OH kinase. Nature 1992; 358(6382):157-59.

33. Leslie NR, Biondi RM, Alessi DR. Pi-regulated kinases and PI phosphatases. Chem Rev 2001; 101(8):2365-80.
34. Vanhaesebroeck B, Leevers SJ, Ahmadi K et al. Synthesis and function of 3-phosphorylated inositol lipids. Annu Rev Biochem 2001; 70:535-602.
35. Sutherland C, O'Brien RM, Granner DK. Phosphatidylinositol 3-kinase, but not p70/p85 ribosomal S6 protein kinase, is required for the regulation of Phosphoenolpyruvate Carboxykinase gene expression by insulin. J Biol Chem 1995; 270(26):15501-06.
36. Gabbay RA, Sutherland C, Gnudi L et al. Insulin regulation of PEPCK gene expression does not require activation of the Ras/MAP kinase signaling pathway. J Biol Chem 1996; 271(4):1890-97.
37. Sutherland C, Waltner-Law M, Gnudi L et al. Activation of the Ras-MAP Kinase-RSK pathway is not required for the repression of PEPCK gene transcription by insulin. J Biol Chem 1998; 273(6):3198-204.
38. Alessi DR, Downes CP. The role of PI3-kinase in insulin action. Biochim Biophys Acta 1998; 1436:151-64.
39. Williams MR, Arthur JS, Balendran A et al. The role of 3-phosphoinositide-dependent protein kinase 1 in activating AGC kinases defined in embryonic stem cells. Curr Biol 2000; 10(8):439-48.
40. Chung J, Kuo CJ, Crabtree GR et al. Rapamycin-FKBP specifically blocks growth-dependent activation of and signaling by the 70kDa S6 protein kinases. Cell 1992; 69:1227-36.
41. Sturgill TW, Ray LB, Erikson E et al. Insulin-stimulated MAP-2 kinase phosphorylates and activates ribosomal protein S6 kinase II. Nature 1988; 334:715-18.
42. Pawson T, Scott JD. Signaling through scaffold, anchoring, and adaptor proteins. Science 1997; 278(5346):2075-80.
43. Moodie SA, Willumsen BM, Weber MJ et al. Complexes of Ras.GTP with Raf-1 and mitogen-activated protein kinase kinase. Science 1993; 260(5114):1658-61.
44. Stokoe D, Macdonald SG, Cadwallader K et al. Activation of Raf as a result of recruitment to the plasma membrane. Science 1994; 264(5164):1463-7.
45. Kyriakis JM, App H, Zhang XF et al. Raf-1 activates MAP kinase-kinase. Nature 1992; 358(6385):417-21.
46. Dent P, Haser W, Haystead TA et al. Activation of mitogen-activated protein kinase kinase by v-Raf in NIH 3T3 cells and in vitro. Science 1992; 257(5075):1404-7.
47. Zheng CF, Guan KL. Activation of MEK family kinases requires phosphorylation of two conserved Ser/Thr residues. EMBO J 1994; 13(5):1123-31.
48. Chang L, Karin M. Mammalian MAP kinase signalling cascades. Nature 2001; 410:37-40.
49. Cohen P. The search for physiological substrates of MAP and SAP kinases in mammalian cells. Trends in Cell Biology 1997; 7:353-61.
50. Lee YH, Giraud J, Davis RJ et al. c-Jun N-terminal kinase (JNK) mediates feedback inhibition of the insulin signaling cascade. J Biol Chem 2003; 278(5):2896-902.
51. Hirosumi J, Tuncman G, Chang L et al. A central role for JNK in obesity and insulin resistance. Nature 2002; 420(6913):333-36.
52. Pearson G, Robinson F, Beers-Gibson T et al. Mitogen-activated protein (MAP) kinase pathways: Regulation and physiological functions. Endocrin Rev 2001; 22:153-83.
53. Parekh DB, Ziegler W, Parker PJ. Multiple pathways control PKC phosphorylation. EMBO J 2000; 19(4):496-503.
54. Messina JL, Weinstock RS. Evidence for diverse roles of PKC in the inhibition of gene expression by insulin. Endocrinology 1994; 135(6):2327-34.
55. Toker A, Meyer M, Reddy KK et al. Activation of PKC family members by the novel polyphosphoinositides PI-3,4-P2 and PI-3,4,5-P3. J Biol Chem 1994; 269(51):32358-67.
56. Bandyopadhay G, Standaert ML, Zhao L et al. Activation of Protein Kinase C (α, β and ζ) by Insulin in 3T3/L1 cells. J Biol Chem 1997; 272(4):2551-58.
57. Le Good JA, Ziegler WH, Parekh DB et al. PKC isotypes controlled by PI3-kinase through the protein kinase PDK1. Science 1998; 281:2042-45.
58. Parekh D, Ziegler W, Yonezawa K et al. Mammalian TOR controls one of two kinase pathways acting upon nPKCd and nPKCe. J Biol Chem 1999; 274:34758-64.
59. Krieger-Brauer HI, Kather H. Human fat cells possess a plasma membrane-bound H2O2-generating system that is activated by insulin via a mechanism bypassing the receptor kinase. J Clin Invest 1992; 89(3):1006-13.
60. Prasad RK, Ismail-Beigi F. Mechanism of stimulation of glucose transport by H2O2: Role of PLC. Arch Biochem Biophys 1999; 362(1):113-22.
61. Mahadev K, Wu X, Zilbering A et al. Hydrogen peroxide generated during cellular insulin stimulation is integral to activation of the distal insulin signaling cascade in 3T3-L1 adipocytes. J Biol Chem 2001; 276(52):48662-69.

62. Mahadev K, Zilbering A, Zhu L et al. Insulin-stimulated hydrogen peroxide reversibly inhibits protein-tyrosine phosphatase 1b in vivo and enhances the early insulin action cascade. J Biol Chem 2001; 276(24):21938-42.
63. Heffetz D, Rutter WJ, Zick Y. The insulinomimetic agents H2O2 and vanadate stimulate tyrosine phosphorylation of potential target proteins for the insulin receptor kinase in intact cells. Biochem J 1992; 288(Pt 2):631-35.
64. Sutherland C, Tebbey PW, Granner DK. Oxidative and chemical stress mimic insulin by selectively inhibiting the expression of phosphoenolpyruvate carboxykinase in hepatoma cells. Diabetes 1997; 46(1):17-22.
65. Morel Y, Barouki R. Repression of gene expression by oxidative stress. Biochem J 1999; 342:481-96.
66. Elchebly M, Payette P, Michaliszyn E et al. Increased insulin sensitivity and obesity resistance in mice lacking the protein tyrosine phosphatase-1B gene. Science 1999; 283(5407):1423-25.
67. Czech MP. Lipid Rafts and insulin action. Nature 2000; 407:147-48.
68. Karin M. Signal transduction from the cell surface to the nucleus through the phosphory-lation of transcription factors. Curr Opin Cell Biol 1994; 6:415-24.
69. Angel P, Imagawa M, Chiu R et al. Phorbol ester-inducible genes contain a common cis element recognised by a TPA-modulated trans-acting factor. Cell 1987; 49:729-39.
70. Mohn KL, Laz TM, Melby AE et al. Immediate-early gene expression differs between regenerating liver, insulin-stimulated H-35 cells, and mitogen-stimulated Balb/c 3T3 cells. J Biol Chem 1990; 265:21914-21.
71. Chen R, Sarnecki C, Blenis J. Nuclear localisation and regulation of erk- and rsk-encoded protein kinases. Mol Cell Biol 1992; 12(3):915-27.
72. Deng T, Karin M. c-Fos transcriptional activity stimulated by H-Ras-activated protein kinase distinct from JNK and ERK. Nature 1994; 371(6493):171-75.
73. Morton S, Davis RJ, McLaren A et al. A reinvestigation of the multisite phosphorylation of the transcription factor c-Jun. EMBO J 2003; 22(15):3876-86.
74. Denton RM, Tavare JM. Does mitogen-activated-protein kinase have a role in insulin action? The cases for and against. Eur J Biochem 1995; 227(3):597-611.
75. Treisman R. Regulation of transcription by MAP kinase cascades. Curr Opin in Cell Biology 1996; 8:205-15.
76. Griffiths MR, Black EJ, Culbert AA et al. Insulin-stimulated expression of c-fos, fra1 and c-jun accompanies the activation of the activator protein-1 (AP-1) transcriptional complex. Biochem J 1998; 335(1):19-26.
77. Hurd TW, Culbert AA, Webster KJ et al. Dual role for mitogen-activated protein kinase (Erk) in insulin-dependent regulation of Fra-1 (fos-related antigen-1) transcription and phosphorylation. Biochem J 2002; 368(2):573-80.
78. Lin K, Dorman JB, Rodan A et al. daf-16: An HNF3/forkhead family member that can function to double the life-span of C. elegans. Science 1997; 278:1319-22.
79. Ogg S, Paradis S, Gottlieb S et al. The forkhead transcription factor DAF-16 transduces insulin-like metabolic and longevity signals in C.elegans. Nature 1997; 389:994-99.
80. Biggs 3rd WH, Meisenhelder J, Hunter T et al. Protein kinase B/Akt-mediated phosphorylation promotes nuclear exclusion of the winged helix transcription factor FKHR1. Proc Natl Acad Sci USA 1999; 96(13):7421-6.
81. Brunet A, Bonni A, Zigmond MJ et al. Akt promotes cell survival by phosphorylating and inhibiting a forkhead transcription factor. Cell 1999; 96:857-68.
82. Kops GJPL, de Ruiter ND, De Vries-Smits AMM et al. Direct control of the forkhead transcription factor AFX by protein kinase B. Nature 1999; 398:630-34.
83. Rena G, Guo S, Cichy SC et al. Phosphorylation of the transcription factor forkhead family member FKHR by PKB. J Biol Chem 1999; 274(24):17179-83.
84. Tang ED, Nunez G, Barr FG et al. Negative regulation of the forkhead transcription factor FKHR by Akt. J Biol Chem 1999; 274(24):16741-46.
85. Rena G, Woods YL, Prescott AR et al. Two novel phosphorylation sites on FKHR that are critical for its nuclear exclusion. EMBO J 2002; 21(9):2263-71.
86. Guo S, Rena G, Cichy S et al. Phosphorylation of Ser 256 by PKB disrupts transactivation by FKHR and mediates effects of insulin on IGFBP-1 promoter activity through a conserved insulin response sequence. J Biol Chem 1999; 274(24):17184-92.
87. Nakae J, Barr V, Accili D. Differential regulation of gene expression by insulin and IGF-1 receptors correlates with phosphorylation of a single amino acid residue in the forkhead transcription factor FKHR. EMBO J 2000; 19(5):989-96.
88. Hall RK, Yamasaki T, Kucera T et al. Regulation of phosphoenolpyruvate carboxykinase and insulin-like growth factor-binding protein-1 gene expression by insulin. The Role Of Winged Helix/Forkhead Proteins. J Biol Chem 2000; 275:30169-75.

89. Kops GJPL, Burgering BMT. Forkhead transcription factors: New insights into protein kinase B (c-akt) signaling. J Mol Med 1999; 77:656-65.
90. Woods YL, Rena G. Effect of multiple phosphorylation events on the transcription factors FKHR, FKHRL1 and AFX. Biochem Soc Trans 2002; 30(4):391-7.
91. Durham SK, Suwanichkul A, Scheimann AO et al. FKHR binds the insulin response element in the insulin-like growth factor binding protein-1 promoter. Endocrinology 1999; 140(7):3140-46.
92. Ayala JE, Streeper RS, Desgrosellier JS et al. Conservation of an insulin response unit between mouse and human glucose-6-phosphatase catalytic subunit gene promoters: Transcription factor FKHR binds the insulin response sequence. Diabetes 1999; 48(9):1885-89.
93. Sudhof TC RD, Brown MS, Goldstein JL. 42 bp element from LDL receptor gene confers end-product repression by sterols when inserted into viral TK promoter. Cell 1987; 48(6):1061-69.
94. Brown MS, Goldstein JL. Sterol regulatory element binding proteins (SREBPs): Controllers of lipid synthesis and cellular uptake. Nutr Rev 1998; 56(2):S1-3.
95. Foufelle F, Ferre P. New perspectives in the regulation of hepatic glycolytic and lipogenic genes by insulin and glucose: A role for the transcription factor sterol regulatory element binding protein-1c. Biochem J 2002; 366(2):377-91.
96. Horton JD, Goldstein JL, Brown MS. SREBPs: Activators of the complete program of cholesterol and fatty acid synthesis in the liver. J Clin Invest 2002; 109(9):1125-31.
97. Hua X, Yokoyama C, Wu J et al. SREBP-2, a second basic-helix-loop-helix-leucine zipper protein that stimulates transcription by binding to a sterol regulatory element. Proc Natl Acad Sci USA 1993; 90:11603-07.
98. Loewen CJ LT. Cholesterol homeostasis: Not until the SCAP lady INSIGs. Current Biology 2002; 12(22):R779-81.
99. Yabe D KR, Liang G, Goldstein JL et al. Liver-specific mRNA for Insig-2 down-regulated by insulin: Implications for fatty acid synthesis. Proc Natl Acad Sci USA 2003; 100(6):155-60.
100. Tontonoz P, Kim JB, Graves RA et al. ADD1: A novel helix-loop-helix transcription factor associated with adipocyte determination and differentiation. Mol Cell Biol 1993; 13:4753-59.
101. Kim JB SG, Halvorsen YD, Shih HM et al. Dual DNA binding specificity of ADD1/SREBP1 controlled by a single amino acid in the basic helix-loop-helix domain. Mol Cell Biol 1995; 15(5):2582-88.
102. Shimomura I, Bashmakov Y, Horton JD. Increased levels of nuclear SREBP-1c associated with fatty livers in two mouse models of diabetes mellitus. J Biol Chem 1999; 274(42):30028-32.
103. Shimano H, Horton JD, Shimomura I et al. Isoform 1c of sterol regulatory element binding protein is less active than isoform 1a in livers of transgenic mice and in cultured cells. J Clin Invest 1997; 99(5):846-54.
104. Horton JD, Bashmakov Y, Shimomura I et al. Regulation of sterol regulatory element binding proteins in livers of fasted and refed mice. Proc Natl Acad Sci USA 1998; 95:5987-92.
105. Liang G, Yang J, Horton JD et al. Diminished hepatic response to fasting/refeeding and liver X receptor agonists in mice with selective deficiency of sterol regulatory element-binding protein-1c. J Biol Chem 2002; 277(11):9520-28.
106. Matsumoto M, Ogawa W, Teshigawara K et al. Role of the insulin receptor substrate 1 and phosphatidylinositol 3-kinase signaling pathway in insulin-induced expression of sterol regulatory element binding protein 1c and glucokinase genes in rat hepatocytes. Diabetes 2002; 51(6):1672-80.
107. Roth G, Kotzka J, Kremer L et al. MAP kinases Erk1/2 phosphorylate sterol regulatory element-binding protein (SREBP)-1a at serine 117 in vitro. J Biol Chem 2000; 275(43):33302-07.
108. Giandomenico V SM, Gronroos E, Ericsson J. Coactivator-dependent acetylation stabilizes members of the SREBP family of transcription factors. Mol Cell Biol 2003; 23(7):2587-99.
109. Paulauskis JD, Sul HS. Hormonal regulation of mouse fatty acid synthase gene transcription in liver. J Biol Chem 1989; 264(1):574-77.
110. Latasa MJ, Griffin MJ, Moon YS et al. Occupancy and function of the -150 SRE and -65 E-box in nutritional rgulation of the FAS gene in living animals. Mol Cell Biol 2003; 16:5896-907.
111. Chakravarty K, Leahy P, Becard D et al. Sterol regulatory element-binding protein-1c mimics the negative effect of insulin on phosphoenolpyruvate carboxykinase (GTP) gene transcription. J Biol Chem 2001; 276(37):34816-23.
112. Becard D, Hainault I, Azzout-Marniche D et al. Adenovirus-mediated overexpression of sterol regulatory element binding protein-1c mimics insulin effects on hepatic gene expression and glucose homeostasis in diabetic mice. Diabetes 2001; 50(11):2425-30.
113. Suske G. The Sp-family of transcription factors. Gene 1999; 238(2):291-300.
114. Bouwman P, Philipsen S. Regulation of the activity of Sp1-related transcription factors. Mol Cell Endocrinol 2002; 195:27-38.

115. Banfi C, Eriksson P, Giandomenico G et al. Transcriptional regulation of PAI-1 gene by insulin:insights into the signalling pathway. Diabetes 2001; 50(7):1522-30.
116. Murao K, Wada Y, Nakamura T et al. Effects of glucose and insulin on rat apolipoprotein A-I gene expression. J Biol Chem 1998; 273(30):18959-65.
117. Samson SL, Wong NC. Role of Sp1 in insulin regulation of gene expression. J Mol Endocrinol 2002; 29(3):265-79.
118. Fry CJ, Farnham PJ. Context-dependent transcriptional regulation. J Biol Chem 1999; 274(42):29583-86.
119. Alexander M, Curtis G, Avruch J et al. Insulin regulation of protein biosynthesis in differentiated 3T3 adipocytes. J Biol Chem 1985; 260:11978-82.
120. Alexander MC, Lomanto M, Nasrin N et al. Insulin stimulates glyceraldehyde-3-phosphate dehydrogenase gene expression through cis-acting DNA sequences. Proc Natl Acad Sci USA 1988; 85:5092-95.
121. Nasrin N, Ercolani L, Denaro M et al. An insulin response element in the GAPDH gene binds a nuclear protein induced by insulin in cultured cells and by nutritional manipulations in vivo. Proc Natl Acad Sci 1990; 87:5273-77.
122. Zheng XL, Matsubara S, Diao C et al. Epidermal growth factor induction of apolipoprotein A-I is mediated by the Ras-MAP kinase cascade and Sp1. J Biol Chem 2001; 276(17):13822-29.
123. Zheng XL, Matsubara S, Diao C et al. Activation of apolipoprotein AI gene expression by protein kinase A and kinase C through transcription factor, Sp1. J Biol Chem 2000; 275(41):31747-54.
124. Pan X, Solomon S, Borromeo DM et al. Insulin deprivation leads to deficiency of Sp1 transcription factor in H-411E hepatoma cells and in streptozotocin-induced diabetic ketoacidosis in the rat. Endocrinology 2001; 142(4):1635-42.
125. Streeper RS, Chapman SC, Ayala JE et al. A phorbol ester-insensitive AP-1 motif mediates the stimulatory effect of insulin on rat malic enzyme gene transcription. Mol Endocrinol 1998; 12(11):1778-91.
126. Barroso I, Santisteban P. Insulin-induced early growth response gene (Egr-1) mediates a short term repression of rat malic enzyme gene transcription. J Biol Chem 1999; 274(25):17997-8004.
127. Messina JL. Insulin's regulation of c-fos gene transcription in hepatoma cells. J Biol Chem 1990; 265:11700-05.
128. Burgering BM, Medema RH, Maassen JA et al. Insulin stimulation of gene expression mediated by p21ras activation. EMBO J 1991; 10:1103-09.
129. Medema RH, Wubbolts R, Bos JL. Two dominant inhibitory mutants of p21ras interfere with insulin-induced gene expression. Mol Cell Biol 1991; 11(12):5963-67.
130. Yamauchi K, Holt K, Pessin J. PI3-kinase functions upstream of ras and raf in mediating insulin stimulation of c-fos transcription. J Biol Chem 1993; 268(20):14597-600.
131. Stumpo DJ, Stewart TN, Gilman MZ et al. Identification of c-fos sequences involved in induction by insulin and phorbol esters. J Biol Chem 1988; 263(4):1611-14.
132. Treisman R. The Serum Response Element. TIBS 1992; 17:423-26.
133. Malik RK, Roe MW, Blackshear PJ. Epidermal growth factor and other mitogens induce binding of a protein complex to the c-fos serum response element in human astrocytoma and other cells. J Biol Chem 1991; 266:8576-82.
134. Thompson MJ, Roe MW, Malik RK et al. Insulin and other growth factors induce binding of the ternary complex and a novel protein complex to the c-fos serum response element. J Biol Chem 1994; 269:21127-35.
135. Gille H, Sharrocks AD, Shaw PE. Phosphorylation of transcription factor p62TCF by MAP kinase stimulates ternary complex formation at c-fos promoter. Nature 1992; 358:414-17.
136. Marais R, Wynne J, Treisman R. The SRF accessory protein Elk-1 contains a growth factor-regulated transcriptional activation domain. Cell 1993; 73:381-87.
137. Santisteban P, Kohn LD, Di Lauro R. Thyroglobulin gene expression is regulated by insulin and IGF-1, as well as thyrotropin, in FRTL-5 thyroid cells. J Biol Chem 1987; 262(9):4048-52.
138. Santisteban P, Acebron A, Polycarpou-Schwarz M et al. Insulin and IGFBP-1 regulate a thyroid-specific nuclear protein that binds to the thyroglobulin promoter. Mol Endocrinology 1992; 6:1310-17.
139. Ortiz L, Zannini M, Di Lauro R et al. Transcriptional control of the forkhead thyroid transcritpion factor TTF-2 by thyrotropin, insulin, and IGF-1. J Biol Chem 1997; 272(37):23334-39.
140. Aza-Blanc P, Di Lauro R, Santisteban P. Identification of a cis-regulatory element and a thyroid-specific nuclear factor mediating the hormonal regulation of rat thyroid peroxidase promoter activity. Mol Endocrinol 1993; 7(10):1297-306.

141. Ortiz L, Aza-Blanc P, Zannini M et al. The interaction between the forkhead thyroid transcription factor TTF-2 and the constitutive factor CTF/NF-1 is required for efficient hormonal regulation of the thyroperoxidase gene transcription. J Biol Chem 1999; 274(21):15213-21.
142. Francis-Lang H, Price M, Polycarpou-Schwarz M et al. Cell-type-specific expression of the rat thyroperoxidase promoter indicates common mechanisms for thyroid-specific gene expression. Mol Cell Biol 1992; 12(2):576-88.
143. Zannini M, Avantaggiato V, Biffali E et al. TTF-2, a new forkhead protein, shows a temporal expression in the developing thyroid which is consistent with a role in controlling the onset of differentiation. EMBO J 1997; 16(11):3185-97.
144. Hanson RW, Reshef L. Regulation of PEPCK gene expression. Annu Rev Biochem 1997; 66:581-611.
145. Sasaki K, Granner DK. Regulation of PEPCK gene transcription by insulin and cAMP: Reciprocal actions on initiation and elongation. Proc Natl Acad Sci 1988; 85:2954-58.
146. Imai E, Miner JN, Mitchell JA et al. Glucocorticoid receptor-cAMP response element-binding protein interaction and the response of the PEPCK gene to glucocorticoids. J Biol Chem 1993; 268(8):5353-56.
147. O'Brien RM, Printz RL, Halmi N et al. Structural and functional analysis of the human phosphoenolpyruvate carboxykinase gene promoter. Biochem Biophys Acta 1995; 1264:284-88.
148. Wang J-C, Stromstedt P-E, O'Brien RM et al. Hepatic Nuclear Factor 3 is an accessory factor required for the stimulation of PEPCK gene transcription by glucocorticoids. Mol Endo 1996; 10(7):794-800.
149. Scott DK, Mitchell JA, Granner DK. Identification and characterization of a second retinoic acid response element in the PEPCK gene promoter. J Biol Chem 1996; 271(11):6260-64.
150. Wang J-C, Stafford JM, Scott DK et al. The molecular physiology of HNF3 in the regulation of gluconeogenesis. J Biol Chem 2000; 275(19):14717-21.
151. Waltner-Law M, Duong DT, Daniels MC et al. Elements of the glucocorticoid and retinoic acid response units are involved in cAMP-mediated expression of the PEPCK gene. J Biol Chem 2003; 278(12):10427-35.
152. Lucas PC, Granner DK. Hormone response domains in gene transcription. Ann Rev Biochem 1992; 61:1131-73.
153. O'Brien RM, Lucas PC, Forest CD et al. Identification of a sequence in the PEPCK gene that mediates a negative effect of insulin on transcription. Science 1990; 249:533-37.
154. O'Brien RM, Noisin EL, Suwanichkul A et al. HNF3 and hormone-regulated expression of the PEPCK and insulin-like growth factor-binding protein 1 genes. Mol Cell Biol 1995; 15(3):1747-58.
155. Nitsch D, Boshart M, Schutz G. Activation of the tyrosine aminotransferase gene is dependent on synergy between liver-specific and hormone-responsive elements. Proc Natl Acad Sci 1993; 90:5479-83.
156. Unterman TG, Fareeduddin A, Harris MA et al. HNF-3 binds to the insulin response sequence in the IGFBP-1 promoter and enhances promoter function. Biochem Biophys Res Comm 1994; 203(3):1835-41.
157. Faber S, O'Brien RM, Imai E et al Dynamic aspects of DNA/protein interactions in the transcriptional initiation complex and the hormone-responsive domains of the PEPCK promoter in vivo. J Biol Chem 1993; 268(33):24976-85.
158. O'Brien RM, Noisin EL, Granner DK. Comparison of the effects of insulin and okadaic acid on PEPCK gene expression. Biochem J 1994; 303:737-42.
159. Liao J, Barthel A, Nakatani K et al. Activation of PKB/Akt is sufficient to repress the glucocorticoid and cAMP induction of PEPCK gene. J Biol Chem 1998; 273(42):27320-24.
160. Lochhead PA, Coghlan MP, Rice SQJ et al. Inhibition of GSK3 selectively reduces G6Pase and PEPCK gene expression. Diabetes 2001; 50:937-47.
161. Duong DT, Waltner-Law ME, Sears R et al. Insulin inhibits hepatocellular glucose production by utilizing LIP to disrupt the association of CBP and RNA polymerase II with the PEPCK gene promoter. J Biol Chem 2002; 277(35):32234-42.
162. Lochhead PA, Salt IP, Walker KS et al. AICAR mimics the effects of insulin on the expression of 2 key gluconeogenic genes PEPCK and G6Pase. Diabetes 2000; 49(6):896-903.
163. Chu DTW, Granner DK. The effect of phorbol esters and diacylglycerols on expression of the PEPCK (GTP) gene in rat hepatoma H4IIE cells. J Biol Chem 1986; 261(36):16848-53.
164. Waltner-Law ME, Wang XL, Law BK et al. Epigallocatechin gallate, a constituent of green tea, represses hepatic glucose production. J Biol Chem 2002; 277(38):34933-40.
165. Hwa V, Oh Y, Rosenfeld RG. The insulin-like growth factor-binding protein (IGFBP) superfamily. Endocr Rev 1999; 20(6):761-87.

166. Lee PDK, Giudice LC, Conover CA et al. IGFBP-1: Recent findings and new directions. Proc Soc Exp Biol Med 1997; 216:319-57.
167. Suwanichkul A, Morris SL, Powell DR. Identification of an IRE in the promoter of the human gene for IGFBP-1. J Biol Chem 1993; 268(23):17063-68.
168. Goswami R, Lacson R, Yang E et al. Functional analysis of glucocorticoid and insulin response sequences in the rat IGFBP-1 promoter. Endocrinology 1994; 134(2):736-43.
169. Schweizer-Groyer G, Jibard N, Neau E et al. The glucocorticoid response element II is functionally homologous in rat and human insulin-like growth factor-binding protein-1 promoters. J Biol Chem 1999; 274(17):11679-86.
170. Suwanichkul A, Allander SV, Morris SL et al. Glucocorticoids and insulin regulate expression of the human gene for IGFBP-1 through proximal promoter elements. J Biol Chem 1994; 269(49):30835-41.
171. Mithieux G. New knowledge regarding glucose-6-phosphatase gene and protein and their roles in the regulation of glucose metabolism. Eur J Endocrin 1997; 136:137-45.
172. Foster JD, Pederson BA, Nordlie RC. Glucose-6-phosphatase structure, regulation, and function: An update. Proc Soc Exp Biol Med 1997; 215(4):314-32.
173. van de Werve G, Lange A, Newgard C et al. New lessons in the regulation of glucose metabolism taught by the glucose 6-phosphatase system. Eur J Biochem 2000; 267(6):1533-49.
174. Van Schaftingen E, Gerin I. The Glucose-6-Phosphatase System. Biochem J 2002; 362(3):513-32.
175. Chou JY, Matern D, Mansfield BC et al. Type I glycogen storage diseases: Disorders of the glucose-6-phosphatase complex. Curr Mol Med 2002; 2(2):121-43.
176. Efendic S, Karlander S, Vranic M. Mild type 2 diabetes markedly increases glucose cycling in the postabsorptive state during glucose infusion irrespective of obesity. J Clin Invest 1988; 81(6):1953-61.
177. Barzilai N, Rossetti L. Role of glucokinase and glucose-6-phosphatase in the acute and chronic regulation of hepatic glucose fluxes by insulin. J Biol Chem 1993; 268(33):25019-23.
178. Seoane J, Trinh K, O'Doherty RM et al. Metabolic impact of adenovirus-mediated overexpression of the G-6-Pase catalytic subunit in hepatocytes. J Biol Chem 1997; 272(43):26972-77.
179. Daniele N, Rajas F, Payrastre B et al. Phosphatidylinositol 3-kinase translocates onto liver endoplasmic reticulum and may account for the inhibition of glucose-6-phosphatase during refeeding. J Biol Chem 1999; 274(6):3597-601.
180. Streeper RS, Svitek CA, Chapman S et al. A multicomponent IRS mediates a strong repression of mouse G-6-Pase gene transcription by insulin. J Biol Chem 1997; 272:11698-701.
181. Dickens M, Svitek CA, Culbert AA et al. Central Role for PI3-kinase in the repression of glucose-6-phosphatase gene transcription by insulin. J Biol Chem 1998; 273:20144-49.
182. Streeper RS, Eaton EM, Ebert DH et al. HNF-1 acts as an accessory factor to enhance the inhibitory action of insulin on mouse G-6-Pase gene transcription. Proc Natl Acad Sci 1998; 95:9208-13.
183. Massillon D, Barzilai N, Chen W et al. Glucose regulates in vivo glucose-6-phosphatase gene expression in the liver of diabetic rats. J Biol Chem 1996; 271(17):9871-74.
184. Vander Kooi BT, Streeper RS, Svitek CA et al. The Three insulin response sequences in the glucose-6-phosphatase catalytic subunit gene promoter are functionally distinct. J Biol Chem 2003,(published on-line as M212570200).
185. Ganss R, Weih F, Schutz G. The cAMP and the glucocorticoid dependent enhancers are targets for insulin repression of tyrosine aminotransferase gene transcription. Molec Endocrinology 1994; 8:895-903.
186. Nitsch D, Schutz G. The distal enhancer implicated in the developmental regulation of the tyrosine aminotransferase gene is bound by liver-specific and ubiquitous factors. Mol Cell Biol 1993; 13(8):4494-504.
187. Grange T, Cappabianca L, Flavin M et al. In vivo analysis of the model tyrosine aminotransferase gene reveals multiple sequential steps in glucocorticoid receptor action. Oncogene 2003; 20(24):3028-38.
188. Carmichael D, Koontz J. Insulin-resonsive TAT transcription requires multiple promoter regions. Biochem Biophys Res Commun 1992; 187(2):778-82.
189. Rondinone CM, Carvalho E, Rahn T et al. Phosphorylation of PDE3B by phosphatidylinositol 3-kinase associated with the insulin receptor. J Biol Chem 2000; 275(14):10093-98.
190. Patel S, Lipina C, Sutherland C. Different mechanisms are used by insulin to repress three genes that contain a homologous thymine-rich insulin response element. FEBS Lett 2003; 549:72-76.
191. Nakae J, Biggs WH, Kitamura T et al. Regulation of insulin action and pancreatic beta-cell function by mutated alleles of the gene encoding forkhead transcription factor Foxo1. Nat Genet 2002; 32(2):245-53.
192. Band CJ, Posner BI. PI3-kinase and p70S6 kinase are required for insulin but not bisperoxovanadium 1,10-phenanthroline inhibition of IGFBP gene expression. J Biol Chem 1997; 272(1):138-45.

193. Patel S, Lochhead PA, Rena G et al. Insulin regulation of IGF-binding protein-1 gene expression is dependent on mammalian target of rapamycin (mTOR), but independent of S6K activity. J Biol Chem 2002; 277(12):9889-95.
194. Patel S, van der Kaay J, Sutherland C. Insulin regulation of IGFBP-1 gene expression is impaired by the presence of hydrogen peroxide due to a reduction in mTOR signalling. Biochem J 2002; 265(2):537-45.
195. Barthel A, Schmoll D, Kruger KD et al. Differential regulation of endogenous glucose-6-phosphatase and phosphoenolpyruvate carboxykinase gene expression by the forkhead transcription factor FKHR in H4IIE-hepatoma cells. Biochem Biophys Res Commun 2001; 285(4):897-902.
196. Velho G, Froguel P. Maturity-onset diabetes of the young (MODY), MODY genes and NIDDM. Diabetes Metab 1997; 23(Suppl 2):34-37.
197. Yki-Jarvinen H. MODY genes and mutations in hepatocyte nuclear factors. Lancet 1997; 349:516-17.
198. Woodcroft KJ, Hafner MS, Novak RF. Insulin signalling in the transcriptional and posttranscriptional regulation of CYP2E1 expression. Hepatology 2002; 35(2):263-73.
199. Au WS, Kung HF, Lin MC. Regulation of MTP gene by insulin in HepG2 cells: Roles of MAPK and p38MAPK. Diabetes 2003; 52(5):1073-80.
200. Miakotina OL, Goss KL, Snyder JM. Insulin utilises the PI s-kinase pathway to inhibit SP-A gene expression in lung epithelial cells. Respir Res 2002; 3(1):27.
201. Wang D, Sul H-S. Insulin stimulation of the fatty acid synthase promoter is mediated by the PI3-kinase pathway. J Biol Chem 1998; 273(39):25420-26.
202. Keeton AB, Amsler MO, Venable DY et al. Insulin signal transduction pathways and insulin-induced gene expression. J Biol Chem 2002; 277(50):48565-73.
203. Yamada K, Kawata H, Shou Z et al. Insulin induces the expression of the SHARP-2/Stra13/DEC1 gene via a PI3-kinase pathway. J Biol Chem 2003; Epub.
204. Osawa H, Sutherland C, Robey RB et al. Analysis of the signaling pathway involved in the regulation of hexokinase II gene transcription by insulin. J Biol Chem 1996; 271(28):16690-94.
205. Roques M, Vidal H. A phosphatidylinositol 3-kinase/p70 ribosomal S6 protein kinase pathway is required for the regulation by insulin of the p85 alpha regulatory subunit of phosphatidylinositol 3-kinase gene expression in human muscle cells. J Biol Chem 1999; 274(48):34005-10.
206. Buyse M, Veingchareun S, Bado A et al. Insulin and glucocorticoids differentially regulate leptin transcription and secretion in brown adipocytes. FASEB J 2001; 15(8):1357-66.
207. Hayakawa J, Ohmichi M, Tasaka K et al. Regulation of the PRL promoter by Akt through cAMP response element binding protein. Endocrinology 2002; 143(1):13-22.
208. Teruel T, Valverde AM, Navarro P et al. Inhibition of PI3-kinase and Ras blocks IGF-1 and insulin-induced UCP1 gene expression in brown adipocytes. J Cell Physiol 1998; 176(1):99-109.
209. Leibiger B, Leibiger IB, Moede T et al. Selective insulin signalling through A and B insulin receptors regulates transcription of insulin and glucokinase genes in pancreatic beta cells. Mol Cell 2001; 7(3):559-70.
210. Kremerskothen J, Wendholt D, Teber I et al. Insulin-induced expression of the activity-regulated cytoskeleton-associated gene (ARC) in human neuroblastoma cells requires p21(ras), MAP kinase and src tyrosine kinases but is PKC-independent. Neurosci Lett 2002; 321(3):153-56.
211. Iida KT, Shimano H, Kawakami Y et al. Insulin up-regulates TNF-alpha production in macrophages through an erk-dependent pathway. J Biol Chem 2001; 276(35):32531-37.
212. Meile C, Rochford JJ, Filippa N et al. Insulin and IGF-1 induce vEGF mRNA expression via different signalling pathways. J Biol Chem 2000; 275(28):21695-702.
213. Sekar N, Veldhuis JD. Concerted transcriptional activation of the ldl receptor gene by insulin and LH in cultured porcine granulosa-luteal cells. Endocrinology 2001; 142(7):2921-28.

CHAPTER 7

# Insulin Action in the Islet β-Cell

Rohit N. Kulkarni*

## Summary

The techniques that allow spatio-temporal control of gene deletion or gene expression in transgenic and knockout animals have been useful to directly evaluate the roles of the insulin and IGF-1 receptors and proteins in their signaling pathway in islet cells. While a functional role for insulin signaling in islet β-cells is now well established, the precise pathways and proteins that mediate β-cell growth and function during development and adulthood are continuing to be unraveled. This chapter will focus on recent studies derived from using transgenic, knockout and siRNA techniques to *directly* examine the role of insulin and IGF-I in the regulation of development, growth and function of islet β-cells.

## Introduction

Insulin and insulin-like-growth-factor-I (IGF-I) constitute two members of the growth factor family, which play important roles in the regulation of metabolism and growth of virtually all tissues in mammals. The receptors for insulin and IGF-I are expressed ubiquitously and mediate the growth and metabolic effects of the hormones.[1-3] Most of the information we currently know regarding insulin/IGF-I signaling pathways is derived from studies underlying the defects in insulin and IGF-I action in type 2 diabetes.[2,4,5] Insulin and IGF-1 bind to distinct receptors that in turn transmit signals by phosphorylating insulin receptor substrates (IRS) including the four IRS proteins, Shc, Gab-1, FAK, Cbl, and potentially other substrates.[2,5-10] These insulin receptor substrates play different but crucial roles in cellular processes that are important for the metabolism and growth of tissues including glucose transport and utilization, protein synthesis, cell growth, proliferation and anti-apoptosis. Several reviews provide an update on these signaling networks.[2,5,7,8,10] Over the last decade, several laboratories have created global and tissue-specific knockouts of genes that code for protein(s), which are considered potentially important in regulating the effects of insulin and/or IGF-I. The pleiotropic signaling effects of insulin and IGF-I family of growth factors have been studied in great detail in classic insulin sensitive tissues including skeletal muscle, liver and adipose.[6,10,11] While IGFs have been studied for their contributions to islet development, a role for insulin during growth of the endocrine cells in the embryonic and post-natal periods is not fully understood. Figure 1 shows a schematic of the pathways and proteins implicated in insulin/IGF-I signaling in islet β-cells.

Although the presence of functional insulin receptors in β-cells is now indisputable (reviewed in ref. 12), it has been a challenge to study the signaling pathways activated by insulin in β-cells for several reasons. First, the precise localization of receptors on apical and/or basolateral

---

*Rohit N. Kulkarni—Division of Cell and Molecular Physiology, Joslin Diabetes Center and Department of Medicine, Harvard Medical School, Boston Massachusetts 02215, U.S.A. Email: Rohit.Kulkarni@joslin.harvard.edu

*Mechanisms of Insulin Action*, edited by Alan Saltiel and Jeffery Pessin.
©2007 Landes Bioscience and Springer Science+Business Media.

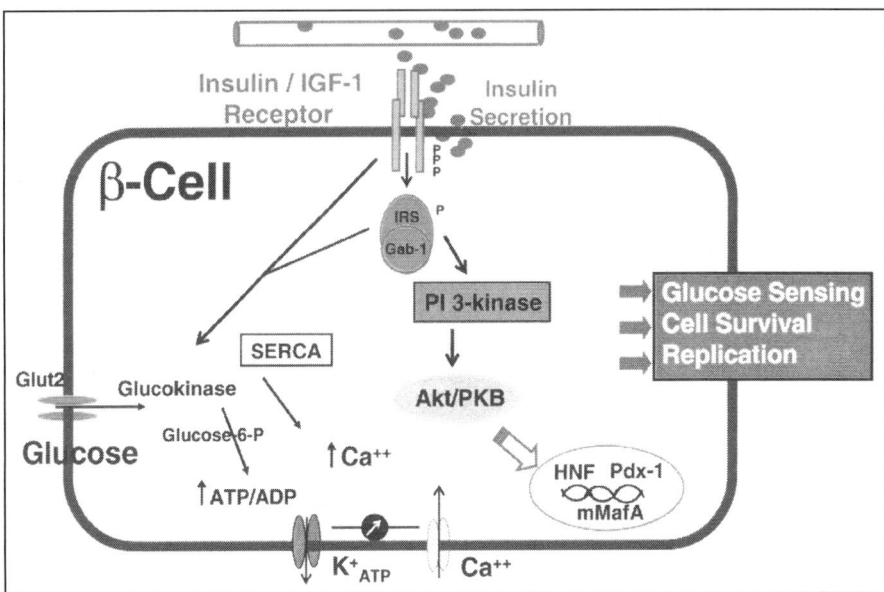

Figure 1. Schematic of insulin/IGF-I signaling pathways in the islet β-cell. Figure adapted with permission form Kulkarni RN. IJBCB 2004; 36:365-371.[174]

surfaces of different islet cells using immunohistochemistry, has been limited due to lack of a robust anti-insulin receptor antibody. Second, the continuous secretion of insulin by β-cells via the regulated and constitutive pathways allows potential internalization and downregulation of insulin receptors and confounds the effects of added ligand. Thus, several studies have used experimental protocols wherein islets/beta cells are either treated with exogenous insulin.[13,14] or with inhibitors of regulated insulin secretion such as somatostatin or diazoxide followed by examining the consequences on insulin secretion or synthesis due to direct or indirect effects. A confounding factor in the latter approach is the inability to completely inhibit insulin secretion since most secretory cells possess regulated and constitutive secretion pathways.[15,16] Thus, very small amounts of insulin secreted by the constitutive pathway likely maintain downregulation of insulin receptors in the presence of inhibitors of regulated secretion such as somatostatin and diazoxide, which in turn, can lead to erroneous interpretation of data. Finally, the high degree of homology between insulin and IGF-1 receptors[17] and the activation of common signaling proteins precludes accurate assessment of functional end points in response to activation of individual receptors. Thus, insulin, depending on the concentrations used, can also activate the IGF-1 receptors leading to inhibitory effects,[18] in contrast to reports of the effects of insulin in single murine[19] or human β-cells[20,21] demonstrating an increase in intracellular $Ca^{++}$ flux or alterations in gene expression.[22] Recent studies in humans provide further evidence for a role for insulin action in the β-cell in vivo.[23,24,24a,24b] The development of powerful genetic engineering techniques has circumvented several disadvantages discussed above and allow for disruption of the gene(s) coding for a given protein and enable direct evaluation of the targeted protein(s) in mouse models. This chapter will focus mostly on *direct* evidence provided by these techniques for a role for insulin and IGF-I during early growth and development of islets and in the maintenance of adult β-cell mass.

## Embryonic and Early Post-Natal Development of the Endocrine Pancreas

The pancreas develops from the fusion of two diverticula of primordial gut tissue to form the distinct endocrine and exocrine components observed in adulthood.[25] Recent reviews provide insight into the development of the endocrine cells[26,27] and the role of numerous transcription factors that are considered essential for the development of the different endocrine cell types.[28,29] Several elegant studies provide compelling evidence to disprove previously held dogmas. For example, based on irreversible tagging of progeny through the activity of *Cre* recombinase it is now accepted that α- and β-cell lineages raise independently from a common precursor expressing the pancreatic homeodomain protein, PDX-1, and not from glucagon-expressing progenitors as was originally suggested.[26] Furthermore, direct lineage tracing studies indicate that NGN3+ cells are islet precursors and are distinct from ductal precursors.[30] The role(s) of insulin and IGF-I signaling during embryonic development of islet cells has not yet been explored using similar approaches.

## Global and Conditional Knockouts of Insulin, IGF-I, IGF-II and Proteins in Their Signaling Pathways

The significance of IGF-I and insulin during early development and growth of islet cells has been a major focus of study for several decades and several important insights have emerged from these experiments.[31-35] The fetal pancreas expresses IGF-I, IGF-II, and IGF-binding protein 3 during late gestation.[31,36,37] IGF-II levels are higher than IGF-I during fetal development and IGF-II has been localized to islets and duct epithelial cells by immunohistochemistry and in situ hybridization techniques.[31,36,38] Together, these studies "indirectly" implicate a role for IGF-I and IGF-II during the post-natal development of the endocrine pancreas.[39]

The development of genetic engineering techniques over the last decade, to create gain of function or loss of function mutations[40,41] has been used successfully to examine the function of specific proteins in different islet cell. Further, the ability to "turn off" a gene encoding for a particular protein in a time-dependent manner provides a tool to simulate the gradual dysfunction that is usually observed in chronic diseases.[41] While adaptation to the creation of a genetic mutation during embryonic life is a natural consequence of this method the information obtained in studies in multiple biological disciplines has been extremely useful to unravel potentially novel and unexpected functions of proteins. These observations are comparable to those made from humans bearing naturally occurring genetic mutations, who of necessity adapt to the mutation, but nevertheless provide important clues to understand the function of the proteins encoded by the gene(s). A partial list of global and conditional knockouts/transgenics of insulin, IGFs and proteins in their signaling pathways is provided in Tables 1 and 2. Unfortunately, many references could not be cited due to space limitations.

To "directly" evaluate the role of growth factors, several investigators have utilized homologous recombination in mice. Thus, global knockout of genes coding for the insulin gene in mice leads to growth retardation, and death due to diabetes mellitus with ketoacidosis and liver steatosis.[42,43] Interestingly, pancreas examination during the post-natal period revealed large islets and prompted the authors to suggest that insulin is a negative regulator of islet growth.[42,43] However, since IGF-II levels are reported to be elevated during the immediate post-natal period,[39] it is possible that lack of insulin allows for unopposed action of IGF-II at both insulin and/or IGF-1 receptors to promote islet hyperplasia. Alternatively, the enhanced vascularization in the absence of insulin may lead to an increase in local concentrations of morphogens, derived from the circulation and/or endothelial cells, to promote islet cell growth.[44-46] A recent study in which the IGF-I gene was inactivated in islets using the PDX-1 promoter described hyperplastic islets that are resistant to streptozotocin-induced diabetes.[47] The increase in the size of

## Table 1. Global knockouts/transgenics

| Protein | Phenotype | Reference |
|---|---|---|
| **Insulin** | Intrauterine growth retardation, | Duvillie et al, 1997[42] |
| | Neonatal lethality, ketoacidosis, | Duvillie et al, 2002[154] |
| | Liver steatosis | |
| **IGF-1** | Dwarfism, | Liu et al, 1993[155] |
| | Variable survival | Powell-Braxton et al, 1993[52] |
| **IGF-II** (overexpressor) | Islet hyperplasia, organ overgrowth | Petrik et al, 1999[48] |
| **Insulin receptor** | Neonatal lethality, | Accili et al, 1996[156] |
| | Ketoacidosis | Joshi et al, 1996[56] |
| **IGF-1 receptor** | Dwarfism, neonatal lethality | Liu et al, 1993[155] |
| **IRS-1** | Post-natal growth retardation | Araki et al, 1994[67] |
| | Insulin resistance | Tamemoto et al, 1994[68] |
| | Islet hyperplasia | Kulkarni et al, 1999[69] |
| | Insulin secretory defect | Aspinwall et al, 2000[157] |
| | | Kubota et al, 2000[70] |
| | | Kulkarni et al, 2004[72] |
| | | Hennige et al, 2005[73] |
| **IRS-2** | Insulin resistance | Withers et al, 1998[57] |
| | Diabetes | Kubota et al, 2000[70] |
| | Islet hypoplasia | |
| **IRS-3** | Relatively normal | Liu et al, 1999[84] |
| **IRS-4** | Mild glucose intolerance | Fantin et al, 2000[85] |
| **PI 3-kinase isoforms** | | |
| p85α | Increased insulin sensitivity, | Terauchi et al, 1999[87] |
| | Hypoglycemia | Mauvais-Jarvis et al, 2002[88] |
| p85β | Increased insulin sensitivity, | Ueki et al, 2002[86] |
| | Hypoglycemia | |
| p50α/p55α | Increased insulin sensitivity | Chen et al, 2004[173] |
| **Akt1** | Growth retardation, Increased apoptosis, | Chen et al, 2001[158] |
| | Normal glucose tolerance. | |
| | Insulin resistance in liver and muscle, | Cho et al, 2001[159] |
| | Increased islet mass. | |
| **P70S6kinase** | Hypoinsulinemia, Glucose | Pende et al, 2000[97] |
| | intolerance, and reduced beta-cell size. | |
| **Insulin receptor-related receptor** | | |
| | Normal phenotype | Kitamura et al, 2001[160] |

Partial list of phenotypes and references of global knockout/transgenics of insulin and IGF-1 genes and proteins in the insulin receptor/IGF-1 signaling pathway.

islets was disproportionate to the mild hyperglycemia suggesting that IGF-II or insulin acting via insulin and/or IGF-1 receptors enhanced islet growth in the absence of locally produced IGF-I. In this context, it is worth noting that overexpression of IGF-II in β-cells has also been reported to lead to hyperplastic islets[48] and intriguingly the mice develop diabetes.[49] While the islet growth effects induced by IGF-II could be mediated via the insulin receptor, the creation of a model of IGF-II overexpression in a mouse lacking insulin receptors in β-cells will directly address whether the IGF-II/insulin receptor pathway is indeed critical for growth.

Not surprisingly, mice with null mutations of the IGF-I and IGF-II genes show similar but milder defects compared to mice lacking the insulin gene. IGF-I null mice show growth defects similar to IGF-II null mutants[17,43,51] and depending on the genetic background, some of the

### Table 2. Tissue-specific knockouts/transgenics

| Protein/Tissue | Phenotype | References |
|---|---|---|
| **IGF-1** | | |
| Liver | Normal growth and development, Muscle insulin sensitivity. | Yakar et al, 1999[161] Liu et al, 2000[162] Yakar et al, 2001[163] |
| Islets | Islet hyperplasia, resistance to diabetes | Lu et al, 2004[47] |
| **IGF-II** | | |
| β-cell (overexpressor) | Islet hyperplasia, diabetes | Devedjian et al, 2000[164] |
| **Insulin Receptor** | | |
| Muscle (transgenic) | Increased adiposity, Dyslipidemia and Glucose intolerance. | Moller et al, 1996[165] |
| | Normal glucose tolerance, Elevated triglyceride and FFA levels | Bruning et al, 1998[166] |
| β-cell | Glucose intolerance, Loss of acute phase insulin secretion, Reduced β-cell mass. | Kulkarni et al, 1999[59] Mauvais-Jarvis et al.[167] |
| | Overt diabetes | Kulkarni, Kahn, 2001[64] |
| Liver | Severe glucose intolerance, Hepatic dysfunction | Michael et al, 2000[151] |
| Brain | Increased food intake and obesity in female Impaired spermatogenesis and ovarian follicle maturation | Bruning et al, 2000[168] |
| Adipose | Protect against obesity and glucose intolerance Heterogeneity in white adipose cell size | Bluher et al, 2002[169] |
| **Insulin Receptor** | | |
| Muscle + Adipose | Impaired glucose tolerance Insulin resistance | Lauro et al, 1998[170] |
| **IGF-1 receptor** | | |
| β-cell | Impaired glucose tolerance Reduced glucose-stimulated insulin secretion | Kulkarni et al, 2002[60] Xuan et al, 2002[61] |
| **IRS-2** | | |
| β-cell | Mild diabetes, obesity | Kubota et al, 2004[76] Lin et al, 2005[75] Choudhury et al, 2005[171] |
| **Insulin Receptor + IGF-1 Receptor** | | |
| Muscle | Insulin resistance, β-cell dysfunction and diabetes | Fernandez et al, 2001[172] |
| **Akt** | | |
| β-cell (overexpressor) | Islet hyperplasia, hyperinsulinemia | Bernal-Mizrachi et al, 2001[95] Bernal-Mizrachi et al, 2004[96] Tuttle et al, 2001[94] |

Partial list of phenotypes and references of tissue-specific knockout/transgenics of insulin and IGF-1 genes and proteins in the insulin receptor/IGF-1 signaling pathway.

IGF-I knockouts die, while others survive into adulthood.[50,51] Mice lacking the IGF-I gene exhibit postnatal lethality, growth retardation, infertility, and defective development of bone and muscle.[51,52] Similar findings were reported in a human with homozygous partial deletion of the IGF-I gene.[53] Taken together these global knockouts underscore the crucial importance of insulin and IGF-I and their cognate receptors in the overlapping regulatory functions of metabolism and growth in mice and humans.

Insulin and IGF-I mediate their effects via the insulin and IGF-1 receptors respectively. Considering the high degree of homology between the insulin and IGF-1 receptors it is likely that the ligands can also act via their cognate receptors.[17] Thus, one would predict either similar phenotypes when either of the receptor is lacking or alternatively one receptor could compensate for the absence of the other receptor in an effort to maintain normal signaling in target tissues. Mice homozygous for a null mutation of the insulin receptor show normal intrauterine growth but die within 48 to 72 h after birth due to severe hyperglycemia and diabetic ketoacidosis.[54-56] On the other hand, IGF-1 receptor null mutants show severe growth deficiency and die at birth due to respiratory failure[51] and manifest a phenotype similar to the IGF-1 null mutants.[50] Although both mutants die early, these studies clearly indicate the mice are born with β-cells. While β-cells in insulin receptor null mutants show degranulation, which likely occurs due to severe hyperglycemia, mice lacking functional IGF-1 receptors show small[57] or relatively normal islet/β-cell mass.[58] These studies provide evidence that neither receptor is critical for the early development and formation of β-cells. Furthermore, these findings have been confirmed in conditional knockouts of insulin or IGF-1 receptors. Thus, β-cell-specific insulin receptor knockouts (βIRKO)[59] or β-cell specific IGF-1 receptor nulls[60,61] are both born with a normal complement of islets/β-cells. Considering the overlapping signaling pathways shared by insulin and IGF-1, both these knockouts develop secretory defects characterized by blunted glucose-stimulated insulin secretory responses secondary to poor glucose sensing.[22,59-62] Studies using insulin receptor siRNA treatment of β-cells report similar defects in glucose sensing.[63] However, one notable difference between the two mutants is the effect on maintenance of β-cell mass in adults. Thus, follow up studies indicate an increased susceptibility of βIRKO mice to develop age-dependent diabetes consequent to a reduced β-cell mass.[62,64] In contrast, β-cell mass in 12 month-old βIGFRKO mice is relatively normal[60,61] (R.N. Kulkarni unpublished observations). Together, these studies point to similar phenotypes in regard to secretory function of β-cells, when the insulin or IGF-1 receptors are disrupted selectively in β-cells, but suggest a prominent role for insulin signaling in maintenance of adult β-cell mass (see below).

Humans bearing mutations of the insulin receptor (leprechaunism), however, manifest quite a different phenotype, compared to mice lacking insulin receptors, and are characterized by intrauterine growth retardation and only mild hyperglycemia and display large islets.[65,66] In these patients, it is unclear whether the insulin receptor is devoid of all signaling capability or whether the mutated protein continues to transmit some signals that allows selective growth pathways to be active.[66] Thus, it is possible that the islet hyperplasia in humans with leprechaunism is due to selective activation of proteins critical for mediating growth effects in the β-cells. The lack of reports describing mutations in insulin or IGF-1 receptors that are restricted only to β-cells in humans makes it difficult to directly resolve the issue of whether insulin signaling plays a role in modulating β-cell growth and function in vivo. The creation of mouse models bearing mutations in insulin receptors in β-cells, similar to mutations that occur in leprechaunism in humans, perhaps using a knock-in strategy, is one way to gain insight into this question. Furthermore, recent studies in vivo in humans indicate a potential role for insulin in modulating β-cell function.[24a,24b]

Gene deletion of proteins downstream to the insulin and IGF-1 receptor, including the IRS proteins, leads to different phenotypes compared to those observed in the receptor mutants. Thus, IRS-1 knockouts exhibit post-natal growth retardation and hyperinsulinemia but a relatively normal lifespan[67,68] and interestingly, the IRS-1 null mice show hyperplastic but dysfunctional islets.[69-73] While the islet hyperplasia in IRS-1 knockouts may be secondary to upregulation of expression of IRS-2 in β-cells,[73] the altered islet function in IRS-1 global mutants has been linked to reduced expression of sarcoplasmic endoplasmic reticulum calcium ATPase (SERCA) proteins.[72] In contrast to IRS-1 knockouts, IRS-2 null mice show only mild growth retardation, and depending on the genetic background of the founder mice either manifest a mild phenotype[70] or develop β-cell hypoplasia leading to overt diabetes.[57] Transgenic mice overexpressing IRS-2 exhibited protection against development of diabetes and transplantation of islets overexpressing IRS-2 promoted glucose tolerance in the recipient mice better than transplantation of islets from wild type mice.[74] It must be noted, however that β-cell-specific loss of IRS-2 does not prevent development and growth of β-cells during the embryonic and early post-natal periods.[75-77] Notwithstanding the observation that some β-cells "escape" *Cre* recombination, it is intriguing that lack of IRS-2 in a "majority" of β-cells during the early post-natal period, when *Cre* expression is maximal and "escape" is minimal, does not lead to a phenotype as severe as that observed in global mutants despite being created on similar genetic backgrounds. This indicates that signals independent of IRS-2 are likely necessary during early growth and development of β-cells. One important role for IRS-2 in β-cell plasticity may be its anti-apoptotic effects linked to cAMP and mediated by the cAMP response binding protein (CREB).[78] A reduced expression of IRS-2 has been reported to promote increased apoptosis.[57,70,79,80] Several studies to rescue the defects in b-cell growth and function due to loss of IRS-2 have been performed. For example, deficiency of protein tyrosine phosphatase 1b (PTP1B) delayed the onset of diabetes in mice lacking IRS-2[81] and transgenic expression of PDX-1 in IRS-2 null mice restored β-cell mass and promoted glucose tolerance.[82] Recently, IRS-2 null mice crossed with mice haploinsufficient for the 3'-lipid-phosphatase Pten also prevented glucose intolerance and the mice survived without diabetes before succumbing to lymphoproliferative disease.[83] Intriguingly, increasing the expression of IRS-4 in β-cell lines that express reduced IRS-2 prevented apoptotic effects.[79] These reports indicate that several signaling proteins and transcription factors can restore the defects in β-cell growth and function due to loss of IRS-2.

By contrast to the defects in β-cell function and growth due to loss of function of IRS-1 and IRS-2, IRS-3 null mice develop normally and have normal glucose tolerance.[84] IRS-4 deficient mice manifest mild growth defects and glucose intolerance and this is evident only in males because the IRS-4 gene is located on the X chromosome.[85] Knockout of the p85 regulatory subunit of PI 3-kinase in mice leads to increased insulin sensitivity and hypoglycemia.[86-88] Islets of mice deficient in the p85 sub-unit of PI 3-kinase exhibited reduced insulin content and mass of endoplasmic reticulum while manifesting an increase in insulin secretory response.[89] Studies using PI 3-kinase inhibitors report an amplification of glucose-stimulated insulin secretion in islets from lean but not obese mice[89-91] and has been suggested to occur at a level distal to mechanisms that alter cytosolic [Ca$^{++}$].[89] While the effects of PI 3-kinase suppression, due to insulin resistance, in classical insulin sensitive tissues leads to a decrease in insulin sensitivity,[2] the opposite effect in islet β-cells has been suggested to promote increased insulin secretion to maintain glucose homeostasis.[91] These studies, largely based on using PI 3-kinase inhibitors, should be interpreted with caution since wortmannin, at concentrations widely used to inhibit PI 3-kinase, has been reported to exert effects on other proteins in mammalian cells.[92]

Two independent groups have highlighted the role of Akt/PKB in the regulation of islet mass. Birnbaum and colleagues reported that null mutants for Akt-1 (PKB-β) manifest insulin resistance in muscle and liver and an increased islet mass, while mice lacking the Akt2 enzyme develop insulin resistance and diabetes-like syndrome.[93,94] Parallel studies in Permutt's lab reported that constitutive activation of Akt1/PKBalpha increased β-cell size, enhanced islet mass, induced hyperinsulinemia and protected against experimental diabetes.[95] Conversely, mice expressing a kinase-dead mutant of Akt manifest defective insulin secretion and increased susceptibility to development of diabetes.[96] Together, these experiments highlight an important role for Akt/PKB in the mainenanace of islet mass. Mutants for p70S6 kinase show reduced β-cell size, lower insulin secretion and reduced pancreatic insulin content.[97]

The lack of a defect in the early growth and development of islet/β-cells in mice lacking insulin or IGF-1 receptors indicates that other growth factors are likely important for the development of the insulin-secreting cells (Fig. 2). Indeed, several reports indicate a role for placental lactogen and GH during the early development of the pancreas (reviewed in refs.

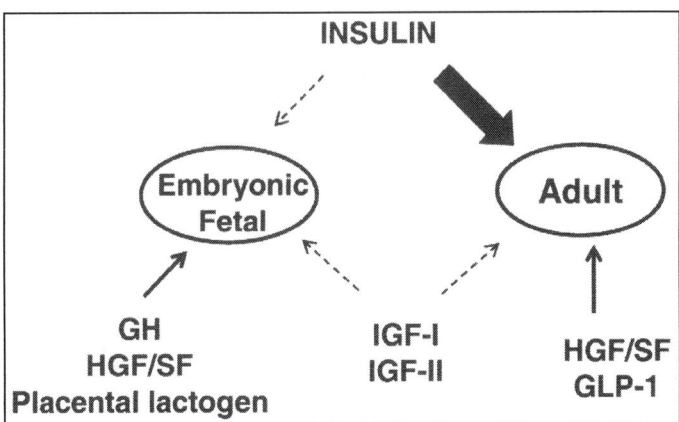

Figure 2. Schematic showing relative significance of growth factors on growth of β-cells during embryonic and adult periods. Data supported by direct evidence is indicated by arrows with solid lines and indirect evidence by dotted lines. Figure reproduced with permission from Kulkarni RN, Rev Endo Metab Disord 2005; 6(3):199-210.[175]

98,99) Overexpression of placental lactogen,[100] PThRP,[101] and HGF[102] individually in β-cells using the rat insulin promoter lead to an increase in β-cell mass and resistance to strepotozotocin-induced β-cell death. Surprisingly, mice with a β-cell-specific knockout of c-met, the receptor mediating the actions of hepatocyte growth factor, developed defects in glucose-stimulated insulin secretion due to reduced glut2 expression but showed little[103] or no effects[104] on β-cell mass. Further, mice doubly transgenic for parathyroid hormone-related protein and placental lactogen showed defects in β-cell function and growth that was similar in severity to single transgenic mice indicating potential saturation of signaling and common downstream partners for the two hormones[105] and also suggest an important role for placental lactogen in β-cell survival. It will be useful to examine potential cross-talk between signaling pathways activated by the receptor tyrosine kinases for c-met, PThRP and placental lactogen and the insulin/IGF-I system to unravel novel interactions underlying the growth and apoptosis of islet cells.

### Maintenance of Adult β-Cell Mass

The mechanisms and signaling pathways that maintain adult β-cell mass are currently intense areas of research in type 1 and type 2 diabetes. Several mechanisms have been proposed to influence adult β-cell mass including neogenesis from ductal cells[106,107] and apoptosis.[108,109] Recent studies using lineage trace analysis provide compelling evidence for β-cell replication as a major pathway for the renewal of adult β-cells in mice.[110] Whether a similar mechanism is operative in humans is not known and impossible to prove by lineage trace analysis. Some evidence for DNA duplication is available from two recent studies suggesting de-differentiation and differentiation of human islet precursor cells.[111,112] Nevertheless, it is conceivable that all three processes are occurring to maintain an appropriate number of β-cells for glucose homeostasis, and identifying the major pathway that contributes to β-cell regeneration will be key to plan strategies to intervene therapeutically. Therefore, until lineage trace analyses or a similar definitive technique can conclusively prove that neogenesis from duct or periductal cells is also a significant source of β-cells in rodents, or until studies in humans can conclusively demonstrate that β-cell regeneration does not involve mitosis, all efforts must be targeted to understand the basic mechanisms underlying β-cell replication. Some immediate questions include – what are the signals that promote β-cell replication, what are the pathways that mediate the mitotic response and how many times can a single β-cell divide over its life span? Answers to these questions, though not trivial, will likely provide crucial information to focus efforts to develop therapeutic targets to enhance β-cell regeneration.

Previous studies have reported a clear role for cyclin dependent kinase 4 (CDK4) as an essential regulator of β-cell cycle control. Rane and coworkers[113] created mice lacking CDK4 and observed the mutants were infertile and developed insulin-deficient diabetes as a consequence of reduced β-cells. Conversely, mice with an activating mutation of CDK4, that inhibited binding to the cell cycle inhibitor P16INK4a, developed pancreatic hyperplasia but did not lead to tumors.[113] Recent studies examining the role of cyclin D1 and D2 support the replication hypothesis.[114,115] Mice lacking cyclin D2 showed a selective decrease in β-cell expansion while maintaining normal ductal cells suggesting that the cell cycle protein is important for proliferation of β-cells independent of influencing duct cells.

One potential mechanism that can contribute to β-cell expansion, includes the replication process and involves a well-recognized pathway that has been studied in considerable detail in organogenesis and in cancer.[116,117] Epithelial-to-mesenchymal transition or EMT occurs in epithelial cells expressing tyrosine kinase receptors and involves disappearance of differentiated junctions, reorganization of cytoskeleton and redistribution of organelles, together transforming epithelial into mesenchymal cells.[117-119] After transformation, the mesenchymal cells may eventually regain a fully differentiated epithelial phenotype via a mesenchyme-to-epithelial transition (MET) or reverse EMT.[120] A characteristic feature of EMT is repression of epithelial markers including E-cadherin and α- and γ-catenins and induction of mesenchymal markers including vimentin, fibronectin and N-cadherin.[117,121] Although the term EMT has been mostly associated either with early development or neoplasia, it is possible that this process is occurring, albeit modified, in normal cells responding to physiological demands that require cell/tissue expansion. Indeed, the E-cadherin/catenin family of proteins can also act as master regulatory and signaling molecules for differentiation, proliferation and apoptosis[117,122] indicating that these proteins have the capacity to regulate growth in normal tissues.[122] Cell-cell adhesion, as mediated by the cadherin-catenin system, is a prerequisite for normal cell function and for the preservation of tissue integrity in most tissues including islet cells. Both E-cadherin and β-catenin and several other members of the cadherin/catenin family are under the control of growth factors including epidermal growth factor (EGF), hepatocyte growth factor/scatter factor (HGF/SF) and insulin like-growth factors (IGF-I and IGF-II).[117,120] Receptors for these growth factors are expressed in β-cells, and proteins in their signaling pathways have been reported to play functional roles in β-cell growth and hormone secretion. Intriguingly, treatment of mouse embryonic stem cells or rat bladder carcinoma cells with IGF-II induces EMT and the withdrawal of IGF-II allows a reversal of the phenotype to an epithelial cell.[120] Direct association between insulin/IGF-1 receptors with the E-cadherin-catenin system, forming a multi-element complex, has been recently suggested based on colocalization of IGF-1 receptors with E-cadherin and β-and α-catenins at points of cell contacts.[117] The presence of IGF-1 receptors and its substrate proteins insulin receptor substrate-1 and SHC in the same complex with E-cadherin indicates potential cross talk between growth factor and catenin-cadherin signaling pathways.[117] Further evidence for a role for cadherin-catenin complex in islet growth is provided by mouse experiments in which dominant-negative expression of E-cadherin on the rat insulin promoter perturbed islet formation without increasing the incidence of tumor formation.[117,123] Together, these data provide a basis for a link between growth factor signaling and the potential for EMT in islet/β-cell growth.

Recently, features suggestive of EMT were described to occur in vivo, for the first time to our knowledge, in a mouse model of insulin resistance manifesting robust islet hyperplasia.[124] The presence of PCNA+ cells within the islets, which also showed down regulation of E-cadherin, and an increased localization of β-catenin to the nucleus, provided evidence for alterations in adhesion properties and an ability of the cells to replicate (Fig. 3A).[124] Furthermore, the lack of close association of replicating cells with lectin (a ductal marker), in multiple pancreas sections, suggests the cells are independent of pancreatic ducts and are likely replicating β-cells that have undergone metaplastic changes. Downregulation of β-catenin, another adhesion protein, in

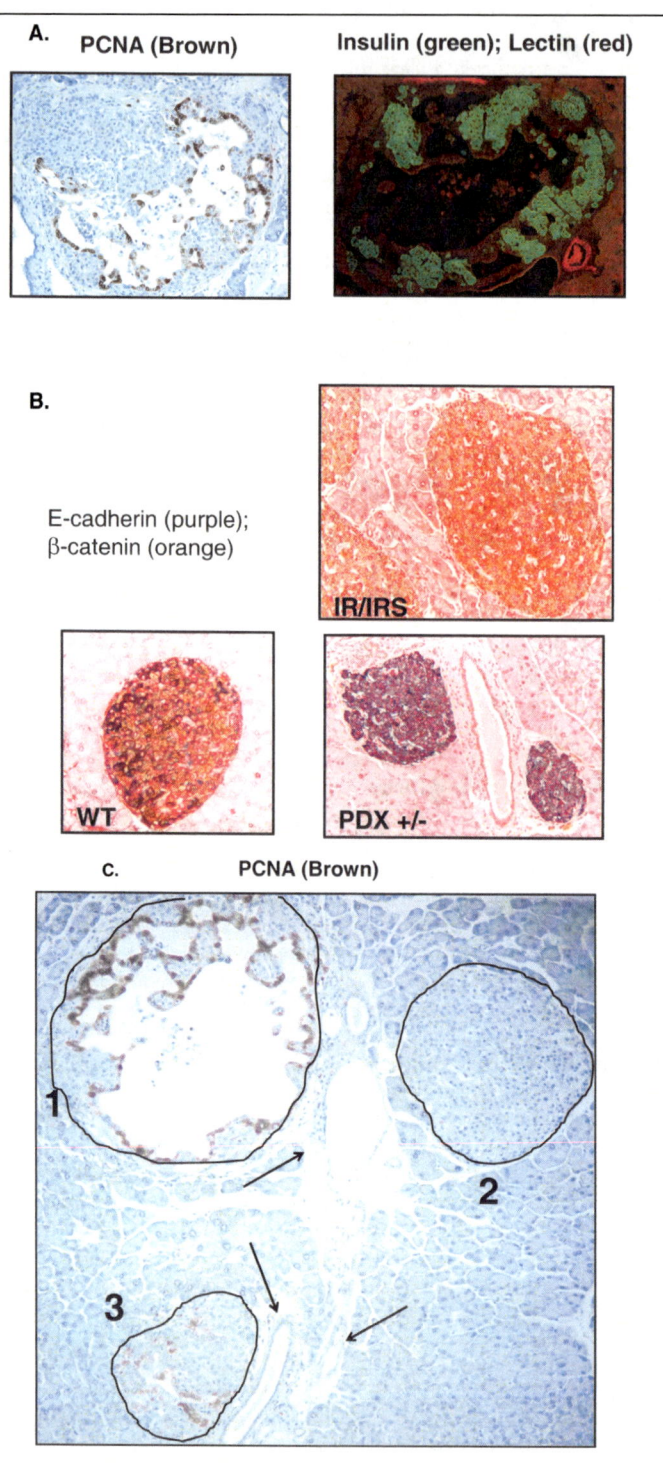

Figure 3. A) Representative islet from a mouse model of islet hyperplasia (insulin receptor/insulin receptor substrate-1 (IR/IRS-1) double heterozygous (DH) mouse) showing multiple PCNA+ cells. A serial section from the same pancreas shows that PCNA+ cells are independent of a pancreatic ductal marker (lectin). B) Pancreas sections from wild-type (WT), DH and PDX-1 haploinsufficient mice (PDX-1+/-). The WT islet shows immunostainng for both E-cadherin and β-catenin, while E-cadherin is down-regulated in DH islet and β-catenin is down-regulated in PDX+/- islet. C) Pancreas sections from a IR/IRS-1 double heterozygous mouse, different from that shown in Figure 3A, showing varying numbers of PCNA+ cells in three sister islets. Note the absence of PCNA+ cells in islet #2 and absence of PCNA+ cells in ducts in the same section. The islets are outlined for ease of comprehension and ducts are indicated by arrows. Figures 3A and 3B reproduced with permission from Kulkarni et al, J Clin Invest 114(6):828-836, 2004.[175]

islet cells from a mouse which is haploinsufficient for PDX-1, indicates a potential role for the homeodomain protein in the EMT process (Fig. 3B). In fact, growth factor signaling has been linked to PDX-1-mediated regulation of β-cell growth[125] providing additional evidence for a role for PDX-1 in β-cell regeneration. Interestingly, the PCNA+ cells are detectable in some, if not all, of the islets. For example, Figure 3C shows three islets in a pancreas section from a mouse that is doubly heterozygous for insulin receptor and insulin receptor substrate-1. PCNA+ cells are clearly evident within the core of two of the islets but not the third suggesting that the proliferation response occurs in susceptible islets likely in a regulated manner. The mechanisms and signals that allow some cells, but not others, to begin the process of replication in response to stimuli such as insulin resistance require further investigation. Thus, it is possible that EMT or an EMT-like process, may promote β-cell expansion under the appropriate stimulatory conditions. The report that EMT also occurs in human islet cell precursor cells[111,112] suggests that this process is a common response across species. Whether this indeed occurs in vivo in humans in early stages of diabetes is an important and timely question to address.

It has been recognized for over a decade that β-cells compensate in order to overcome the ambient hyperinsulinemia.[64,126] However, few studies have examined the pathways and proteins underlying the islet hyperplastic process. Since circulating insulin levels are significantly elevated in insulin resistance, an obvious candidate for β-cell proliferation is insulin itself[33,127] (Fig. 2). In fact, insulin has been shown to enhance islet β-cell replication in neonatal rat monolayer cultures[128] and to increase the regenerative ability of β-cells in transplantation models of fetal rat pancreas.[129,130] A role for insulin as a growth factor is also supported by studies in βIRKO mice, which display an age-dependent decrease in β-cell mass[59,62] and by reports that treatment of MIN6 β-cells with insulin receptor siRNA leads to altered expression of cell cycle proteins and proliferation.[131] Further support for a proliferative role for insulin arises from our recent studies using a transplantation model wherein BrdU+ β-cells show a direct correlation with circulating levels of insulin.[73,132] In addition to its nutrient role, glucose has been shown to increase β-cell mass in several models (reviewed in ref. 98).[133] Whether the effects of glucose are mediated by the secreted insulin acting in an autocrine manner to promote growth and/or prevent apoptosis in β-cells requires further study in models lacking insulin receptors in β-cells. Similarly, it is possible that the effects of GLP-1 on β-cell proliferation[29,134] are mediated, in part, by secreted insulin acting in an autocrine manner.

Other mechanisms that have been reported to contribute to regeneration of β-cells include budding of cells from pancreatic duct tissues,[106] transdifferentiation from acinar cells[135,136] and the ability of transcription factors to induce hepatocytes to differentiate into insulin-secreting cells.[137,138] Figure 4 shows a schematic of current concepts on potential pathways of β-cell regeneration.

## Growth and Development of Islet α-Cells

Insulin and IGF-1 receptors are also expressed in islet α-cells[139,140] and their role in early development and growth of glucagon-producing cells is not fully explored. However, a relative increase in α-cell number is a recognized feature in adult patients with type 2 diabetes.[141] Whether potential stimulation of α-cell proliferation is in fact mediated by high circulating

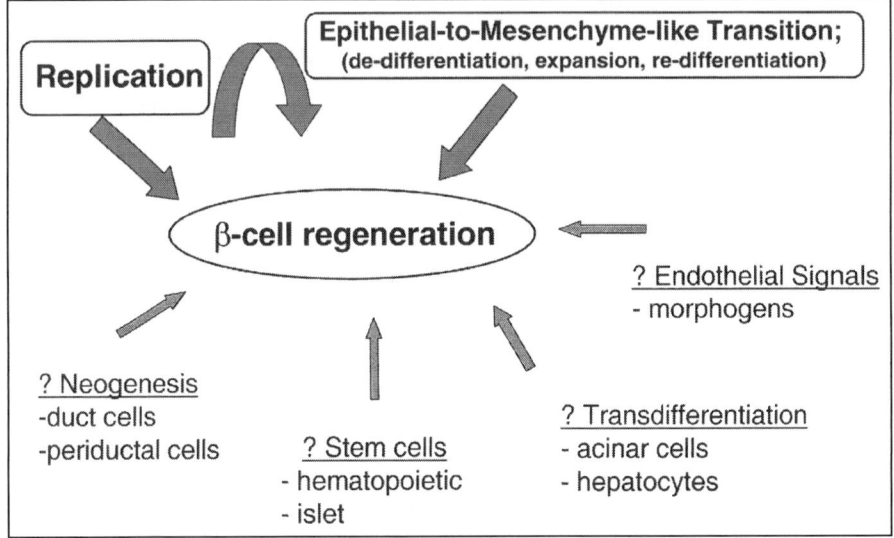

Figure 4. A schematic of potential mechanisms contributing to β-cell regeneration. Figure reproduced with permission from Kulkarni RN, Rev Endo Metab Disord 2005; 6(3):199-210.

levels of insulin in established cases of type 2 diabetes is not clear. Several studies implicate a role for intra-islet insulin to suppress glucagon release as an important factor in poor recovery from hypoglycemia in patients with long-standing type 1 diabetes and in advanced stages of type 2 diabetes on insulin therapy.[142,143] Indeed, a report suggests that glucose-stimulated glucagon secretion in αTC6 cells requires expression of insulin receptors in α-cells.[144] The signaling proteins and pathways that mediate these effects are not fully defined.

A recent study, in which mice were treated with glucagon receptor inhibitors, reported a significant increase in α-cell hyperplasia providing evidence for therapeutic intervention at the receptor level in modulating the ability of islet cells to grow.[145] The potential influence of interactions between β- and α-cells on the modulation of growth of islets during development and for maintenance of mass during adulthood in the face of altering pathophysiological states is worth exploring.

## The Liver-Pancreas Connection

The liver and the pancreas are known to share a common developmental pathway[146,147] and express several common transcription factors, which are essential for their growth.[147] Therefore, it is not surprising that even in the adult organism there is evidence suggestive of communication between the two metabolic tissues. For example, mice with a hepatic glucokinase knock-out manifest impaired insulin secretion in response to glucose suggesting that loss of hepatic glucose sensing impacts on islet function.[148] Furthermore, the liver has long been recognized as a source of circulating growth factors including IGF and HGF/SF, both of which are known to especially influence islet growth (reviewed in ref. 64).[39,102,149,150] In this context, it is interesting that mice lacking insulin receptors in hepatocytes develop large islets.[151] One interpretation of these observations is that in pathophysiological states, the insulin-resistant liver may potentially transmit signals to the islets via secreted growth factors to allow for β-cell compensation.[64] Although studies in hepatocytes implicate a role for insulin in regulating the hepatocyte nuclear transcription factor 3β (Foxa2),[152] and PDX-1 has been linked to growth factor signaling in the context of β-cell growth,[125] further studies to define the proteins linking upstream signals such as insulin with transcription factors in islets/β-cells are worth exploring.[153]

## Future Insights

Genetic engineering techniques have revolutionized the understanding of the role of insulin and IGFs in the development and maintenance of β-cell mass. Understanding the link(s) between insulin/IGF-I/IRS/FoxO1 signaling and PDX-1 with molecules that regulate β-cell cycle control will be crucial for the development of therapeutic strategies aimed at promoting islet cell regeneration.

## *Acknowledgements*

The author thanks Julie Marr and Kezia Frayjo for excellent secretarial assistance. R.N.K. is the recipient of the K08 Clinical Scientist Award (NIH DK 02885) and acknowledges support from NIH R01 DK67536, R01 DK68721, R03 DK66207, the Harvard Stem Cell Institute, the ADA and JDRF Center for Islet Transplantation at Harvard Medical School.

## References

1. Blakesley VA, Butler AA, Koval AP et al. In: Rosenfeld R, Roberts Jr C, eds. The IGF system. 1999:143-164.
2. Cheatham B, Kahn CR. Insulin action and the insulin signaling network. Endocr Rev 1995; 16:117-142.
3. De Meyts P et al. The insulin-like growth factor-1 receptor. Structure, ligand-binding mechanism and signal transduction. Horm Res 1994; 42:152-169.
4. Bell GI, Polonsky KS. Diabetes mellitus and genetically programmed defects in beta-cell function. Nature 2001; 414:788-791.
5. Saltiel AR, Kahn CR. Insulin signalling and the regulation of glucose and lipid metabolism. Nature 2001; 414:799-806.
6. Kahn CR. Diabetogenes and the cause of type II diabetes (Banting Lecture). Diabetes 1994; 1066-1084.
7. White MF. IRS proteins and the common path to diabetes. Am J Physiol Endocrinol Metab 2002; 283:E413-E422.
8. Accili D. Lilly lecture 2003. The struggle for mastery in insulin action: From triumvirate to republic. Diabetes 2004; 53(7):1633-42.
9. White MF, Yenush L. The IRS-signaling system: A network of docking proteins that mediate insulin and cytokine action. Curr Top Microbiol Immunol 1998; 228:179-208.
10. Virkamaki A, Ueki K, Kahn CR. Protein-protein interaction in insulin signaling and the molecular mechanisms of insulin resistance. J Clin Invest 1999; 103:931-943.
11. DeFronzo RA. Lilly Lecture 1987. The triumvirate: β-cell, muscle, liver: A collusion responsible for NIDDM. Diabetes 1988; 37:667-687.
12. Kulkarni RN. Receptors for insulin and insulin-like growth factor-1 and insulin receptor substrate-1 mediate pathways that regulate islet function. Biochem Soc Trans 2002; 30:317-322.
13. Leibowitz G et al. Insulin does not mediate glucose stimulation of proinsulin biosynthesis. Diabetes 2003; 52:998-1003.
14. Wicksteed B, Alarcon C, Briaud I et al. Glucose-induced translational control of proinsulin biosynthesis is proportional to preproinsulin mRNA levels in islet beta-cells but not regulated via a positive feedback of secreted insulin. J Biol Chem 2003; 278:42080-42090.
15. Irminger JC, Vollenweider FM, Neerman-Arbez M et al. Human proinsulin conversion in the regulated and the constitutive pathways of transfected AtT20 cells. J Biochem 1994; 269:1756-1762.
16. Halban PA. Proinsulin processing in the regulated and the constitutive secretory pathway. Diabetologia 1994; 37(Suppl 2):S65-S72.
17. Nakae J, Kido Y, Accili D. Distinct and overlapping functions of insulin and IGF-1 receptors. Endo Reviews 2001; 22:818-835.
18. Persaud SJ, AsareAnane H, Jones PM. Insulin receptor activation inhibits insulin secretion from human islets of Langerhans. Growth Regul 2002; 510:225-228.
19. Aspinwall CA, Lakey JRT, Kennedy RT. Insulin-stimulated insulin secretion in single pancreatic beta cells. The Journal of Biological Chemistry 1998; 274:6360-6365.
20. Johnson JD, Misler S. Nicotinic acid-adenine dinucleotide phosphate-sensitive calcium stores initiate insulin signaling in human beta cells. Proc Natl Acad Sci USA 2002; 99:14566-14571.
21. Luciani DS, Johnson JD. Acute effects of insulin on beta-cells from transplantable human islets. Mol Cell Endocrinol 2005; 241:88-98.

22. Leibiger B, Leibiger IB, Moede T et al. Selective insulin signaling through A and B insulin receptors regulates transcription of insulin and glucokinase genes in pancreatic beta cells. Mol Cell 2001; 7:559-570.
23. Bouche C, Kulkarni RN, Kahn CR et al. Exogenous insulin enhances glucose-stimulated insulin secretion in healthy humans. Diabetes 54 2005, (Ref Type: Generic).
24. Rasouli N, Hale T, Kahn SE et al. Effects of short term experimental insulin resistance and family history of diabetes on pancreatic {beta}-cell function in nondiabetic individuals. J Clin Endocrinol Metab 2005; 90(10):5825-33.
24a. Gunton JE, Kulkarni RN, Yim S et al. Loss of ARNT/HIF1beta mediates altered gene expression and pancreatic-islet dysfunction in human type 2 diabetes. Cell 2005; 122(3):337-49.
24b. Hribal ML, Perego L, Lovari S et al. Chronic hyperglycemia impairs insulin secretion by affecting insulin receptor expression, splicing, and signaling in RIN beta cell line and human islets of Langerhans. FASEB J 2003; 17(10):1340-2.
25. Pictet R, Rutter WJ. In: Steiner DF, Freinkel M, eds. Handbook of Physiology. Washington, DC: American Physiological Society, 1972:25-66.
26. Herrera PL. Defining the cell lineages of the islets of Langerhans using transgenic mice. Int J Dev Biol 2002; 46:97-103.
27. Murtaugh LC, Melton DA. Genes, signals, and lineages in pancreas development. Annu Rev Cell Dev Biol 2003; 19:71-89.
28. Sander M, German MS. The beta cell transcription factors and development of the pancreas. J Mol Med 1997; 75:327-340.
29. Habener JF, Kemp DM, Thomas MK. Minireview: Transcriptional regulation in pancreatic development. Endocrinology 2005; 146:1025-1034.
30. GuG, Dubauskaite J, Melton DA. Direct evidence for the pancreatic lineage: NGN3+ cells are islet progenitors and are distinct from duct progenitors. Development 2002; 129:2447-2457.
31. Fowden AL, Hill DJ. Intra-uterine programming of the endocrine pancreas. Brit Med Bull 2001:123-142.
32. Dupont J, Holzenberger M. Biology of insulin-like growth factors in development. Birth Defects Res C Embryo Today 2003; 69:257-271.
33. Hill DJ, Milner RD. Insulin as a growth factor. Pediatr Res 1985; 19:879-886.
34. Milner RD, Hill DJ. Fetal growth control: The role of insulin and related peptides. Clin Endocrinol (Oxf.) 1984; 21:415-433.
35. Hill DJ. Development of the endocrine pancreas. Rev Endocr Metab Disord 2005; 6:229-238.
36. Hogg J, Han VKM, Clemmons DR et al. Interactions of glucose, insulin-like growth factors (IGFs) and IGF binding proteins in the regulation of DNA synthesis by isolated fetal rat islets of langerhans. J Endocrinol 1993:401-412.
37. Fehmann HC, Jehle P, Goke B. IGF-I and IGF-II: Expression and function in the endocrine pancreas. Exp Clin Endocrinol Diabetes 1995; 103(Suppl 2):37-41.
38. Hill DJ, Hogg J. Growth factor control of pancreatic B cell hyperplasia. Baillieres. Clin Endocrinol Metab 1991; 5:689-698.
39. Hill DJ, Petrik J, Arany E. Growth factors and the regulation of fetal growth. Diabetes Care 1998; (Suppl 2):B60-B69.
40. Orban PC, Chui D, Marth JD. Tissue- and site-specific DNA recombination in transgenic mice. Proc Natl Acad Sci USA 1992; 89:6861-6865.
41. Ryding AD, Sharp MG, Mullins JJ. Conditional transgenic technologies. J Endocrinol 2001; 171:1-14.
42. Duvillie B et al. Increased islet cell proliferation, decreased apoptosis, and greater vascularization leading to beta-cell hyperplasia in mutant mice lacking insulin. Endocrinology 2002; 143:1530-1537.
43. Duvillie B et al. Phenotypic alterations in insulin-deficient mutant mice. Proc Natl Acad Sci USA 1997; 94:5137-5140.
44. Lammert E, Cleaver O, Melton D. Induction of pancreatic differentiation by siganls from blood vessels. Science 2001:1-10.
45. Lammert E, Cleaver O, Melton D. Role of endothelial cells in early pancreas and liver development. Mech Dev 2003; 120:59-64.
46. Cleaver O, Melton DA. Endothelial signaling during development. Nat Med 2003; 9:661-668.
47. Lu Y et al. Pancreatic-specific inactivation of IGF-I gene causes enlarged pancreatic islets and significant resistance to diabetes. Diabetes 2004; 53:3131-3141.
48. Petrik J et al. Overexpression of insulin-like growth factor-II in transgenic mice is associated with pancreatic islet cell hyperplasia. Endocrinology 1999; 140:2353-2363.
49. Devedjian JC et al. Transgenic mice overexpressing insulin-like growth factor-II in B cells develop type 2 diabetes. J Clin Invest 2000; 105:731-740.

50. Efstratiadis A. Genetics of mouse growth. Int J Dev Biol 1998; 955-976.
51. Liu JP, Baker J, Perkins JA et al. Mice carrying null mutations of the genes encoding insulin-like growth factor I (Igf-1) and type 1 IGF receptor (Igf1r). Cell 1993; 75:59-72.
52. Powell-Braxton L et al. IGF-I is required for normal embryonic growth in mice. Genes & Development 1993; 7:2609-2617.
53. Woods KA, Camacho-HubnerC, Savage MO et al. Intrauterine growth retardation and postnatal growth failure associated with deletion of the insulin-like growth factor I gene. N Engl J Med 1996; 335:1363-1367.
54. Accili D et al. Early neonatal death in mice homozygous for a null allele of the insulin receptor gene. Nat Genet 1996; 12:106-109.
55. Joshi RL et al. Targeted disruption of the insulin receptor gene in the mouse results in neonatal lethality. EMBO J 1996; 15:1542-1547.
56. Joshi RL et al. Targeted disruption of the insulin receptor gene in the mouse results in neonatal lethality. EMBO J 1996; 15:1542-1547.
57. Withers DJ et al. Disruption of IRS-2 causes type 2 diabetes in mice. Nature 1998; 391:900-904.
58. Kido Y, Nakae J, Xuan S et al. Beta cell development in mice lacking insulin and type I IGF receptors. Diabetes 2000; 49(Suppl 1).
59. Kulkarni RN et al. Tissue-specific knockout of the insulin receptor in pancreatic β cells creates an insulin secretory defect similar to that in Type 2 diabetes. Cell 1999; 96:329-339.
60. Kulkarni RN et al. beta-cell-specific deletion of the Igf1 receptor leads to hyperinsulinemia and glucose intolerance but does not alter beta-cell mass. Nat Genet 2002; 31:111-115.
61. Xuan S et al. Defective insulin secretion in pancreatic beta cells lacking type 1 IGF receptor. J Clin Invest 2002; 110:1011-1019.
62. Otani K et al. Reduced beta-cell mass and altered glucose sensing impair insulin-secretory function in betaIRKO mice. Am J Physiol Endocrinol Metab 2004; 286:E41-E49.
63. Da Silva X, Qian Q, Cullen PJ et al. Distinct roles for insulin and insulin-like growth factor-1 receptors in pancreatic beta-cell glucose sensing revealed by RNA silencing. Biochem J 2004; 377:149-158.
64. Kulkarni RN, Kahn CR. Molecular basis of pancreas development and function. In: Habener JF, Hussain M, eds. New York City: Kluwer Academic Publishers, 2001:299-323.
65. Elders MJ et al. Endocrine-metabolic relationships in patients with leprechaunism. J Natl Med Assoc 1982; 74:1195-1210.
66. Taylor SI. Lilly Lecture: Molecular mechanisms of insulin resistance-Lessons from patients with mutations in the insulin receptor gene. Diabetes 1992; 41:1473-1490.
67. Araki E et al. Alternative pathway of insulin signalling in mice with targeted disruption of the IRS-1 gene. Nature 1994; 372:186-190.
68. Tamemoto H et al. Insulin resistance and growth retardation in mice lacking insulin receptor substrate-1. Nature 1994; 372:182-186.
69. Kulkarni RN et al. Altered function of insulin receptor substrate-1-deficient mouse islets and cultured beta-cell lines. J Clin Invest 1999; 104:R69-R75.
70. Kubota N et al. Disruption of insulin receptor substrate 2 causes type 2 diabetes because of liver insulin resistance and lack of compensatory β-cell hyperplasia. Diabetes 2000; 49:1880-1889.
71. Aspinwall CA et al. Roles of insulin receptor substrate-1, phosphatidylinositol 3-kinase, and release of intracellular Ca2+ stores in insulin-stimulated insulin secretion in beta -cells. J Biol Chem 2000; 275:22331-22338.
72. Kulkarni RN, Roper M, Dahlgren GM.et.al. Insulin secretory defect in IRS-1 null mice is linked with reduced calcium signaling and altered expression of SERCA-2b and-3. Diabetes 2004; 53:1517-1525.
73. Hennige AM et al. Alterations in growth and apoptosis of insulin receptor substrate-1 deficient beta-cells. Am J Physiol Endocrinol Metab 2005; 289:E337-E346.
74. Hennige AM et al. Upregulation of insulin receptor substrate-2 in pancreatic beta cells prevents diabetes. J Clin Invest 2003; 112:1521-1532.
75. Lin X et al. Dysregulation of insulin receptor substrate 2 in beta cells and brain causes obesity and diabetes. J Clin Invest 2004; 114:908-916.
76. Kubota N et al. Insulin receptor substrate 2 plays a crucial role in beta cells and the hypothalamus. J Clin Invest 2004; 114:917-927.
77. Choudhury AI et.al. The role of insulin receptor substrate 2 in hypothalamic and beta-cell function. Journal of Clinical Investigation 2005; 115:940-950.
78. Jhala US et al. cAMP promotes pancreatic beta cell survival via CREB mediated induction of IRS2. Genes and Development 2003; 17:1575-1580.

79. Lingohr MK et al. Decreasing IRS-2 expression in pancreatic beta-cells (INS-1) promotes apoptosis, which can be compensated for by introduction of IRS-4 expression. Mol Cell Endocrinol 2003; 209:17-31.
80. Briaud I et al. Insulin receptor substrate-2 proteasomal degradation mediated by a mammalian target of rapamycin (mTOR)-induced negative feedback downregulates protein kinase B-mediated signaling pathway in beta-cells. J Biol Chem 2005; 280:2282-2293.
81. Kushner JA et al. Islet-sparing effects of protein tyrosine phosphatase-1b deficiency delays onset of diabetes in IRS2 knockout mice. Diabetes 2004; 53:61-66.
82. Kushner JA et al. Pdx1 restores beta cell function in Irs2 knockout mice. J Clin Invest 2002; 109:1193-1201.
83. Kushner JA et al. Pten regulation of islet growth and glucose homeostasis. J Biol Chem 2005.
84. Liu SC, Wang Q, Lienhard GE et al. Insulin receptor substrate 3 is not essential for growth or glucose homeostasis. J Biol Chem 1999; 274:18093-18099.
85. Fantin VR, Wang GE, Lienhard GE et al. Mice lacking insulin receptor substrate 4 exhibit mild defects in growth, reproduction, and glucose homostasis. Am. J Physiol 2000; 278:E127-133.
86. Ueki K et al. Increased insulin sensitivity in mice lacking p85beta subunit of phosphoinositide 3-kinase. Proc Natl Acad Sci USA 2002; 99:419-424.
87. Terauchi Y et al. Increased insulin sensitivity and hypoglycaemia in mice lacking the p85 alpha subunit of phosphoinositide 3-kinase. Nat Genet 1999; 21:230-235.
88. Mauvais-Jarvis F et al. Reduced expression of the murine p85a subunit of phosphoinositide 3-kinase improves insulin signaling and ameliorates diabetes. J Clin Invest 2002; 109:141-149.
89. Eto K et al. Phosphatidylinositol 3-kinase suppresses glucose-stimulated insulin secretion by affecting post-cytosolic [Ca(2+)] elevation signals. Diabetes 2002; 51:87-97.
90. Zawalich WS, Tesz GJ, Zawalich KC. Inhibitors of phosphatidylinositol 3-kinase amplify insulin release from islets of lean but not obese mice. J Endocrinol 2002; 174:247-258.
91. Zawalich WS, Zawalich KC. A link between insulin resistance and hyperinsulinemia: Inhibitors of phosphatidylinositol 3-kinase augment flucose-induced insulin secretion from islets of lean, but not obese, rats. Endocrinology 2000; 141:3287-3295.
92. Liu Y et al. Wortmannin, a widely used phosphoinositide 3-kinase inhibitor, also potently inhibits mammalian polo-like kinase. Chem Biol 2005; 12:99-107.
93. Cho H et al. Insulin resistance and a diabetes mellitus-like syndrome in mice lacking the protein kinase Akt2 (PKBβ). Science 2001; 292:1728-1731.
94. Tuttle RL et al. Regulation of pancreatic beta-cell growth and survival by the serine/threonine protein kinase Akt1/PKBalpha. Nat Med 2001; 7:1133-1137.
95. Bernal-Mizrachi E, Wen W, Stahlhut S et al. Islet beta cell expression of constitutively active Akt1/PKB alpha induces striking hypertrophy, hyperplasia, and hyperinsulinemia. J Clin Invest 2001; 108:1631-1638.
96. Bernal-Mizrachi E et al. Defective insulin secretion and increased susceptibility to experimental diabetes are induced by reduced Akt activity in pancreatic islet beta cells. J Clin Invest 2004; 114:928-936.
97. Pende M et al. Hypoinsulinaemia,glucose intolerance and diminished beta-cell size in S6K1-deficient mice. Nature 2000; 408:994-997.
98. Bonner-Weir S, Smith FE. Islet cell growth and the growth factors involved. TEM 1994; 5:60-64.
99. Bonner-Weir S. Regulation of pancreatic β-cell mass in vivo. Recent Prog Horm Res 1994; 49:91-104.
100. Vasavada RC et al. Targeted expression of placental lactogen in the beta cells of transgenic mice results in beta cell proliferation, islet mass augmentation, and hypoglycemia. J Biol Chem 2000; 275:15399-15406.
101. Vasavada RC et al. Overexpression of parathyroid hormone-related protein in the pancreatic islets of transgenic mice causes islet hyperplasia, hyperinsulinemia, and hypoglycemia. J Biol Chem 1996; 271:1200-1208.
102. Garcia-Ocana A et al. Hepatocyte growth factor overexpression in the islet of transgenic mice increases beta cell proliferation, enhances islet mass, and induces mild hypoglycemia. J Biol Chem 2000; 275:1226-1232.
103. Dai C, Huh CG, Thorgeirsson SS et al. Beta-cell-specific ablation of the hepatocyte growth factor receptor results in reduced islet size, impaired insulin secretion, and glucose intolerance. Am J Pathol 2005; 167:429-436.
104. Roccisana J et al. Targeted inactivation of hepatocyte growth factor receptor c-met in beta-cells leads to defective insulin secretion and GLUT-2 downregulation without alteration of beta-cell mass. Diabetes 2005; 54:2090-2102.

105. Fujinaka Y, Sipula D, Garcia-Ocana A et al. Characterization of mice doubly transgenic for parathyroid hormone-related protein and murine placental lactogen: A novel role for placental lactogen in pancreatic beta-cell survival. Diabetes 2004; 53:3120-3130.
106. Bonner-Weir S et al. In vitro cultivation of human islets from expanded ductal tissue. Proc Natl Acad Sci USA 2000; 97:7999-8004.
107. Suarez-Pinzon WL, Lakey JR, Brand SJ et al. Combination therapy with epidermal growth factor and gastrin induces neogenesis of human islet {beta}-cells from pancreatic duct cells and an increase in functional {beta}-cell mass*. J Clin Endocrinol Metab 2005.
108. Yoon KH et al. Selective beta-Cell loss and alpha-cell expansion in patients with type 2 diabetes mellitus in korea. J Clin Endocrinol Metab 2003; 88:2300-2308.
109. Butler AE et al. Beta-cell deficit and increased beta-cell apoptosis in humans with type 2 diabetes. Diabetes 2003; 52:102-110.
110. Dor Y, Brown J, Martinez OI et al. Adult pancreatic beta-cells are formed by self-duplication rather than stem-cell differentiation. Nature 2004; 429:41-46.
111. Gershengorn MC et al. Epithelial-to-mesenchymal transition generates proliferative human islet precursor cells. Science 2004; 306:2261-2264.
112. Lechner A, Nolan AL, Blacken RA et al. Redifferentiation of insulin-secreting cells after in vitro expansion of adult human pancreatic islet tissue. Biochem Biophys Res Commun 2005; 327:581-588.
113. Rane SG et al. Loss of cyclin-dependent kinase (Cdk4) expression causes insulin-deficient diabetes and cdk4 activation results in β-islet cell hyperplasia. Nat Genet 1999; 22:44-52.
114. Georgia S, Bhushan A. Beta cell replication is the primary mechanism for maintaining postnatal beta cell mass. J Clin Invest 2004; 114:963-968.
115. Kushner JA, Ciemerych MA, Sicinska E et al. Cyclins D2 and D1 are essential for postnatal pancreatic beta-cell growth. Mol Cell Biol 2005; 25:3752-3762.
116. Thiery JP. Epithelial-mesenchymal transitions in development and pathologies. Curr Opin Cell Biol 2003; 15:740-746.
117. Potter E, Bergwitz C, Brabant G. The cadherin-catenin system: Implications for growth and differentiation of endocrine tissues. Endocr Rev 1999; 20:207-239.
118. Thiery JP. Epithelial-mesenchymal transitions in tumour progression. Nat Rev Cancer 2002; 2:442-454.
119. Savagner P. Leaving the neighborhood: Molecular mechanisms involved during epithelial-mesenchymal transition. Bio Essays 2001; 23:912-923.
120. Morali OG et al. IGF-II induces rapid beta-catenin relocation to the nucleus during epithelium to mesenchyme transition. Oncogene 2001; 20:4942-4950.
121. Kang Y, Massague J. Epithelial-mesenchymal transitions: Twist in development and metastasis. Cell 2004; 118:277-279.
122. El Bahrawy MA, Pignatelli M. E-cadherin and catenins: Molecules with versatile roles in normal and neoplastic epithelial cell biology. Microsc Res Tech 1998; 43:224-232.
123. Dahl U, Sjodin A, Semb H. Cadherins regulate aggregation of pancreatic beta-cells in vivo. Development 1996; 122:2895-2902.
124. Kulkarni RN et al. PDX-1 haploinsufficiency limits the compensatory islet hyperplasia that occurs in response to insulin resistance. J Clin Invest 2004; 114:828-836.
125. Kitamura T et al. The forkhead transcription factor Foxo1 links insulin signaling to Pdx1 regulation of pancreatic beta cell growth. J Clin Invest 2002; 110:1839-1847.
126. Bonner-Weir S, Scaglia L, Montana E et al. Diabetes 1994. In: Baba S, Kaneko T, eds. Excerta Medica International Congress, 1995:179-228.
127. Accili D. A kinase in the life of the beta cell. J Clin Invest 2001; 108:1575-1576.
128. Rabinovitch A, Quigley C, Russell T et al. Insulin and multiplication stimulating activity (and insulin-like growth factor) stimulate islet beta-cell replication in neonatal rat pancreatic monolayer cultures. Diabetes 1982:160-164.
129. McEvoy RC, Schmitt RV, Hegre OD. Syngeneic transplantation of fetal rat pancreas. I. Effect of insulin treatment of the reversal of alloxan diabetes. Diabetes 1978; 27:982-987.
130. Movassat J, Saulnier C, Portha B. Insulin administration enhances growth of the beta-cell mass in streptozotocin-treated newborn rats. Diabetes 1997; 46:1445-1452.
131. Ohsugi M et al. Reduced expression of the insulin receptor in mouse insulinoma (MIN6) cells reveals multiple roles of insulin signaling in gene expression, proliferation, insulin content, and secretion. J Biol Chem 2005; 280:4992-5003.
132. Flier SN, Kulkarni RN, Kahn CR. Evidence for a circulating islet cell growth factor in insulin- resistant states. Proc Natl Acad Sci USA 2001; 98:7475-7480.
133. Bonner-Weir S, Deery D, Leahy JL et al. Compensatory growth of pancreatic β-cells in adult rats after short-term glucose infusion. Diabetes 1989; 38:49-53.

134. Stoffers DA. The development of beta-cell mass: Recent progress and potential role of GLP-1. Horm Metab Res 2004; 36:811-821.
135. Song KH et al. In vitro transdifferentiation of adult pancreatic acinar cells into insulin-expressing cells. Biochem Biophys Res Commun 2004; 316:1094-1100.
136. Paris M, Tourrel-Cuzin C, Plachot C et al. Review: Pancreatic beta-cell neogenesis revisited. Exp Diabesity Res 2004; 5:111-121.
137. Ferber S et al. Pancreatic and duodenal homeobox gene 1 induces expression of insulin genes in liver and ameliorates streptozotocin-induced hyperglycemia. Nature Medicine 2000; 6:568-572.
138. Kojima H et al. NeuroD-betacellulin gene therapy induces islet neogenesis in the liver and reverses diabetes in mice. Nat Med 2003; 9:596-603.
139. Kaneko K et al. Insulin inhibits glucagon secretion by the activation of PI3-kinase in In-R1-G9 cells. Diabetes Res Clin Pract 1999; 44:83-92.
140. Van Schravendijk CF, Foriers A, Van den Brande JL et al. Evidence for the presence of type I insulin-like growth factor receptors on rat pancreatic A and B cells. Endocrinology 1987; 121:1784-1788.
141. Unger RH. Glucagon physiology and pathophysiology in the light of new advances. Diabetologia 1985; 28:574-578.
142. Harvel PJ, Veith RC, Dunning BE et al. Role for autonomic nervous system to increase pancreatic glucagon secretion durning marked insulin-induced hypoglycemia in dogs. Diabetes 1991; 40:1107-1114.
143. Cryer PE. Banting Lecture. Hypoglycemia: The limiting factor in the management of IDDM. Diabetes 1994; 43:1378-1389.
144. Diao J, Asghar Z, Chan CB et al. Glucose-regulated glucagon secretion requires insulin receptor expression in pancreatic alpha-cells. J Biol Chem 2005; 280:33487-33496.
145. Sloop KW et al. Hepatic and glucagon-like peptide-1-mediated reversal of diabetes by glucagon receptor antisense oligonucleotide inhibitors. J Clin Invest 2004; 113:1571-1581.
146. Duncan SA, Navas MA, DufortD et al. Regulation of a transcription factor network required for differentiation and metabolism. Science 1998; 281:692-695.
147. Odom DT et al. Control of pancreas and liver gene expression by HNF transcription factors. Science 2004; 303:1378-1381.
148. Postic C et al. Dual roles for glucokinase in flucose homeostasis as determined by liver and pancreatic ß cell-specific gene knock-outs using cre recombinase. The Journal of Biological Chemistry 1998:1-11.
149. Scharf JG, Ramadori G, Braulke T et al. Cellular localization and hormonal regulation of biosynthesis of insulin-like growth factor binding proteins and of the acid-labile subunit within rat liver. Prog Growth Factor Res 1995; 6:175-180.
150. Funakoshi H, Nakamura T. Hepatocyte growth factor: From diagnosis to clinical applications. Clin Chim. Acta 2003; 327:1-23.
151. Michael MD et al. Loss of insulin signaling in hepatocytes leads to severe insulin resistance and progressive hepatic dysfunction. Mol Cell 2000; 6:87-97.
152. Wolfrum C, Besser D, Luca E et al. Insulin regulates the activity of forkhead transcription factor Hnf-3beta/Foxa-2 by Akt-mediated phosphorylation and nuclear/cytosolic localization. Proc Natl Acad Sci USA 2003; 100:11624-11629.
153. Kulkarni RN, Kahn CR. Molecular biology. HNFs—linking the liver and pancreatic islets in diabetes. Science 2004; 303:1311-1312.
154. Duvillie B et al. Phenotypic alterations in insulin-deficient mutant mice. Proc Natl Acad Sci USA 1997; 94:5137-5140.
155. Liu JP, Baker J, Perkins AS et al. Mice carrying null mutations of the genes encoding insulin-like growth factor I (Igf-1) and type 1 IGF receptor (Igf1r). Cell 1993; 75:59-72.
156. Accili D et al. Early neonatal death in mice homozygous for a null allele of the insulin receptor gene. Nat Genet 1996; 12:106-109.
157. Aspinwall CA et al. Roles of insulin receptor substrate-1, phosphatidylinositol 3-kinase, and release of intracellular Ca2+ stores in insulin-stimulated insulin secretion in beta -cells. J Biol Chem 2000; 275:22331-22338.
158. Chen WS et al. Growth retardation and increased apoptosis in mice with homozygous disruption of the Akt1 gene. Genes Dev 2001; 15:2203-2208.
159. Cho H, Thorvaldsen JL, Chu Q et al. Akt1/PKBalpha is required for normal growth but dispensable for maintenance of glucose homeostasis in mice. J Biol Chem 2001; 276:38349-38352.
160. Kitamura T et al. Preserved pancreatic beta-cell development and function in mice lacking the insulin receptor-related receptor. Mol Cell Biol 2001; 21:5624-5630.

161. Yakar S et al. Normal growth and development in the absence of hepatic insulin-like growth factor I. Proc Natl Acad Sci USA 1999; 96:7324-7329.
162. Liu JL, Yakar S, LeRoith D. Mice deficient in liver production of insulin-like growth factor I display sexual dimorphism in growth hormone-stimulated postnatal growth. Endocrinology 2000; 141:4436-4441.
163. Yakar S et al. Liver-specific igf-1 gene deletion leads to muscle insulin insensitivity. Diabetes 2001; 50:1110-1118.
164. Devedjian JC et al. Transgenic mice overexpressing insulin-like growth factor-II in beta cells develop type 2 diabetes. The Journal of Clinical Investigation 2000; 105:731-740.
165. Moller DE et al. Transgenic mice with muscle-specific insulin resistance develop increased adiposity, impaired glucose tolerance, and dyslipidemia. Endocrinology 1996; 137:2397-2405.
166. Bruning JC et al. A muscle-specific insulin receptor knockout exhibits features of the metabolic syndrome of NIDDM without altering glucose tolerance. Mol Cell 1998; 2:559-569.
167. Mauvais-Jarvis F et al. A model to explore the interaction between muscle insulin resistance and beta-cell dysfunction in the development of type 2 diabetes. Diabetes 2000; 49:2126-2134.
168. Bruning JC et al. Role of brain insulin receptor in control of body weight and reproduction. Science 2000; 289:2122-2125.
169. Bluher M et al. Adipose tissue selective insulin receptor knockout protects against obesity and obesity-related glucose intolerance. Dev Cell 2002; 3:25-38.
170. Lauro D, Kido Y, Hayashi H et al. Transgenic knock-out mice with a targeted impairment of insulin action in skeletal muscle and adipose tissue. Diabetes 1998; 47(Supp.1):A45.
171. Choudhury AI et al. The role of insulin receptor substrate 2 in hypothalamic and beta cell function. J Clin Invest 2005; 115:940-950.
172. Fernandez AM et al. Functional inactivation of the IGF-I and insulin receptors in skeletal muscle causes type 2 diabetes. Genes Dev 2001; 15:1926-1934.
173. Chen D, Mauvais-Jarvis F, Bluher M et al. p50alpha/p55alpha phosphoinositide 3-kinase knock-out mice exhibit enhanced insulin sensitivity. Mol Cell Biol 2004; 24:324-329.
174. Kulkarni RN. The islet beta-cell. Int J Biochem Cell Biol 2004; 36(3):365-71.
175. Kulkarni RN. New insights into the roles of Insulin/IGF-I in the development and maintenance of beta-cell mass. Rev Endocr Metab Disord 2005; 6(3):199-210.

## CHAPTER 8

# Central Regulation of Insulin Sensitivity

### Silvana Obici and Luciano Rossetti*

"...Hence we could say that in a diabetic individual the liver secretes too much. The matter which produces sugar cannot be transformed into a product with a more complex organization. The dis-assimilation has become prevalent. Therefore we can consider diabetes as a disease of the nervous system caused by excessive activation of the disassimilator nerve of the liver, which drives the premature disassimilation of matter that would otherwise be used for nutrition... ...Hence the treatment of diabetes should address the nervous system. Stimulating the sympathetic nerve could be a valuable tool. But, in order to achieve a treatment with a rationale based on physiology, we should answer many questions, which are still awaiting a solution from the science of physiology".

*Claude Bernard* in *"Leçons sur les phénomènes de la vie"*
Cours de physiologies Generale du Museum d'Histoire Naturelle, 1859

### Introduction

Insulin rapidly lowers blood glucose levels via inhibition of endogenous glucose production and stimulation of glucose uptake. The mechanisms by which insulin modulates hepatic glucose production involve either activation of insulin signaling in hepatocytes (direct effects) or activation of insulin receptors in extra-hepatic sites (indirect effects), which in turn leads to inhibition of glucose production via neural and/or humoral mediators. The direct effects can be further divided into acute insulin action leading to rapid decrease in glucose production and chronic insulin action modulating the gene expression of rate-limiting enzymes within the biochemical pathways leading to glucose production. Short-term effects of insulin on hepatic glucose fluxes may be divided into three major components: (a) direct effects on the liver, mostly leading to rapid inhibition of glycogenolysis;[1,2] (b) indirect effects mediated via peripheral actions of insulin, mostly modulating lipolysis;[3-6] (c) indirect effects mediated via activation of hypothalamic insulin signaling.[7,8] Furthermore, insulin exerts potent long-term effects on liver gene expression and function.[9] These more chronic actions of insulin can in turn markedly affect the acute responses to an increase in circulating insulin levels.[10,11]

Type 2 diabetes is a complex metabolic disorder characterized by insulin resistance and impaired β-cell function.[12] Fasting hyperglycemia is the hallmark of diabetes mellitus and an increase in the rates of GP is its major determinant in the great majority of patients with diabetes mellitus.[13] Patients with type 2 diabetes (DM2) exhibit an elevation in the rates of glucose production despite increased circulating levels of insulin and glucose. This increase is largely accounted for by a marked enhancement in the rate of gluconeogenesis.[14-17]

The worldwide epidemic of type 2 diabetes mellitus is driven by a marked increase in the prevalence of obesity.[18-21] Indeed, epidemiological and metabolic evidence tightly links obesity to type 2 diabetes mellitus (DM2) with insulin resistance providing the strongest etiological

*Corresponding Author: Luciano Rossetti—Albert Einstein College of Medicine,1300 Morris Park Avenue, Bronx, New York, 10461 U.S.A. Email: rossetti@aecom.yu.edu

*Mechanisms of Insulin Action*, edited by Alan R. Saltiel and Jeffrey E. Pessin.
©2007 Landes Bioscience and Springer Science+Business Media.

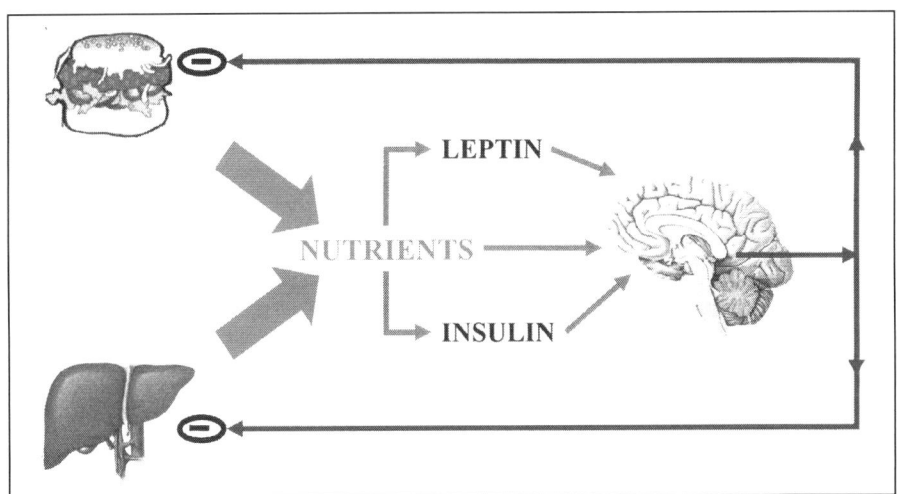

Figure 1. Scheme of homeostasis for circulating nutrients. Nutrients in the systemic circulation originate from exogenous (food) and endogenous sources (hepatic lipid and glucose production). The elevation of circulating nutrients activates nutrient-sensing pathways that stimulate the secretion and biosynthesis of counter-regulatory hormones such as insulin and leptin. These hormones act on the hypothalamus to restrain further ingestion of food (exogenous sources) and hepatic glucose production (endogenous sources). Nutrients themselves may directly activate nutrient sensing pathways in the hypothalamus and trigger counter-regulatory responses which ultimately inhibit further food intake and glucose production. The nutrient-induced hormonal signals and the nutrient sensing pathways in the hypothalamus are likely to be integrated at the neuronal or cellular level.

thread.[19,22,23] This association justifies the quest for basic mechanisms coupling energy balance with glucose homeostasis.[24-26] This rapid epidemic is likely the consequence of complex interactions between genes and environment. Consumption of high calorie diets and sedentary lifestyles are deemed to be the main environmental triggers.[22,27] Thus, it is keen to better our understanding of the basic mechanisms by which environmental factors can lead to insulin resistance. In this regard, adipose tissue is the main endogenous source of circulating lipids but it is also the site of production and secretion of several hormones and cytokines.[28-30] These adipose-derived signaling molecules exert potent metabolic effects in distant organs and they are likely to play a key role in the complex inter-organ communication network, which appears to modulate intermediate metabolism and energy balance. Hypothalamic centers sense the availability of peripheral nutrients partly via redundant nutrient-induced peripheral signals such as leptin and insulin and via direct metabolic signaling.[24,31-34] Indeed, hormonal and nutritional cues are integrated within selective hypothalamic regions, which have been postulated to function as a primary site for biochemical sensing of nutrient availability (Fig. 1). Responding to nutrient availability, these hypothalamic regions in turn exert a negative feedback not only on food intake[32] but also on endogenous glucose production.[7,24,35]

Since obesity is tightly associated to insulin resistance, the notion that the same hypothalamic regions may be involved in the regulation of both energy balance and insulin action leads to envision two ways by which these centers operate (Fig. 2): in a "sequential" model, hypothalamic energy centers alter insulin sensitivity largely via changes in fat mass and distribution. On the other hand, overlapping hypothalamic pathways may operate simultaneously to regulate energy balance and insulin action in response to changes in nutrient intake (parallel model). As discussed below, these models are not mutually exclusive.

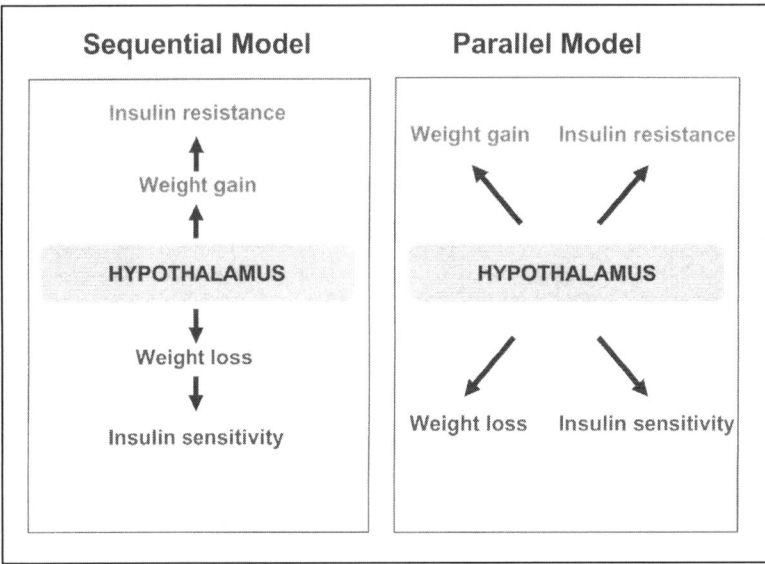

Figure 2. Scheme of the hypothalamic control over systemic insulin action. A) Sequential model: hypothalamic neural pathways regulate feeding behavior, energy expenditure and body composition, which secondarily affect insulin sensitivity. B) Parallel model: hypothalamic neural pathways concurrently modulate energy balance, body composition and insulin action. Sequential and parallel models do not exclude each other and may coexist.

## Insulin Action in the Hypothalamus

Over the past several years, multiple roles of insulin action in the brain have emerged. Insulin gains access to the hypothalamus via a saturable, receptor-mediated process and via diffusion from the eminence,[36] which is partly situated outside of the blood-brain barrier. Micro-dialysis has confirmed that insulin levels in the extracellular fluid of hypothalamic nuclei are regulated during meal absorption.[37,38] The expression of insulin receptors, insulin-receptor substrates (IRS-1, IRS-2, and IRS-4), phosphoinositide-3-kinase (PI3K) and other downstream components of the insulin signaling pathway has been demonstrated in the hypothalamus and specifically in its arcuate nuclei.[39-41] Early work by Porte and Woods[42,43] and by Cherrington and colleagues[44,45] provide solid support for the notion that insulin can rapidly act in the hypothalamus. These 'central' effects of insulin include modulation of feeding behavior,[43,46,47] neuropeptide Y expression,[47-49] hypoglycemia counter-regulation,[44] and autonomic outflow.[45,50,51]

As illustrated in Figure 1, it is plausible that the activation of insulin signaling in the hypothalamus (and perhaps in other areas of the CNS) initiates a negative feedback designed to control the availability of nutrients via regulation of both energy intake and glucose production. In this regard, the role of central insulin in the regulation of energy balance is supported by the observation that insulin deficient animals are markedly hyperphagic and the administration of insulin in a cerebral ventricle normalizes their food intake.[47,51] Furthermore, ICV administration of insulin[43] or insulin mimetics[52] decreases food intake and body weight while intrahypothalamic infusion of anti-insulin antibodies increases food intake and body weight in normal animals.[53] It is also noteworthy that prolonged or life-long impairment in CNS insulin signaling also leads to hyperphagia, increased plasma insulin levels, and decreased insulin sensitivity.[54,55] However, it is important to discern whether these metabolic consequences are directly due to the manipulation of central insulin action or are secondary to the concomitant changes in energy balance and fat mass.

## Hypothalamic Insulin Action: 'Gain of Function' Studies

A novel role of CNS insulin receptors in the regulation of glucose homeostasis has been recently discovered.[7] The stimulation of hypothalamic insulin signaling, in the presence of basal circulating insulin concentrations was sufficient to inhibit glucose production. This observation suggests that the activation of central insulin signaling can inhibit glucose production independently of changes in circulating insulin levels. These potent metabolic effects of intracerobroventricular insulin could be mediated by the activation of insulin signal transduction in regions of the CNS distant from the hypothalamic nuclei.[56] However, intrahypothalamic infusion of insulin entirely reproduced the potent effects of ICV insulin on blood glucose concentration and on GP.[8] Thus, activation of insulin signaling within the mediobasal hypothalamus is sufficient to decrease blood glucose levels via suppression of glucose production. These effects were largely due to a marked inhibition of hepatic gluconeogenesis and were associated with decreases in the hepatic expression of PEPCK and glucose-6-phosphatase.

Insulin activates ATP-sensitive potassium ($K_{ATP}$) channels in selective hypothalamic neurons.[57,58] However, the functional significance of insulin's activation of hypothalamic $K_{ATP}$ channels is poorly understood. Recent evidence has demonstrated that direct activation of hypothalamic $K_{ATP}$ channels with diazoxide is sufficient to reproduce the potent effects of central insulin on glucose homeostasis.[8] Central administration of diazoxide lowered blood glucose levels. During clamp studies, central administration of diazoxide, in the presence of basal circulating insulin levels markedly increased the rate of glucose infusion required to prevent hypoglycemia. This effect was due to a marked decrease in glucose production, while the rate of glucose uptake was not significantly affected. Importantly, the decrease in GP was largely accounted for by marked inhibition of gluconeogenesis, while the rate of glycogenolysis was not significantly decreased. ICV diazoxide markedly decreased liver G6Pase and PEPCK mRNA levels. Thus, direct activation of central $K_{ATP}$ channels is sufficient to recapitulate the action of central insulin on glucose levels, hepatic glucose production and gluconeogenesis, and on the hepatic expression of G6Pase and PEPCK. Thus, activation of either $K_{ATP}$ channels or insulin signalling within the mediobasal hypothalamus is sufficient to decrease blood glucose levels via suppression of GP, hepatic gluconeogenesis, and PEPCK and G6Pase expression. These 'gain-of-function' experiments suggest that modulation of $K_{ATP}$ channels activity within the arcuate nuclei of the hypothalamus can have a major impact on liver glucose homeostasis.

## Hypothalamic Insulin Action: 'Loss of Function' Studies

As in other cell types, insulin action in neurons requires binding to membrane receptors leading to a cascade of phosphorylation events within the neurons. The latter are initiated by the autophosphorylation of the β-subunit of the insulin receptor. The two main downstream pathways of insulin signaling involve activation of MAP kinases (ERK1 and ERK2) and of Phosphoinositide3-Kinase (PI3K) (Fig. 3). An important role of CNS insulin signaling is supported by genetic ablation studies. Inactivation of Insr in nestin-positive neurons results in obesity, hyperinsulinemia, and decreased female fertility.[55] In C.Elegans, the dauer phenotype caused by mutations in the Insr ortholog daf-2 can be rescued by selective reexpression of daf-2 in the brain.[59] Indeed, neural and reproductive inputs have shown to control lifespan of C. elegans.[59,60]

Insulin can activate hypothalamic $K_{ATP}$ channels and the activation of these channels is per se sufficient to lower blood glucose via inhibition of hepatic gluconeogenesis. Sulfonylureas ($K_{ATP}$ blockers) abolish the activation of hypothalamic $K_{ATP}$ channels by insulin and leptin.[57,58] The role of $K_{ATP}$ channels in hypothalamic insulin action was recently tested in rats receiving ICV infusions of vehicle, insulin, insulin and $K_{ATP}$ blocker, or $K_{ATP}$ blocker alone (Table 1). In the presence of near basal circulating insulin levels, glucose infusion was required to prevent hypoglycemia following central administration of insulin. ICV co infusion of the $K_{ATP}$ blocker glibenclamide abolished the effect of ICV insulin.[8] Central administration of $K_{ATP}$ blocker alone did not modify the rate of glucose infusion compared with ICV vehicle. In the presence of basal insulin levels, ICV insulin markedly and significantly decreased GP. This decrease was

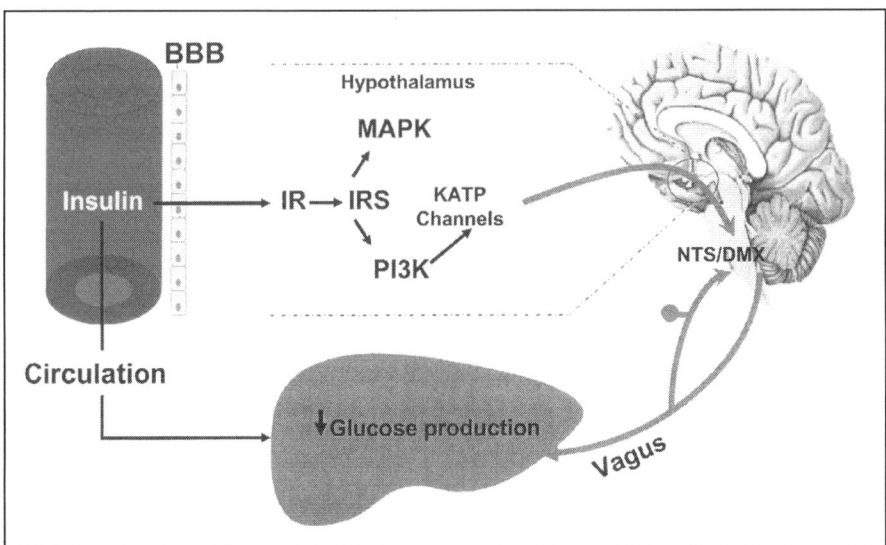

Figure 3. Hypothalamic insulin action and its effect on the regulation of glucose production. Insulin acutely suppresses glucose production via a direct effect on the liver (mainly suppressing glycogenolysis) and via stimulation of neural circuits originating in the arcuate nucleus. Circulating insulin crosses the blood-brain barrier through a saturable transport system and activates insulin receptors located in the arcuate nucleus. The insulin signaling cascade, including the activation of IRS and PI3Kinase, leads ultimately to the opening of KATP channels and neuronal hyperpolarization. The opening of the KATP channels is a step necessary and sufficient to activate descending vagal inputs to the liver, which in turn suppress glucose production by mainly inhibiting gluconeogenesis.

negated by the ICV coinfusion of $K_{ATP}$ blocker and was largely accounted for by a marked inhibition of gluconeogenesis. Importantly, in the presence of ICV glibenclamide, central insulin did not modify hepatic glucose fluxes. Central stimulation of insulin action also resulted in marked $K_{ATP}$ channels- dependent decreases in liver G6Pase and PEPCK mRNA levels. Thus, the ICV infusion of $K_{ATP}$ blocker completely prevented the central effects of insulin on liver glucose homeostasis. This work established that the activation of hypothalamic $K_{ATP}$

Table 1. Role of hypothalamic ATP-dependent potassium channels (KATP) in the modulation of insulin sensitivity. Summary of the acute effects of intrahypothalamic injection of KATP blockers or opener during euglycemic clamp studies on glucose fluxes.

|  | $K_{ATP}$ Blocker | Insulin | $K_{ATP}$ Blocker + Insulin | $K_{ATP}$ Opener |
|---|---|---|---|---|
| Glucose disposal | – | – | – | – |
| Glucose production | – | – | – | – |
| Glycogenolysis | – | – | – | – |
| Gluconeogenesis | – | – | – | – |
| PEPCK expression | – | – | – | – |
| G6Pase expression | – | – | – | – |

channels is required for the potent effects of central insulin on liver glucose homeostasis. How does the activation of $K_{ATP}$ channels in the hypothalamus result in dramatic changes in biochemical and molecular events in the liver?

Hypothalamic centers participate in the short-term regulation of ingestive behavior via descending neural connections to the caudal brainstem leading to activation of vagal input to the gastrointestinal tract.[32,56] Since autonomic neural input to the liver can also rapidly modulate liver metabolism,[61] it is plausible that the central administration of insulin decreases liver glucose production and the expression of G6Pase and PEPCK via activation of hepatic efferent vagal fibers. Indeed, hepatic branch vagotomy (HV) abolished the effects of the central administration of insulin on liver glucose homeostasis.[8] In the presence of ~basal circulating insulin levels, ICV insulin markedly and significantly decreased glucose production in sham-operated rats but not in rats with hepatic branch vagotomy. Hepatic branch vagotomy also negated the marked decreases in gluconeogenesis and in the hepatic expression of G6Pase and PEPCK following central administration of insulin. Further work established that the efferent vagal input to the liver is required for the inhibition of glucose production following central administration of insulin while the afferent input from the hepatic branch of the vagus nerve to the brainstem is not required. These studies identify key components of a neural circuit linking hypothalamic insulin action to the regulation of liver glucose homeostasis. Is this brain-liver cross-talk required for the physiological regulation of glucose homeostasis?

To address this question it was necessary to examine the effects of physiological increases in the circulating insulin levels on glucose homeostasis in the presence of selective loss-of-functions within this insulin-dependent brain-liver neural circuit. Results from several independent models demonstrate that hypothalamic insulin action is required for suppression of glucose production by hyperinsulinemia in conscious rats and mice. The ICV administration of insulin antagonists markedly diminished insulin's ability to inhibit glucose production during insulin clamp procedures.[7] In the presence of physiologic hyperinsulinemia, the ICV administration of insulin antibodies or treatment with antisense oligonucleotides specific for the insulin receptor led to impaired hepatic insulin action. Similarly, ICV infusion of inhibitors of PI3Kinase markedly decreased the ability of circulating insulin to suppress glucose production. Thus, antagonism of central insulin signaling induces hepatic insulin resistance in the presence of physiological elevations in circulating insulin concentrations. It is particularly noteworthy that the downregulation of insulin receptor expression with intracerebroventricular infusion of antisense oligonucleotide was mainly restricted to the the medial aspect of the Arcuate nuclei, but did not significantly alter insulin receptors in the ventrolateral aspect of the Arcuate nuclei or in other areas of the CNS.[54] This indicates that a defect of insulin action limited to the medial aspect of the Arcuate nucleus, which includes NPY- and Agrp-containing neurons, is sufficient to impair insulin action on glucose production. This is consistent with a negative feedback system designed to monitor and regulate the input of nutrients in the circulation. This restraint on glucose output may be required for the maintenance of metabolic homeostasis and its failure could lead to glucose intolerance.

Additionally, the effects of physiologic increases in plasma insulin levels on glucose production were markedly and similarly impaired by ICV or intra-hypothalamic infusion of a $K_{ATP}$ blocker.[8] Indeed, blocking the effects of insulin on hypothalamic insulin signaling or on hypothalamic $K_{ATP}$ channels results in a loss of ~half of the inhibitory action of circulating insulin on glucose production. $K_{ATP}$ channels are expressed in several tissues and couple energy metabolism to the electrical activity of the cells.[62] SUR1 mRNA is detectable in the mediobasal hypothalamus (arcuate nucleus). Since hypothalamic neurons expressing $K_{ATP}$ channels are targets of insulin[58] and display high sensitivity to diazoxide and sulfonylurea[63] it is likely that SUR1 is a component of these insulin-responsive $K_{ATP}$ channels. Consistent with this postulate, insulin's ability to restrain hepatic gluconeogenesis was selectively impaired in SUR1 null (SUR1KO) mice.[8] Taken together with the results of 'gain-of-function' experiments, these

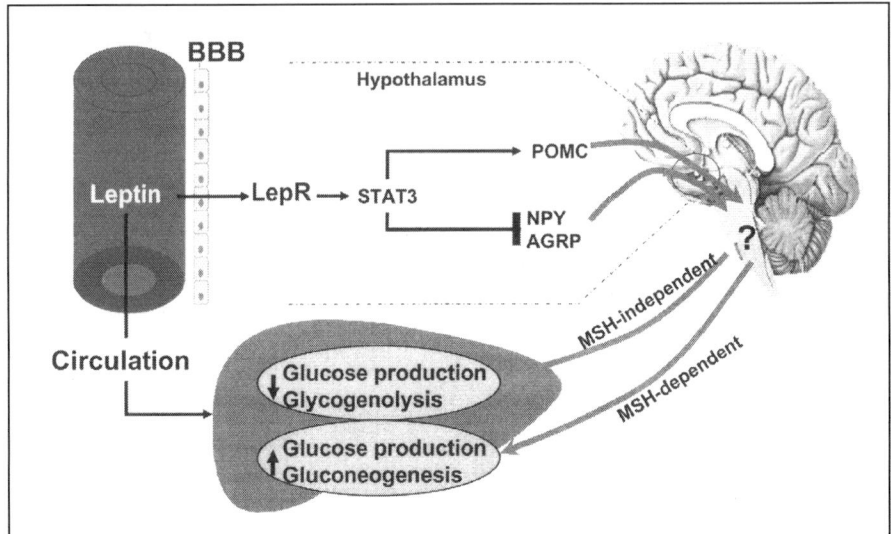

Figure 4. Scheme of the effects of leptin on hepatic glucose metabolism. Elevations of systemic leptin lead to activation of hypothalamic leptin signaling, which in turn increases the hepatic expression of gluconeogenic enzymes, stimulates the rate of gluconeogenesis and decreases the rate of glycogenolysis. The activation of gluconeogenesis is mediated by the leptin-dependent activation of melanocortin pathway since central administration of MSH reproduces these hepatic effects and coadministartion of a melanocortin antagonist with leptin abolishes them. On the other hand, the effect of leptin on suppression of glycogenolysis is independent of activation of the melanocortin pathway.

pharmacologic and genetic 'loss-of-function' experiments in rats and mice respectively indicate that SUR1-containing $K_{ATP}$ channels within the medial hypothalamus play an important role in the regulation of liver glucose homeostasis.

Finally, blocking the effects of insulin on the hepatic branch of the vagus also results in a loss of ~half of the inhibitory action of circulating insulin on glucose production. Hepatic branch vagotomy also interferes with the inhibitory effects of systemic insulin on gluconeogenesis and on the hepatic expression of G6Pase and PEPCK.[8] Overall, the presence of hepatic insulin resistance in (a) rats with acute impairments in either hypothalamic insulin signaling or $K_{ATP}$ channels activation; (b) in SUR1 null mice, and (c) in rats with hepatic branch vagotomy establish a physiological role for this insulin-mediated brain-liver circuit in the modulation of hepatic glucose homeostasis. In fact, the neuronal circuit engaged in response to the activation of hypothalamic insulin signaling plays an important role in restraining hepatic gluconeogenesis in response to physiological increases in the circulating insulin levels. Since increased gluconeogenesis is a main cause of fasting and postprandial hyperglycemia in DM2,[64] improving hypothalamic insulin signaling is likely to become a novel therapeutic strategy for this disease.

## Central Effects of Leptin on Insulin Sensitivity

Leptin, the product of the ob gene, regulates food intake and energy expenditure by modulation of hypothalamic neuropeptinergic pathways.[32] These central pathways can exert "orectic" (e.g., promote food intake and energy efficiency) or "anorectic" (e.g., reduce food intake and energy efficiency) effects, responding to the signals of negative and positive energy balance. Orectic pathways include neuropeptide Y (NPY), melanocyte-concentrating hormone (MCH), and orexins. Anorectic pathways include melanocortins (e.g., α-melanocyte stimulating hormone, α-MSH) and corticotrophin releasing hormone (CRH). Two populations of neurons in the arcuate (ARC) nucleus of the hypothalamus are highly responsive to leptin (Fig. 4). One of these populations responds to leptin by increasing the expression of proopiomelanocortin (POMC), the precursor molecule for α-MSH. The other population of neurons responds to leptin by markedly decreasing the expression of NPY and the agouti-related protein (Agrp). The latter is a natural antagonist of the melanocortin pathway acting on the MC4 (and MC3) receptors.[65,66] Mice over-expressing Agrp develop an obesity syndrome likely mediated via antagonism of the melanocortin pathway.[67] α-MSH is the natural ligand for the central melanocortin receptors (MC3 and MC4). The MC4 receptor is expressed in the appetite centers of the hypothalamus and has been convincingly implicated in the regulation of energy homeostasis.[68] In particular, genetic knockout of the MC4 receptor gene and ICV administration of agonists and antagonists result in dramatic effects on feeding behavior and energy balance.[69-71] Since obesity is tightly associated with insulin resistance, it has been postulated that the leptin pathway may also play a major role in carbohydrate metabolism and insulin action.[35,72-74] Indeed, the marked improvement in glucose tolerance following leptin administration to obese animal models implicates leptin as a major regulator of intermediate metabolism. For example, rodents with a genetic deficiency of leptin function, such as the ob/ob and db/db mice, and the Zucker fa/fa rats, are markedly resistant to insulin action and develop diabetes mellitus later in life. Prolonged leptin administration in ob/ob mice markedly decreases both plasma insulin and glucose concentration.[75,76] Since the decline in plasma glucose and insulin is greater in leptin-treated mice than in pair-fed control mice, leptin may directly improve in vivo insulin action. Studies performed with experimental models of congenital generalized lipodistrophy have demonstrated that the systemic administration of leptin reverses insulin resistance and diabetes mellitus without correcting the defect in adipocyte differentiation.[77]

Leptin regulates food intake and body adiposity partly via activation of melanocortin receptors in the hypothalamus and in other areas within the central nervous system.[56] Indeed, bi-directional modulation of the activity of the central melanocortin pathway leads to significant changes in hepatic and peripheral insulin action.[71] On the other hand, the prolonged administration of either leptin or melanocortin agonists or antagonists also impacts on the distribution of body adiposity and on lipid homeostasis.[75,76] These effects are in turn likely to secondarily influence the in vivo action of insulin on metabolic parameters, since it is well established that changes in body weight, fat mass and/or fat distribution similar to those associated with long-term treatment with either leptin or melanocortin agonists can per se alter insulin action. Therefore, it is important to delineate the acute effects of leptin and melanocortins on metabolic fluxes prior to the onset of changes in energy balance, body composition, and lipid storage.

In this regard, acute administration of leptin to postabsorptive rats causes a marked redistribution of intra-hepatic glucose fluxes, with a marked increase in the relative contribution of gluconeogenesis and a parallel decrease in the contribution of glycogenolysis to hepatic glucose fluxes.[73,78] This effect is seen following either systemic or central administration of the hormone. Which are the mechanism(s) by which leptin stimulates gluconeogenesis and inhibits glycogenolysis? Leptin appears to exert its pleiotropic behavioral, metabolic, and neuro-endocrine actions via multiple and partly divergent central pathways. For example, it has been suggested that leptin controls the hypothalamic melanocortin pathway and energy balance via STAT3-dependent signaling and the NPY pathway, reproductive function, and glucose homeostasis via STAT3-independent signaling.[79] Furthermore, leptin and insulin receptors appear to share mechanisms of signal transduction such as the activation of phosphatidylinositol-3-kinase (PI3K).[80] These observations are intriguing since these two hormones have partly overlapping and partly divergent physiological actions. Of particular interest, the acute ICV administration of insulin markedly inhibits endogenous glucose production (GP) via a PI3K-dependent mechanism, while the acute ICV administration of leptin stimulates gluconeogenesis and liver PEPCK expression.[33,35,78]

Leptin was also shown to regulate glucose tolerance, insulin signaling/action, and lipid metabolism independently of its anorectic effects.[78,81-85] These early observations generated

Table 2. *Summary of the acute effects of leptin administration during hyperinsulinemic-euglycemic clamp studies on glucose fluxes. The melanocortin-dependent effects of leptin on glucose fluxes are abolished by ICV administration of a melanocorting antagonist (SHU9119).*

| Acute Effects of Leptin: | Melanocortin Dependent | Melanocortin Independent |
|---|---|---|
| Glucose disposal | – | – |
| Glucose Production | – | – |
| Glycogenolysis | – | – |
| Gluconeogenesis | – | – |
| PEPCK expression | – | – |
| G6Pase expression | – | – |

great interest in the investigation of the mechanisms by which leptin interferes with metabolic processes. In this regard, leptin activates central melanocortin receptors partly via increased biosynthesis of the physiological ligand α-MSH and via decreased biosynthesis of an antagonist agouti-related protein (Agrp) at the level of the hypothalamus.[65] The activation of the central melanocortin pathway appears to be a required step in leptin action on food intake, energy expenditure, sympathetic nervous system, insulin secretion, and body fat distribution.[72,82,86-90] Conversely, there is also recent evidence for effects of leptin on energy balance and intermediate metabolism that may be independent of the activation of central melanocortin receptors.[35] Specifically, leptin stimulates gluconeogenesis and the hepatic expression of PEPCK and Glc-6-Pase via central activation of melanocortin receptors. On the other hand, the systemic or central administration of leptin also restrains hepatic glycogenolysis via a central melanocortin-independent mechanism.

The acute and central activation of melanocortin receptors stimulates the expression of gluconeogenic enzymes within the liver, markedly increases the rate of gluconeogenesis, and diminishes the suppressive effect of insulin on GP (Table 2). These rapid metabolic consequences of the central modulation of the melanocortin pathway stand in sharp contrast with the insulin sensitizing effects of more prolonged manipulations of the central melanocortin receptors.[71] In fact, week-long activation of this central pathway leads to decreased visceral adiposity and improved insulin action.[71] Furthermore, genetic animal models in which signaling via the central melanocortin receptors is defective, display hyperphagia, adult onset diabetes or glucose intolerance, and hyperinsulinemia.[70] The contrast between acute and chronic effects of central melanocortin modulation are likely due to the dramatic effects of this pathway on body fat mass and distribution, lipid oxidation and storage, and sympathetic nervous system activity. Acutely, the activation of the melanocortin pathway in the CNS is likely to enhance autonomic outflow to peripheral organs in the absence of changes in visceral adiposity and lipid storage.[70,71] In the liver, an increase in adrenergic tone leads to increased expression of Glc-6-Pase and PEPCK and to increased fat oxidation, which in turn can drive up gluconeogenesis.

The effects of leptin on hepatic glucose fluxes appear to be more complex than those of α-MSH. The systemic infusion of leptin during insulin clamp studies similarly increases the hepatic expression of PEPCK and Glc-6-Pase and the rate of gluconeogenesis, but it also markedly suppresses the rate of glycogenolysis so that GP is not increased (Table 2). When the stimulatory effect of systemic leptin on central melanocortin receptors are prevented by the coadministration of a melanocortin antagonist, equal increases in circulating leptin levels fail to alter gluconeogenesis and liver PEPCK/Glc-6-Pase mRNA levels. However, in the presence of central melanocortin blockade, increasing plasma leptin concentration markedly inhibits the rate of glycogenolysis and consequently enhances the inhibition of GP by insulin.[35] These experiments suggest that the effects of a physiological increase in systemic leptin on the hepatic expression of gluconeogenic enzymes as well as on gluconeogenesis are centrally mediated since they are abolished by central antagonism of the melanocortin receptors. Conversely, the insulin-like effects of systemic leptin on hepatic glucose fluxes could be mediated via direct effects on the liver as well as via extra-hepatic (presumably hypothalamic) actions of leptin. In this regard, leptin has also been shown to directly modulate insulin signaling, lipid metabolism, and glycogenolysis in some studies conducted in isolated liver cells[91,92] or perfused liver.[93] However, central leptin administration is sufficient to markedly suppress GP in the presence of central melanocortin blockade. Thus, leptin can inhibit hepatic glycogenolysis and GP via a melanocortin-independent central pathway. This is consistent with the observation that neuron-specific deletion of leptin receptor expression largely recapitulates the behavioral and metabolic defects induced by whole-body loss-of-function in this receptor.[94]

While the CNS pathways and efferent signals mediating the melanocortin-independent effects of leptin on liver glycogenolysis are unknown, it is tempting to speculate that the activation of an 'insulin-like' signaling pathway within the hypothalamus plays a key role. Leptin and insulin can both increase the activity of PI3K in the hypothalamus and this signaling pathway is involved in their anorectic effects.[80,95] Importantly, central activation of insulin signaling markedly suppresses GP and the effect of a physiological increase in circulating insulin on GP requires the central activation of PI3K. The metabolic effects of leptin via a melanocortin-independent signaling pathway closely resemble the potent inhibitory effect of central administration of insulin on GP. These results indicate that leptin can modulate hepatic insulin action via melanocortin-dependent as well as melanocortin-independent effects. In particular, leptin stimulates hepatic gluconeogenesis and PEPCK/Glc-6-Pase expression via central activation of melanocortin receptors, but it also markedly suppresses hepatic glycogenolysis via a central melanocortin-independent mechanism (Table 2). Thus selective signaling (or selective resistance) via either of these two central circuits mediating the rapid actions of leptin on the liver can lead to dramatic changes in hepatic insulin action and glucose metabolism. In this regard, it will be important to discern whether common forms of leptin resistance, such as diet-induced obesity, symmetrically impact on the melanocortin-dependent and melanocortin-independent effects of leptin on liver glucose metabolism.

## Hypothalamic Lipid Sensing

According to the "lipostatic model" for the regulation of energy balance, peripheral signals proportional to the size of the energy stores communicate the energy status to brain centers involved in the regulation of food intake and fuel metabolism.[32,96] Leptin and insulin are ideal candidates for this lipostatic function since their circulating levels are proportional to adiposity and their central administration decreases food intake.[76,96-98] Circulating macronutrients also acutely increase leptin and insulin plasma levels, which in turn activate hypothalamic responses to nutrient abundance.[32] In addition to their anorectic effects, central delivery of insulin or leptin can induce rapid shifts in the metabolic fluxes of peripheral tissues such as liver and skeletal muscle.[73,85] As outlined above, the effect of circulating insulin on the suppression of glucose production is partly due to the activation of hypothalamic insulin receptors located in the arcuate nucleus. Based on these observations a homeostatic model of negative feedback between circulating nutrients and hypothalamic responses has been proposed. According to this model, increased levels of plasma glucose and lipids stimulate insulin and leptin[99] biosynthesis and secretion, which in turn, regulate hypothalamic efferent pathways. The activation of these central responses leads to inhibition of food intake and hepatic production of nutrients. Furthermore, recent findings suggest that hypothalamic neurons are also capable of directly sensing the levels of circulating nutrients.[100] For example, the central administration of the long-chain fatty acid oleic acid inhibits food intake and glucose production. The central effect of oleic acid is not reproduced by fatty acids with medium chain, suggesting that long-chain fatty acids are a specific signal of nutrient abundance, which is not simply mediated by increased production of ATP via fatty acid oxidation. Fatty acids are transported from the circulation

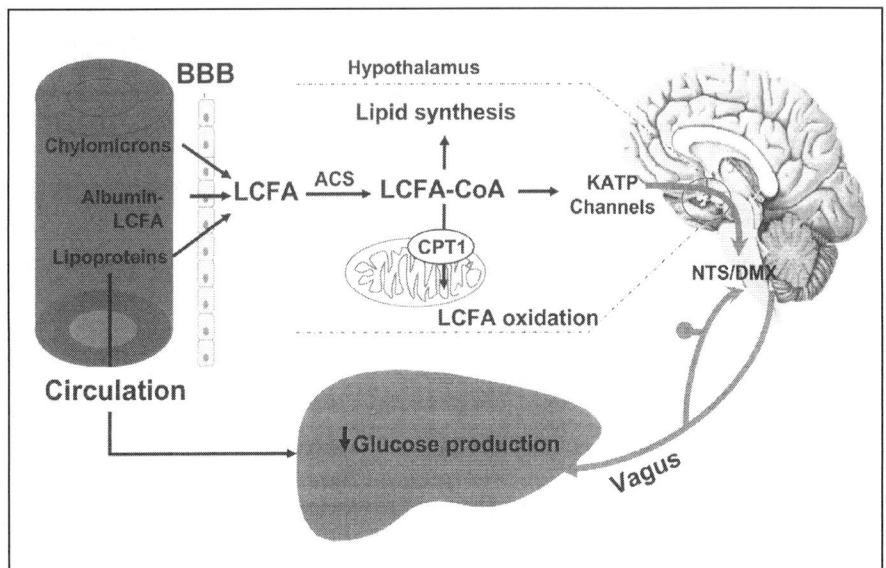

Figure 5. Scheme of the metabolic and neuronal steps involved in hypothalamic lipid sensing and suppression of hepatic glucose production. Circulating lipids may affect hepatic glucose production either directly or via activation of hypothalamic lipid-sensing pathways. The former mechanism increases hepatic glucose production via the stimulation of glucose flux through the enzyme glucose-6-phosphatase. On the other hand, circulating LCFAs levels are continuously sensed in the hypothalamus and their increase activates a neural efferent input that involves the activation of hypothalamic KATP channels and the stimulation of vagal motor neurons innervating the liver. Such descendent pathway counteracts the peripheral effect of LCFAs and restrains hepatic glucose production. Intracellular levels of LCFA-CoAs are an essential signaling component of the hypothalamic lipid sensing since inhibition of the enzyme that forms LCFA-CoAs from LCFAs (acyl-CoA synthetase) abolishes the central effects of LCFAs and inhibition of the enzyme that initiates the oxidization of LCFA-CoAs (carnitine palmitoyl-CoA transferase, CPT1) reproduces the central effects of LCFAs delivery.

to the brain (Fig. 5), converted into LCFA-CoAs, and further metabolized in oxidative (by β-oxidation in mitochondria) or biosynthetic pathways (incorporation in phospholipids). Although the brain largely derives its ATP from the oxidation of glucose, lipid oxidation occurs as well.[101] Studies performed with radiotracers techniques have shown that, although up to 50% of fatty acids delivered to the whole brain are oxidized to acetate, the bulk of palmitate and oleate incorporated into brain lipids is derived from circulating FAs and not from newly synthesized LCFA-CoA. The potent anorectic effects of inhibitors of fatty acid synthase (FAS) have further focused attention on the role of the neuronal lipid metabolism.[102] The effect of FAS inhibitors on food intake requires the accumulation of malonyl-CoA, a potent inhibitor of carnitine palmitoyltransferase-1 (CPT1). This enzyme is necessary for the transport of LCFA-CoA in mitochondria and is the rate-limiting step in the oxidative metabolism of LCFA-CoA (Fig. 5). In peripheral tissues (liver and muscle), malonyl-CoA has been identified as a fuel sensor that regulates the rate of fatty acid oxidation and, consequently determines the intracellular levels of LCFA-CoAs. Recent evidence supports the notion that the accumulation of LCFA-CoAs in hypothalamic neurons represents a signal of nutrient abundance.[103] Inhibition of hypothalamic CPT1 with specific drug inhibitors or antisense technique increases neuronal levels of LCFA-CoA's. This elevation is sufficient to suppress food intake and endogenous glucose production. In physiological conditions, the inhibition of CPT1 activity may occur when levels of malonyl-CoA increase due to an increased flux of glucose-derived carbons.

Therefore, increased availability of LCFAs and carbohydrates may activate a central "lipid-sensing" signal of negative feedback designed to further restrain the entry of nutrients in the circulation.

Can this hypothalamic lipid sensing mechanism detect physiological changes in circulating LCFAs? An elevation in circulating LCFA in the presence of hyperinsulinemia, markedly diminishes the action of insulin on glucose production. However, in the presence of postabsorptive insulin concentrations, the elevation of plasma LCFA concentration via lipid infusions stimulates gluconeogenesis, without altering glucose production in nondiabetic humans and dogs because of a compensatory decrease in hepatic glycogenolysis. This rapid metabolic adaptation has been referred to as hepatic autoregulation. In type 2 diabetes, hepatic autoregulation appears to be impaired since reciprocal changes in glycogenolysis fail to compensate for changes in gluconeogenesis when the plasma LCFA concentrations are experimentally manipulated.[104] The mechanism by which an increase in plasma LCFA levels restrains hepatic glycogenolysis has been recently investigated.[105] Similarly to insulin, the modulatory effects of circulating LCFAs on hepatic glucose production include direct and indirect effects. Indeed, LCFAs can regulate liver glucose homeostasis via their metabolic signaling within the hypothalamus. For example, central administration of oleic acid results in changes in hepatic metabolic fluxes that are virtually opposite to those elicited by direct action of LCFA on the liver.[106,107] Therefore, LCFAs modulate hepatic glucose metabolism via distinct biochemical signaling both in liver as well as in brain. The central action of circulating LCFA is required to counteract LCFA-induced stimulation of gluconeogenesis and to prevent an increase in glucose production, thereby providing an extra-hepatic site for hepatic autoregulation (Fig. 5).

## Conclusion

As originally proposed by Claude Bernard the central nervous system plays an important role in the regulation of insulin action and intermediate metabolism. A fundamental site for this regulation is the hypothalamus where multiple nutritional and endocrine signals are integrated in order to gather information on the body's nutritional status. This information in turn activates complex neural pathways that modulate not only energy intake and expenditure but also intermediate metabolism and insulin action. While recent progress on the biochemical and physiological characterization of some of these neural circuits has shed light on the mechanisms by which hormones such as insulin and leptin and nutrients such as fatty acids regulate glucose homeostasis, it is likely that additional regulatory systems will soon be delineated that couple the sensing of nutrients within hypothalamic energy centers to the regulation of basic metabolic processes. Considering the dominant role that nutrient excess and obesity play in the current worldwide epidemic of type 2 diabetes mellitus, it is particularly important to rapidly translate these novel insights into useful clinical applications.

## References

1. Sindelar DK, Chu CA, Neal DW et al. Interaction of equal increments in arterial and portal vein insulin on hepatic glucose production in the dog. Am J Physiol 1997; 273:E972-E980.
2. Sindelar DK, Chu CA, Venson P et al. Basal hepatic glucose production is regulated by the portal vein insulin concentration. Diabetes 1998; 47:523-529.
3. Boden G, Chen X, Capulong E. et al. Effects of free fatty acids on gluconeogenesis and autoregulation of glucose production in type 2 diabetes. Diabetes 2001; 50:810-816.
4. Lewis GF, Vranic M, Harley P et al. Fatty acids mediate the acute extrahepatic effects of insulin on hepatic glucose production in humans. Diabetes 1997; 46:1111-1119.
5. Rebrin K, Steil GM, Getty L et al. Free fatty acid as a link in the regulation of hepatic glucose output by peripheral insulin. Diabetes 1995; 44:1038-1045.
6. Sindelar DK, Chu CA, Rohlie M et al. The role of fatty acids in mediating the effects of peripheral insulin on hepatic glucose production in the conscious dog. Diabetes 1997; 46:187-196.
7. Obici S, Zhang BB, Karkanias G et al. Hypothalamic insulin signaling is required for inhibition of glucose production. Nat Med 2002; 8:1376-1382.
8. Pocai A, Lam TK, Gutierrez-Juarez R et al. Hypothalamic KATP channels control hepatic glucose production. Nature (In press).

9. Hornbuckle LA, Edgerton DS, Ayala JE et al. Selective tonic inhibition of G-6-Pase catalytic subunit, but not G-6-P transporter, gene expression by insulin in vivo. Am J Physiol Endocrinol Metab 2001; 281:E713-E725.
10. Fisher SJ, Kahn CR. Insulin signaling is required for insulin's direct and indirect action on hepatic glucose production. J Clin Invest 2003; 111:463-468.
11. Michael MD, Kulkarni RN, Postic C et al. Loss of insulin signaling in hepatocytes leads to severe insulin resistance and progressive hepatic dysfunction. Mol Cell 2000; 6:87-97.
12. Saltiel AR, Kahn CR. Insulin signalling and the regulation of glucose and lipid metabolism. Nature 2001; 414:799-806.
13. Cherrington AD. Banting Lecture 1997. Control of glucose uptake and release by the liver in vivo. Diabetes 1999; 48:1198-1214.
14. Consoli A, Nurjhan N, Capani F et al. Predominant role of gluconeogenesis in increased hepatic glucose production in NIDDM. Diabetes 1989; 38:550-557.
15. Ferrannini E, Galvan AQ, Gastaldelli A et al. Insulin: New roles for an ancient hormone. Eur J Clin Invest 1999; 29:842-852.
16. Magnusson I, Rothman DL, Gerard DP et al. Contribution of hepatic glycogenolysis to glucose production in humans in response to a physiological increase in plasma glucagon concentration. Diabetes 1995; 44:185-189.
17. Rothman DL, Magnusson I, Katz LD et al. Quantitation of hepatic glycogenolysis and gluconeogenesis in fasting humans with 13C NMR. Science 1991; 254:573-576.
18. Moller DE. New drug targets for type 2 diabetes and the metabolic syndrome. Nature 2001; 414:821-827.
19. Kahn BB, Flier JS. Obesity and insulin resistance. J Clin Invest 2000; 106:473-481.
20. Kopelman PG. Obesity as a medical problem. Nature 2000; 404:635-643.
21. Friedman JM. Obesity in the new millennium. Nature 2000; 404:632-634.
22. Kopelman PG, Hitman GA. Diabetes. Exploding type II. Lancet 1998; 352(Suppl 4):SIV5.
23. Porte D, Seeley RJ, Woods SC et al. Obesity, diabetes and the central nervous system. Diabetologia 1998; 41:863-881.
24. Obici S, Rossetti L. Minireview: Nutrient sensing and the regulation of insulin action and energy balance. Endocrinology 2003; 144:5172-5178.
25. Flier JS. Diabetes. The missing link with obesity? Nature 2001; 409:292-293.
26. Flier JS. Obesity wars: Molecular progress confronts an expanding epidemic. Cell 2004; 116:337-350.
27. Flegal KM, Carroll MD, Ogden CL et al. Prevalence and trends in obesity among US adults, 1999-2000. JAMA 2002; 288:1723-1727.
28. Combs TP, Berg AH, Obici S et al. Endogenous glucose production is inhibited by the adipose-derived protein Acrp30. J Clin Invest 2001; 108:1875-1881.
29. Rajala MW, Obici S, Scherer PE et al. Adipose-derived resistin and gut-derived resistin-like molecule-beta selectively impair insulin action on glucose production. J Clin Invest 2003; 111:225-230.
30. Rangwala SM, Rich AS, Rhoades B et al. Abnormal glucose homeostasis due to chronic hyperresistinemia. Diabetes 2004; 53(8):1937-41.
31. Porte D, Baskin DG, Schwartz MW. Leptin and insulin action in the central nervous system. Nutr Rev 2002; 60:S20-S29.
32. Schwartz MW, Woods SC, Porte D et al. Central nervous system control of food intake. Nature 2000; 404:661-671.
33. Woods SC, Schwartz MW, Baskin DG et al. Food intake and the regulation of body weight. Annu Rev Psychol 2000; 51:255-277.
34. Ahima RS, Prabakaran D, Mantzoros C et al. Role of leptin in the neuroendocrine response to fasting. Nature 1996; 382:250-252.
35. Gutierrez-Juarez R, Obici S, Rossetti L. Melanocortin-independent effects of leptin on hepatic glucose fluxes. J Biol Chem 2004; 279:49704-49715.
36. Schwartz MW, Sipols A, Kahn SE et al. Kinetics and specificity of insulin uptake from plasma into cerebrospinal fluid. Am J Physiol 1990; 259:E378-E383.
37. Gerozissis K, Rouch C, Nicolaidis S et al. Brain insulin response to feeding in the rat is both macronutrient and area specific. Physiol Behav 1998; 65:271-275.
38. Gerozissis K, Orosco M, Rouch C et al. Insulin responses to a fat meal in hypothalamic microdialysates and plasma. Physiol Behav 1997; 62:767-772.
39. Unger JW, Betz M. Insulin receptors and signal transduction proteins in the hypothalamo-hypophyseal system: A review on morphological findings and functional implications. Histol Histopathol 1998; 13:1215-1224.
40. Marks JL, Porte D, Stahl WL et al. Localization of insulin receptor mRNA in rat brain by in situ hybridization. Endocrinology 1990; 127:3234-3236.

41. Baskin DG, Sipols AJ, Schwartz MW et al. Immunocytochemical detection of insulin receptor substrate-1 (IRS-1) in rat brain: Colocalization with phosphotyrosine. Regul Pept 1993; 48:257-266.
42. Woods SC, Seeley RJ, Porte D et al. Signals that regulate food intake and energy homeostasis. Science 1998; 280:1378-1383.
43. Woods SC, Lotter EC, McKay LD et al. Chronic intracerebroventricular infusion of insulin reduces food intake and body weight of baboons. Nature 1979; 282:503-505.
44. Davis SN, Dunham B, Walmsley K et al. Brain of the conscious dog is sensitive to physiological changes in circulating insulin. Am J Physiol 1997; 272:E567-E575.
45. Davis SN, Colburn C, Dobbins R et al. Evidence that the brain of the conscious dog is insulin sensitive. J Clin Invest 1995; 95:593-602.
46. Richardson RD, Ramsay DS, Lernmark A et al. Weight loss in rats following intraventricular transplants of pancreatic islets. Am J Physiol 1994; 266:R59-R64.
47. Sipols AJ, Baskin DG, Schwartz MW. Effect of intracerebroventricular insulin infusion on diabetic hyperphagia and hypothalamic neuropeptide gene expression. Diabetes 1995; 44:147-151.
48. Schwartz MW, Marks JL, Sipols AJ et al. Central insulin administration reduces neuropeptide Y mRNA expression in the arcuate nucleus of food-deprived lean (Fa/Fa) but not obese (fa/fa) Zucker rats. Endocrinology 1991; 128:2645-2647.
49. Sahu A, Dube MG, Phelps CP et al. Insulin and insulin-like growth factor II suppress neuropeptide Y release from the nerve terminals in the paraventricular nucleus: A putative hypothalamic site for energy homeostasis. Endocrinology 1995; 136:5718-5724.
50. Liang C, Doherty JU, Faillace R et al. Insulin infusion in conscious dogs. Effects on systemic and coronary hemodynamics, regional blood flows, and plasma catecholamines. J Clin Invest 1982; 69:1321-1336.
51. Rowe JW, Young JB, Minaker KL et al. Effect of insulin and glucose infusions on sympathetic nervous system activity in normal man. Diabetes 1981; 30:219-225.
52. Air EL, Strowski MZ, Benoit SC et al. Small molecule insulin mimetics reduce food intake and body weight and prevent development of obesity. Nat Med 2002; 8:179-183.
53. McGowan MK, Andrews KM, Grossman SP. Chronic intrahypothalamic infusions of insulin or insulin antibodies alter body weight and food intake in the rat. Physiol Behav 1992; 51:753-766.
54. Obici S, Feng Z, Karkanias G et al. Decreasing hypothalamic insulin receptors causes hyperphagia and insulin resistance in rats. Nat Neurosci 2002; 5:566-572.
55. Bruning JC, Gautam D, Burks DJ et al. Role of brain insulin receptor in control of body weight and reproduction. Science 2000; 289:2122-2125.
56. Grill HJ, Schwartz MW, Kaplan JM et al. Evidence that the caudal brainstem is a target for the inhibitory effect of leptin on food intake. Endocrinology 2002; 143:239-246.
57. Spanswick D, Smith MA, Groppi VE et al. Leptin inhibits hypothalamic neurons by activation of ATP-sensitive potassium channels. Nature 1997; 390:521-525.
58. Spanswick D, Smith MA, Mirshamsi S et al. Insulin activates ATP-sensitive K+ channels in hypothalamic neurons of lean, but not obese rats. Nat Neurosci 2000; 3:757-758.
59. Wolkow CA, Kimura KD, Lee MS et al. Regulation of C. elegans life-span by insulinlike signaling in the nervous system. Science 2000; 290:147-150.
60. Apfeld J, Kenyon C. Cell nonautonomy of C. elegans daf-2 function in the regulation of diapause and life span. Cell 1998; 95:199-210.
61. Matsuhisa M, Yamasaki Y, Shiba Y et al. Important role of the hepatic vagus nerve in glucose uptake and production by the liver. Metabolism 2000; 49:11-16.
62. Aguilar-Bryan L, Bryan J. Molecular biology of adenosine triphosphate-sensitive potassium channels. Endocr Rev 1999; 20:101-135.
63. Seino S, Miki T. Physiological and pathophysiological roles of ATP-sensitive K+ channels. Prog Biophys Mol Biol 2003; 81:133-176.
64. Magnuson MA. Tissue-specific regulation of glucokinase gene expression. J Cell Biochem 1992; 48:115-121.
65. Cowley MA, Smart JL, Rubinstein M et al. Leptin activates anorexigenic POMC neurons through a neural network in the arcuate nucleus. Nature 2001; 411:480-484.
66. Cowley MA, Pronchuk N, Fan W et al. Integration of NPY, AGRP, and melanocortin signals in the hypothalamic paraventricular nucleus: Evidence of a cellular basis for the adipostat. Neuron 1999; 24:155-163.
67. Klebig ML, Wilkinson JE, Geisler JG et al. Ectopic expression of the agouti gene in transgenic mice causes obesity, features of type II diabetes, and yellow fur. Proc Natl Acad Sci USA 1995; 92:4728-4732.
68. Butler AA, Cone RD. Knockout studies defining different roles for melanocortin receptors in energy homeostasis. Ann NY Acad Sci 2003; 994:240-245.

69. Butler AA, Kesterson RA, Khong K et al. A unique metabolic syndrome causes obesity in the melanocortin-3 receptor-deficient mouse. Endocrinology 2000; 141:3518-3521.
70. Huszar D, Lynch CA, Fairchild-Huntress V et al. Targeted disruption of the melanocortin-4 receptor results in obesity in mice. Cell 1997; 88:131-141.
71. Obici S, Feng Z, Tan J et al. Central melanocortin receptors regulate insulin action. J Clin Invest 2001; 108:1079-1085.
72. Barzilai N, She L, Liu L et al. Decreased visceral adiposity accounts for leptin effect on hepatic but not peripheral insulin action. Am J Physiol 1999; 277:E291-E298.
73. Liu L, Karkanias GB, Morales JC et al. Intracerebroventricular leptin regulates hepatic but not peripheral glucose fluxes. J Biol Chem 1998; 273:31160-31167.
74. Fan W, Dinulescu DM, Butler AA et al. The central melanocortin system can directly regulate serum insulin levels. Endocrinology 2000; 141:3072-3079.
75. Halaas JL, Gajiwala KS, Maffei M et al. Weight-reducing effects of the plasma protein encoded by the obese gene. Science 1995; 269:543-546.
76. Halaas JL, Boozer C, Blair-West J et al. Physiological response to long-term peripheral and central leptin infusion in lean and obese mice. Proc Natl Acad Sci USA 1997; 94:8878-8883.
77. Shimomura I, Hammer RE, Ikemoto S et al. Leptin reverses insulin resistance and diabetes mellitus in mice with congenital lipodystrophy. Nature 1999; 401:73-76.
78. Rossetti L, Massillon D, Barzilai N et al. Short term effects of leptin on hepatic gluconeogenesis and in vivo insulin action. J Biol Chem 1997; 272:27758-27763.
79. Bates SH, Stearns WH, Dundon TA et al. STAT3 signalling is required for leptin regulation of energy balance but not reproduction. Nature 2003; 421:856-859.
80. Niswender KD, Morrison CD, Clegg DJ et al. Insulin activation of phosphatidylinositol 3-kinase in the hypothalamic arcuate nucleus: A key mediator of insulin-induced anorexia. Diabetes 2003; 52:227-231.
81. Schwartz MW, Baskin DG, Bukowski TR et al. Specificity of leptin action on elevated blood glucose levels and hypothalamic neuropeptide Y gene expression in ob/ob mice. Diabetes 1996; 45:531-535.
82. Barzilai N, Wang J, Massilon D et al. Leptin selectively decreases visceral adiposity and enhances insulin action. J Clin Invest 1997; 100:3105-3110.
83. Muoio DM, Dohm GL, Fiedorek FT et al. Leptin directly alters lipid partitioning in skeletal muscle. Diabetes 1997; 46:1360-1363.
84. Muoio DM, Dohm GL, Tapscott EB et al. Leptin opposes insulin's effects on fatty acid partitioning in muscles isolated from obese ob/ob mice. Am J Physiol 1999; 276:E913-E921.
85. Minokoshi Y, Kim YB, Peroni OD et al. Leptin stimulates fatty-acid oxidation by activating AMP-activated protein kinase. Nature 2002; 415:339-343.
86. Balthasar N, Coppari R, McMinn J et al. Leptin receptor signaling in POMC neurons is required for normal body weight homeostasis. Neuron 2004; 42:983-991.
87. da Silva AA, Kuo JJ, Hall JE. Role of hypothalamic melanocortin 3/4-receptors in mediating chronic cardiovascular, renal, and metabolic actions of leptin. Hypertension 2004; 43:1312-1317.
88. Haynes WG, Morgan DA, Djalali A et al. Interactions between the melanocortin system and leptin in control of sympathetic nerve traffic. Hypertension 1999; 33:542-547.
89. Rahmouni K, Haynes WG, Morgan DA et al. Role of melanocortin-4 receptors in mediating renal sympathoactivation to leptin and insulin. J Neurosci 2003; 23:5998-6004.
90. Muzumdar R, Ma X, Yang X et al. Physiologic effect of leptin on insulin secretion is mediated mainly through central mechanisms. FASEB 2003; 17:1130-1132.
91. Zhao AZ, Shinohara MM, Huang D et al. Leptin induces insulin-like signaling that antagonizes cAMP elevation by glucagon in hepatocytes. J Biol Chem 2000; 275:11348-11354.
92. Lam NT, Lewis JT, Cheung AT et al. Leptin increases hepatic insulin sensitivity and protein tyrosine phosphatase 1B expression. Mol Endocrinol 2004; 18:1333-1345.
93. Huang W, Dedousis N, Bhatt BA et al. Impaired activation of phosphatidylinositol 3-kinase by leptin is a novel mechanism of hepatic leptin resistance in diet-induced obesity. J Biol Chem 2004; 279:21695-21700.
94. Cohen P, Zhao C, Cai X et al. Selective deletion of leptin receptor in neurons leads to obesity. J Clin Invest 2001; 108:1113-1121.
95. Niswender KD, Morton GJ, Stearns WH et al. Intracellular signalling. Key enzyme in leptin-induced anorexia. Nature 2001; 413:794-795.
96. Baskin DG, Figlewicz LD, Seeley RJ et al. Insulin and leptin: Dual adiposity signals to the brain for the regulation of food intake and body weight. Brain Res 1999; 848:114-123.
97. Schwartz MW, Figlewicz DP, Baskin DG et al. Insulin in the brain: A hormonal regulator of energy balance. Endocr Rev 1992; 13:387-414.

98. Seeley RJ, van Dijk G, Campfield LA et al. Intraventricular leptin reduces food intake and body weight of lean rats but not obese Zucker rats. Horm Metab Res 1996; 28:664-668.
99. Wang J, Liu R, Hawkins M et al. A nutrient-sensing pathway regulates leptin gene expression in muscle and fat. Nature 1998; 393:684-688.
100. Levin BE, Dunn-Meynell AA, Routh VH. Brain glucose sensing and body energy homeostasis: Role in obesity and diabetes. Am J Physiol 1999; 276:R1223-R1231.
101. Miller JC, Gnaedinger JM, Rapaport SI. Utilization of plasma fatty acids in rat brain: Distribution of 14C-Palmitate between oxidative and synthetic pathways. J Neurochem 1987; 49:1507-1514.
102. Loftus TM, Jaworsky DE, Frehywot GL et al. Reduced food intake and body weight in mice treated with fatty acid synthase inhibitors. Science 2000; 88:2379-2381.
103. Obici S, Feng Z, Arduini A et al. Inhibition of hypothalamic carnitine palmitoyltransferase-1 decreases food intake and glucose production. Nat Med 2003; 9:756-761.
104. Boden G, Chen X, Ruiz J et al. Mechanisms of fatty acid-induced inhibition of glucose uptake. J Clin Invest 1994; 93:2438-2446.
105. Lam TK, Pocai A, Gutierrez-Juarez R et al. Hypothalamic sensing of circulating fatty acids is required for glucose homeostasis. Nat Med 2005; 11:320-327.
106. Obici S, Feng Z, Morgan K et al. Central administration of oleic acid inhibits glucose production and food intake. Diabetes 2002; 51:271-275.
107. Morgan K, Obici S, Rossetti L. Hypothalamic responses to long-chain fatty acids are nutritionally regulated. J Biol Chem 2004; 279:31139-31148.

CHAPTER 9

# Transgenic Models of Impaired Insulin Signaling

Francesco Oriente and Domenico Accili*

## Abstract

Insulin resistance plays a key role in the pathogenesis of several human diseases, including diabetes, obesity, hypertension and cardiovascular diseases. The predisposition to insulin resistance results from genetic and environmental factors. The search for gene variants that predispose to insulin resistance has been thwarted by its genetically heterogeneous pathogenesis. However, using techniques of targeted mutagenesis and transgenesis in rodents, investigators have developed mouse models to test critical hypotheses on the pathogenesis of insulin resistance. Moreover, experimental crosses among mutant mice have shed light onto the polygenic nature of the interactions underlying this complex metabolic condition. This review focuses on targeted mutations affecting the function of genes in the insulin signaling pathway in mice.

## Introduction

In recent years, transgenic mouse models have played an increasingly large part in studies of insulin signalling.[1] The predisposition to insulin resistance and/or β-cell dysfunction is inherited in a nonMendelian fashion.[2] Thus, human genetic studies of the predisposition to insulin resistance are fraught with uncertainties and complicated by the variations of the human gene pool. To dissect the complex genetics of insulin resistance and β-cell dysfunction, investigators have generated transgenic and knockout mice bearing mutations in genes required for insulin action and/or insulin secretion (Table 1).

## Insulin Receptor Knockout

Mice bearing null *Insr* mutations are born with slight growth retardation (~10%).[3] After birth, there is a rapid rise of glucose levels, accompanied by a transient, robust increase in insulin levels up to 1,000-fold above normal. However, β-cell "failure" occurs within few days and is followed by death in diabetic ketoacidosis.[4,5] This phenotype indicates that *Insr* is necessary for postnatal fuel homeostasis, but not for prenatal growth and metabolic control. The *Insr* null mouse differs phenotypically from humans carrying similar mutations.[6] Unlike mice, humans lacking insulin receptors show severe intra-uterine growth-retardation, lipodystrophy and hypoglycemia. To study this apparent discrepancy, we have used mosaic analysis to introduce partial ablations of Insr function in mice. These experiments indicate that, when >80% of somatic cells lose Insr expression, mice develop a phenotype similar to human leprechaunism, with severe post-natal growth retardation and hypoglycemia.[7] The growth retardation appears

*Corresponding Author: Domenico Accili—Berrie Research Pavilion, 1150 St. Nicholas Ave., New York, New York 10032, U.S.A. Email: da230@columbia.edu

*Mechanisms of Insulin Action*, edited by Alan R. Saltiel and Jeffrey E. Pessin.
©2007 Landes Bioscience and Springer Science+Business Media.

*Table 1. Summary of mutant mice described in the text*

| Gene | Phenotype | References |
|---|---|---|
| *Irs1* | Growth retardation, insulin resistance | 60,61 |
| *Irs2* | β-cell dysfunction | 47 |
| *Irs3* | None apparent | 71 |
| *Irs4* | Growth retardation, insulin resistance | 72 |
| *Irs1+Irs2* | Embryonic lethality | 73,74 |
| *Irs1+Irs3* | Lipoatrophy | 75 |
| *Irs1+Irs4* | Growth retardation, insulin resistance | 75 |
| *Pi3kr (p85α)* | Hypoglycemia | 100 |
| *Pan-Pi3kr* | Hypoglycemia | 101 |
| *Glut4* | Cardiac hypertrophy, | 80 |
| *Glut4+/−* | Insulin resistance, diabetes | 81 |
| *Glut4 (heart-specific)* | Cardiac hypertrophy | 86 |
| *Glut4 (muscle-specific)* | Insulin-resistant diabetes | 83 |
| *Glut4 (adipose-specific)* | Insulin-resistant diabetes | 88 |
| *Syntaxin4* | Insulin resistance | 89 |
| *Ptg1* | Insulin resistance | 93 |
| *Ceacam* | Impaired insulin clearance | 96 |
| *Akt2* | Insulin resistance, diabetes | 78,79 |
| *Ptp1b* | Resistance to diet-induced insulin resistance | 97,98 |
| *Ship2* | Hypoglycemia | 103 |
| *Jnk2* | Increased insulin sensitivity | 109 |
| *Ikkβ* | Increased insulin sensitivity | 113 |
| *Foxo1* | Increased insulin sensitivity | 53,118,119 |

to be due to a >50-fold increase in expression of Igfbp1,[8-10] while the hypoglycemia is due to a profound depletion of hepatic glycogen stores. These data are consistent with two important conclusions: that insulin acting through Insr promotes growth, and that it does so by inhibiting hepatic expression of Igfbp1. Thus, the phenotype of human leprechaunism likely reflects residual Insr function, either because of hypomorphic alleles, or –in patients with complete knockouts- because of compensation through Igf1 receptors.

## Conditional *Insulin Receptor* Knockouts

The lethal phenotype of *Insr* knockouts precludes a detailed analysis of insulin receptor function in adult mice. Since an impairment of insulin action in one or more insulin target cells is a likely cause of diabetes, mouse models have been developed to circumvent this limitation (Table 2).

### Hepatocytes

The liver plays a central role in the control of glucose homeostasis. Insulin resistance in the liver is closely correlated with fasting hyperglycemia in type 2 diabetes. Insulin affects hepatic glucose suppression initially by inhibiting glycolysis. With prolonged fasting, however, insulin reverts to the alternate pathway of inhibiting gluconeogenesis. There is considerable debate as to whether this latter phenomenon is secondary to a direct effect of insulin on the liver or rather an indirect effect to decrease the supply of gluconeogenic precursors.[11] Liver-specific *Insr* knockout (LIRKO) mice exhibit elevations in blood glucose levels and glucose intolerance. In addition, they develop marked hyperinsulinemia due to a combination of increased β-cell mass and decreased insulin clearance, as well as failure of insulin to suppress hepatic

Table 2. Summary of tissue-specific insulin receptor knockouts

| Insr Knockout | Phenotype | Ref. |
|---|---|---|
| Constitutive | Diabetic ketoacidosis | 4,5 |
| Muscle/adipose tissue | Impaired glucose tolerance | 21 |
| Liver | Moderate insulin resistance, transient hyperglycemia | 12 |
| Adipocyte | Protection against obesity, longevity | 15,121 |
| Brown adipose tissue | β-cell failure | 17 |
| Muscle | Dyslipidemia | 20 |
| Cardiac muscle | Reduced heart size and performance | 24 |
| Central nervous system | Obesity, sub-fertility | 35 |
| β-cell | Impaired glucose tolerance | 40 |
| Vascular endothelium | Protection from hypoxia-induced neovascularization | 122 |
| Mosaicism | Growth retardation, lipoatrophy, hypoglycemia | 7 |

gluconeogenesis,[12,13] suggesting that hepatocyte insulin signaling is required for both direct and indirect control of hepatic glucose production.

## Adipose Tissue

Insulin has been implicated as an important player in key adipocyte functions, including differentiation, stimulation of glucose uptake and inhibition of lipolysis.[14] Mice with Insr knockout limited to white and brown fat tissues (FIRKO) demonstrate significant defects in insulin-mediated glucose uptake and in the ability of insulin to inhibit isoproterenol-stimulated lipolysis.[15] Although basal glucose uptake remains intact, there is evidence of increased basal lipolysis, associated with a polarization of adipocyte size into large and small cells. Leptin levels in these mice are unexpectedly high for the apparent fat mass. Interestingly, FIRKO mice are resistant to obesity and attendant glucose intolerance when injected with gold-thio-glucose, a hypothalamic toxin. A further interesting phenotype noted in the FIRKO mice is a 20% increase in mean life span, with parallel increases in median and maximum life spans. These data support the notion that a decreased fat mass can affect life span independently of caloric restriction.[15]

## Brown Adipocyte

Brown adipose tissue is thought to play an important role in determining peripheral insulin sensitivity,[16] as well as thermal adaptation. Insr has been inactivated selectively in this tissue (BATIRKO) using the uncoupling protein-1 (UCP1) promoter to drive Cre expression. Although brown adipose tissue develops normally in these mice, it hypotrophies gradually with a 50% reduction at three months, and ~75% by 6-12 months. The loss of brown adipocytes is associated with deterioration of β-cell function and decreased β-cell mass, giving rise to hyperglycemia.[17] This observation suggests that the maintenance of an adequate β-cell mass requires brown adipose tissue. It remains to be determined whether this is an endocrine effect of factors produced in brown adipose tissue, or whether it reflects a broader metabolic change. It is also unclear why this phenotype is not observed in FIRKO mice, which carry Insr mutations in both brown and white adipocytes.

## Skeletal Muscle

Resistance of skeletal muscle to insulin-dependent glucose uptake and phosphorylation is an early step in the development of type 2 diabetes.[18] To examine the role of insulin receptors in muscle metabolism, and how the latter affects the development of diabetes, Insr has been inactivated in this tissue (MIRKO). Conditional Insr knockout in skeletal muscle leads to

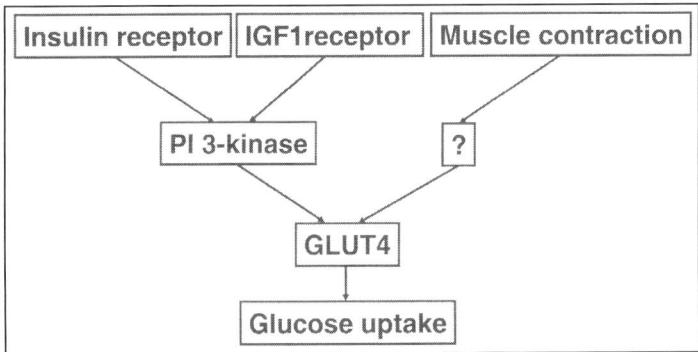

Figure 1. Divergent pathways of glucose transporter translocation. There appear to be at least three different pathways regulating GLUT4 translocation and leading to glucose uptake and utilization in muscle. In addition to activation of Insr, contraction is a powerful trigger to GLUT4 translocation through activation of the AMP-activated kinase. In addition, muscle Igf1r signals through Irs proteins and PI 3-kinase to stimulate GLUT4 translocation via activation of Akt and other inositol-tris-phosphate ($PIP_3$)-dependent kinases, such as PKC isoforms. These pathways explain why an isolated *Insr* knockout does not suffice to cause insulin resistance to glucose uptake, while a muscle-specific *Glut4* knockout causes severe insulin resistance and diabetes.

impaired insulin signaling, decreased insulin-stimulated glucose transport and glycogen synthesis and a concurrent increase in insulin stimulated glucose transport in adipose tissue,[19] without systemic insulin resistance.[20] This is similar to the phenotype of mice carrying a dominant negative *Insr* transgene in muscle and a heterozygous systemic Insr knockout.[21] MIRKO mice develop a metabolic syndrome with increased fat stores and hypertriglyceridemia. It appears that two alternative pathways of glucose uptake compensate for the lack of insulin signaling, including Igf1r[22] and contraction-activated signaling[23] (Fig. 1). Therefore, although MIRKO mice do not demonstrate significant insulin resistance or diabetes, the phenotypes of mice expressing dominant negative *Igf1r* and muscle-specific knockout of *Glut4* confirm the critical role of muscle tissue in glucose metabolism.

## Heart

Mice in which Insr knockout is restricted to the heart have been generated by way of the cardiac-specific α-myosin heavy chain promoter (CIRKO). Ablation of insulin signaling in cardiomyocytes resulted in smaller hearts, due to reduced cardiomyocyte size. Glucose uptake and glycolysis were increased when measured in perfused hearts, probably due to a compensatory increase in Glut4 expression. In contrast, the heart's ability to oxidize fatty acids was decreased, and cardiac performance was moderately impaired.[24,25] The CIRKO hearts are more prone to injury and failure when subjected to pressure overload,[26] indicating that insulin resistance may affect cardiac performance independently of changes in lipid and glucose levels.

## Vascular Endothelium

Vascular endothelium is a target of Insr signaling, and several functions have been ascribed to Insr in this cell type,[27] including stimulation of NO-mediated relaxation of smooth muscle cells and protection from apoptosis. However, it has also been shown to act as a potent stimulator of VEGF and endothelin expression, which would be expected to cause vasoconstriction and potentially lead to hypertension. Moreover, it has been suggested that Insr in vascular endothelium may contribute to delivery of insulin to target cells.[28,29] Finally, there is evidence that insulin is a vasodilator in vivo,[30,31] and that some of its effects on glucose uptake may be secondary to increased tissue blood flow.[32] Mice lacking Insr in vascular endothelium display slight reductions of arterial blood pressure under both low- and high-salt diets, and are mildly insulin-resistant. The failure of these mice to develop hypertension, either basally or in response to a high-salt diet, refutes the notion that the association between insulin resistance and hypertension can be explained by insulin's action on the vascular endothelium. Similarly, the lack of significant insulin resistance indicates that insulin transport across the endothelial barrier is not limiting for peripheral insulin action. Interestingly, when these mice are subjected to chronic hypoxia, they display a decrease in retinal neovascularization, the primary cause of diabetic retinopathy and blindness.[33]

## Central Nervous System

Insulin receptors are expressed at high levels in the brain.[34] The function of neuronal Insr has been studied by generating neuron-specific *Insr* knockout mice (NIRKO).[35] Loss of Insr in nestin-positive cells -which include both neurons and glia- leads to mild insulin resistance, obesity, hypertriglyceridemia and increased food intake (the latter limited to female mice). This suggests that, in addition to glucose metabolism, Insr is important in brain nutrient sensing and regulation of energy expenditure. A further, unexpected result in NIRKO mice is impaired male and female fertility. Initial assessment indicates hypogonadotrophic hypogonadism in relation to defective spermatogenesis and follicular maturation. It is notable that *Irs2* knockout mice, in addition to developing diabetes, also demonstrate abnormalities in food intake, increased obesity despite elevated leptin levels and infertility, particularly in the females.[36]

A similar phenotype results from ICV delivery of reagents that inhibit Insr signaling.[37,38] From these experiments, it is also becoming apparent that Insr in hypothalamic nuclei play an important role in insulin control of hepatic glucose production.

## Pancreatic β-Cell

Multiple defects in insulin secretion and β-cell mass are present in patients with type 2 diabetes and also during the insulin-resistant prediabetic stage.[39] To address the role of insulin signaling in these processes, *Insr* has been inactivated in mature β-cells using a minimal *Insulin2* promoter to drive CRE-dependent recombination (βIRKO). These mice demonstrate a selective impairment in the first phase of glucose-stimulated insulin secretion, a phenotype reminiscent of that seen in type 2 diabetes patients (Fig. 2). This defect leads to age-dependent glucose intolerance and, in some mice, to overt diabetes. These data suggest that the insulin-resistant state in type 2 diabetics may, in part, also be responsible for the defect in insulin secretion that is seen in this disease.[40]

Similar to Insr, Igf1r ablation in β-cells affects insulin secretion and results in fasting hyperinsulinemia with impaired glucose tolerance (Fig. 2). The likeliest cause of these combined abnormalities is that basal insulin secretion is increased, but glucose-stimulated insulin secretion is impaired. These finding are consistent with the known effect of Igf1 to inhibit insulin secretion from β-cells[41] and perfused rat pancreas.[42]

In contrast, the Insulin receptor-related receptor does not appear to be involved in β-cell function.[43]

There is substantial evidence for a role of receptor tyrosine kinases in insulin synthesis and release, as well as β-cell proliferation and survival (Fig. 2). In addition to *Insr*[40] and *Igf1r*,[44,45] inactivation of *Irs1* leads to defective insulin secretion.[46] On the other hand, inactivation of *Irs2* leads to reduced β-cell mass.[47] Over-expression of Akt in β-cells increases neogenesis and results in increased β-cell mass and size, without affecting insulin secretion,[48,49] while ablation of its substrate $p70^{s6k1}$ results in decreased cell size and hypoinsulinemia. In contrast, mutations of the eukaryotic translation initiation factor 2α (eIF2α)[50] and its kinase *Perk*[51] cause diabetes by affecting *Insulin* mRNA translation.

While it is clear that tyrosine kinase signaling is important for β-cell proliferation and survival, evidence that Insr and/or Igf1r are the receptors that promote β-cell growth is, at this point, inconclusive. An alternate explanation is that Insr and/or Igf1r promote β-cell neogenesis through terminal differentiation of β-cell precursors.[52,53]

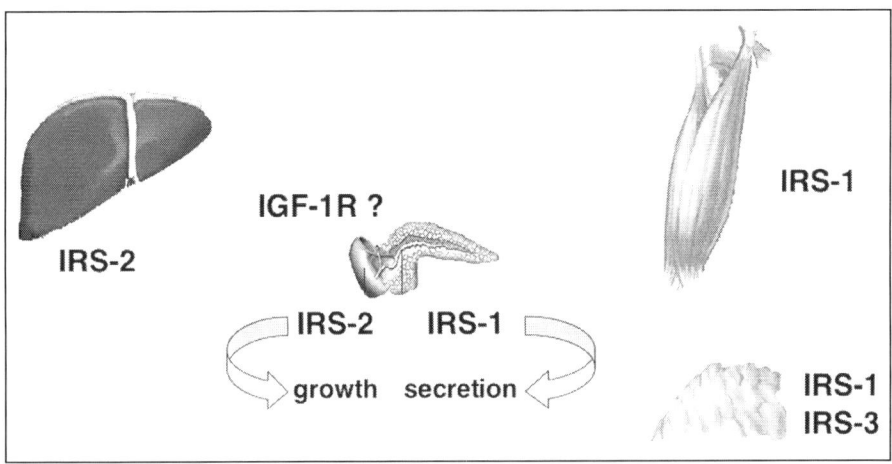

Figure 2. Mutations of *Insr* and *Igf1r* signaling in pancreatic β-cells. *Insr* and *Igf1r* knockouts in β-cells have been obtained using the insulin promoter to drive cre expression. In both instances, insulin secretion is impaired, albeit the mechanisms appear to differ. The effect of *Irs1* knockout on β-cells is similar to that of *Igf1r*, suggesting that Irs1 lies downstream of Igf1r. Alterations of insulin secretion result also from mutations of Pi 3-kinase. Although ablation of Irs2 has a profound effect on β-cell proliferation, neither receptor knockout affects this aspect of β-cell physiology. Thus, it is possible that Irs2 is activated by additional receptors. A prime candidate was Irr, but studies of *Irr* knockouts rule out an effect on β-cell proliferation and glucose-induced insulin secretion.

The role of Pi3k also presents a complex picture (Fig. 2). Pharmacologic inhibition of Pi3k signaling by wortmannin is associated with increased insulin release from purified islets.[54] Indeed, Eto and colleagues have shown that islets from mice lacking the Pi3kr subunit have increased insulin secretion, due to elevation in cytosolic calcium levels.[55] This is contrast to data in *Insr*, *Igf1r* and *Irs1* knockouts, in which insulin secretion appears to be blunted and calcium signaling unaffected.[40,44-46,56-58] This discrepancy may reflect different experimental conditions and different parameters assessed in each study.[55]

## Mutations Affecting Insulin Receptor Signaling

### Insulin Receptor Substrates
Irs proteins mediate insulin, IGF and cytokine actions.[59] The results of gene inactivation experiments are consistent with the view that Irs proteins have different specificities and functions (Fig. 3).

### Irs1
The main consequence of *Irs1* inactivation in mice is intrauterine and post-natal growth retardation, associated with mild insulin resistance.[60,61] Lack of Irs1 has also been shown to impair insulin secretion from β-cells,[46] suggesting that insulin/IGF-1 signaling through Irs1 is required for proper β-cell function. Moreover, combined heterozygosity for *Insr* and *Irs1* null alleles causes a severe impairment of insulin action associated with a steep rise in the incidence of diabetes in the resulting progeny,[62,63] suggesting that Irs1 has an important role on insulin action.

### Irs2
Mice lacking Irs2 develop diabetes due to a combination of insulin deficiency and impaired insulin action. The disease is lethal in some genetic backgrounds[47] and milder in others.[64] The

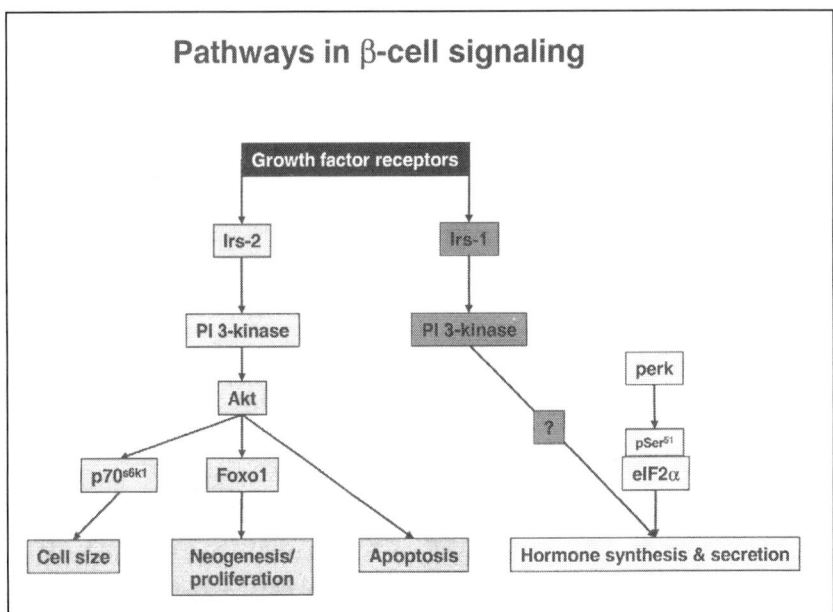

Figure 3. Tissue-specific functions of Irs proteins. Different Irs proteins appear to have specific roles in different tissues. These conclusions are supported by knockout studies. Irs1 appears to be the main mediator of insulin signaling in skeletal muscle, adipose tissue and in β-cell secretory function. Irs2 is important for liver metabolism and β-cell proliferation/protection from apoptosis. Irs3 plays an ancillary role in the fat cell, as demonstrated by the lipoatrophy of double mutant mice lacking both Irs1 and Irs3. The function of Irs4 is elusive. Mice lacking Irs4 have a mild phenotype of growth retardation and hyperinsulinemia, but the phenotype of double mutants Irs1/Irs4 is the same as in Irs1 knockouts.

metabolic syndrome is due to combined insulin resistance in peripheral tissues[63,65] and impaired growth or increased apoptosis of pancreatic β-cells.[53,66] These data are consistent with an important role of Irs2 in fuel homeostasis, but also underline the conclusion that multiple substrates are required to mediate insulin action, since the phenotype due to lack of insulin receptors is substantially more severe than that due to lack of Irs2.

### *Irs3*

Irs3 is the most abundant Irs protein in adipocytes,[67,68] but is also found in other tissues.[69,70] Lack of Irs3 has no apparent effect on mouse development and metabolism, or on insulin-dependent glucose uptake in isolated adipose cells.[71] However, this may be due to compensation by Irs1 in adipose cells, as described below.

### *Irs4*

Ablation of Irs4 results in a mild phenotype with reduced growth, impaired glucose tolerance and reproductive abnormalities.[72] This phenotype is reminiscent of the *Irs1* knockout phenotype, albeit considerably milder.

### *Combined Irs Knockouts*

Mice lacking both Irs1 and Irs2 die prior to implantation on the C57BL x 129SV background,[73] while mutant embryos can be recovered in the DBA x 129SV background,[74] suggesting that the actions of Irs1 and Irs2 are dependent on modifier genes. An additive effect is seen when *Irs1* and *Irs3* knockouts are inter-crossed. Combined deficiency of these two proteins causes lipoatrophy with insulin resistance. Unlike other lipoatrophic models, *Irs1/Irs3* knockouts fail to develop intra-hepatic and intra-muscular deposits of triglycerides. As predicted from other models of lipoatrophy, leptin administration reverses the hyperglycemia and hyperinsulinemia.[75] In contrast, mice with combined ablation of *Irs1* and *Irs4* show the same phenotype as *Irs1* knockouts, with growth retardation and mild insulin resistance, suggesting that Irs4 is not a primary mediator of insulin and Igf action.[75]

## *Akt*

Among the Pdk1 targets implicated in insulin signaling, Akt stands out because of its brisk insulin-dependent activation[76] and insulin-like effects.[77] Two groups, with slightly diverging phenotypes, have recently generated mice lacking Akt2. In one instance, mice show impaired insulin signaling in muscle and adipocytes, which is however insufficient to cause overt diabetes.[78] In another experiment, the phenotype is more pronounced, with growth retardation, reduced fat mass and diabetes.[79] Whether these phenotypic differences are due to compensation by related isoforms remains unclear, but the data are consistent with an important role of Akt in insulin action. Akt appears to be central to β-cell function as well.[48,49]

## *Glut4*

Mice with a complete Glut4 knockout are growth retarded and show cardiac hypertrophy and underdeveloped adipose tissue. They develop moderate insulin resistance, which in male animals is associated with hyperglycemia in the fed state. Female mice do not develop hyperglycemia.[80] Somewhat surprisingly, heterozygous males develop insulin-resistant diabetes, without obesity.[81] It is not clear why heterozygotes are more severely affected than homozygotes, but it could be due to the fact that additional glucose transporters, such as GlutX/Glut8,[82] can be induced in mice lacking Glut4, but not in heterozygous knockouts.

### Muscle-Specific Glut4 Ablation

The phenotype of mice lacking Glut4 indicates that Glut4 is required for insulin-dependent glucose uptake, but not for maintenance of metabolic homeostasis. To better evaluate the contribution of Glut4 to muscle metabolism, mice lacking Glut4 in skeletal muscle have been generated. Unlike total Glut4 knockouts, muscle-specific Glut4 knockouts are insulin resistant and glucose intolerant, consistent with an important role for Glut4 in glucose metabolism.[83] Hyperinsulinemic-euglycemic clamps demonstrate that muscle-specific Glut4 knockout mice have decreased whole body and insulin-stimulated muscle glucose uptake. In a sub-set of mice that develops overt diabetes, there is also insulin resistance in liver and adipose tissue. However, these defects appear to be secondary to glucose toxicity, since they are reversed by phlorizin treatment.[84] Since only some of the muscle Glut4 knockout mice develop diabetes, it appears that there are compensatory mechanisms that prevent the onset of hyperglycemia even when muscle glucose utilization is reduced.

### Heart-Specific Glut4 Ablation

The presence of cardiac abnormalities in Glut4-deficient mice is of great interest, in view of the increased incidence of cardiac disease in insulin-resistant patients.[85] To determine whether the cardiac hypertrophy seen in Glut4-deficient mice is a primary result of Glut4 deficiency in the heart or a consequence of metabolic abnormalities, Abel and coworkers have generated mice lacking Glut4 in the heart. These mice lack insulin-dependent glucose uptake. The metabolic abnormality results in compensated cardiac hypertrophy, increased cardiomyocyte size and preserved contractile function.[86] In addition, recovery from ischemic damage is impaired in the Glut4-deficient heart.[87] The data indicate that Glut4 is important for cardiac function.

### Fat-Specific Glut4 Knockout

Glut4 has been selectively ablated in mature adipocytes using the Ap2 promoter to drive cre expression. Mice lacking adipocyte Glut4 fail to increase glucose uptake in response to insulin

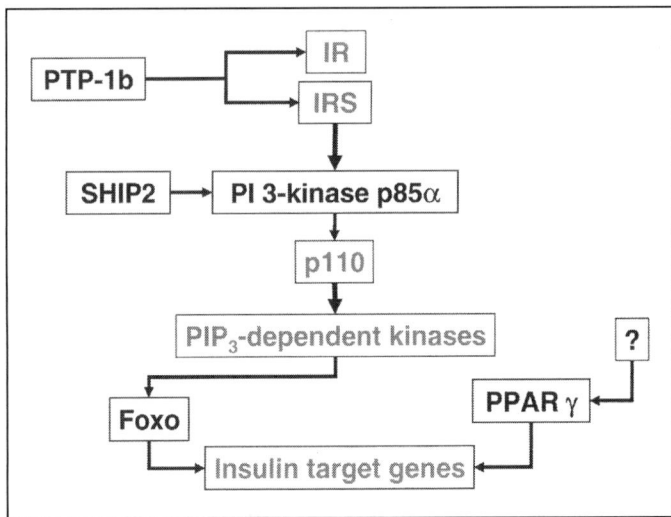

Figure 4. Negative regulators of Insr signaling. Several gene knockouts have insulin-sensitizing effects. Mice lacking the tyrosine phosphatase Ptp1b are refractory to diet-induced insulin resistance, possibly resulting from reduced tyrosine dephosphorylation of Insr and its main substrates. Ablation of the regulatory p85 subunit of Pi 3-kinase restores insulin sensitivity in Insr and Irs1 knockout mice, due to increased activation of Pi 3-kinase-dependent responses. Mutations of the inositol-phosphatase Ship-2 cause neonatal hypoglycemia, and mice heterozygous for the *Ship2* null allele reveal increased glucose tolerance. Pharmacological or genetic inhibition of IKK β kinase protects mice against diet-induced insulin resistance, as do mutations ablating JNK1 function. Haploinsufficiency for the forkhead transcription factor Foxo1, which is normally inhibited by Akt-dependent phosphorylation, results in decreased expression of genes regulating gluconeogenesis.

and develop hyperinsulinemia with hepatic insulin resistance. Since adipocyte-mediated glucose uptake accounts for only ~10% of whole-body glucose disposal, the phenotype indicates that a change in insulin sensitivity in adipocytes can disproportionately affect metabolic control.[88] As observed in mice with muscle-specific Glut4 ablation, the metabolic consequences display individual variations, and only some mice develop diabetes. Of note, insulin-induced suppression of hepatic gluconeogenesis is also impaired in the fat-specific Glut4 knockouts. Therefore, abnormal glucose transport in the adipocyte can lead to insulin resistance and inadequate glucose disposal in other tissues through chronic hyperinsulinemia or an unidentified secondary agent.

### Genes in the Glut4 Translocation Pathway: Syntaxin4

The molecular effectors of Glut4 translocation from the intracellular pool to the plasma membrane are incompletely characterized. Syntaxin 4 plays a role in Glut4 vesicle fusion. While homozygous knockouts are embryonic lethal, heterozygous knockouts develop impaired glucose tolerance and a 50% decrease in Glut4 translocation in response to insulin in skeletal muscle, but not in adipose tissue.[89]

### Genes in the Glycogen Synthesis Pathway: Ptg1

There is substantial evidence that insulin resistance affects nonoxidative pathways of glucose disposal.[90] Nonetheless, the identification of proteins involved in insulin stimulation of this complex process has proven to be as elusive as that of proteins in the Glut4 translocation pathway. Insulin's ability to promote glycogen synthesis is a result of reduced glycogen synthase phosphorylation, which turn is the net effect of reduced glycogen synthase kinase and increased protein phosphatase-1 activities. The relevant roles of these two processes differ in different cell types.[91] The identification of specific proteins targeting protein phosphatase 1α to glycogen synthase has been a key step in our understanding of this process.[92] Protein targeting to glycogen (PTG) is a scaffolding protein that targets protein phosphatase 1 to enzymes involved in glycogen synthesis. Mice lacking Ptg1 die immediately after birth, while heterozygotes develop age-dependent glucose intolerance and insulin resistance. The metabolic syndrome is associated with depletion of tissue glycogen content, accompanied by a compensatory increase in muscle triglyceride content.[93] This picture closely resembles the metabolic abnormalities in the development of type 2 diabetes, and provides a unifying mechanism for the decrease in nonoxidative glucose disposal and increase in tissue triglyceride content that are commonly observed in the disease.

### *Ceacam1 and Insulin Clearance*

Although hyperinsulinemia is generally thought to be a compensatory response to insulin resistance, it can potentially arise as a primary cause of insulin resistance, by triggering a secondary down-regulation of the insulin signaling pathway. The hepatic membrane protein Ceacam has been shown to participate in the process of receptor-mediated insulin internalization and degradation. Ceacam activation is dependent on the phosphorylation of residues in its intracellular domain.[94,95] To test the hypothesis that defects in Ceacam-dependent insulin internalization result in hyperinsulinemia, transgenic mice expressing a dominant negative Ceacam, in which the key phosphorylation site has been mutated, have been generated. These mice do indeed develop hyperinsulinemia and a metabolic syndrome of increased visceral adiposity and increased triglycerides and FFAs. Thus, these findings provide proof-of-principle for the view that alterations in insulin clearance affect peripheral insulin sensitivity.[96]

## Gene Knockouts Associated with Increased Insulin Sensitivity

In addition to mutations affecting insulin action, mutations in genes required to terminate or modulate insulin signaling have proved very informative in examining the pathophysiology of insulin resistance (Fig. 4).

### *Protein Tyrosine Phosphatase (Ptp)1b*

Several tyrosine phosphatases have been implicated in the regulation of insulin signaling. Disruption of the *Ptp1b* gene in mice leads to improved glucose tolerance, decreased glucose levels, and decreased insulin levels, consistent with a state of heightened insulin sensitivity. Increased Insr phosphorylation in muscle and liver is a potential mechanism of enhanced insulin action. Importantly, both homozygous and heterozygous *Ptp1b* mice are refractory to high fat diet-induced obesity and insulin resistance.[97,98] These data suggest that Ptp1b is one of the tyrosine phosphatases required for termination of insulin action.

### *Pi 3-Kinase Regulatory Subunits*

Insulin signaling requires the lipid kinase activity of the enzyme phosphatidylinositol-3-kinase (Pi3k).[99] The gene Pik3r1 encodes three proteins (p85a, p55a and p50a) that serve as regulatory subunits of class 1A Pi3ks. Mice lacking the p85 isoform of this subunit show increased insulin sensitivity and hypoglycemia due to increased glucose transport in skeletal muscle and adipocytes.[100] This unexpected phenotype is associated with a compensatory increase in insulin-dependent binding of Irs proteins to the alternatively spliced products of Pi3kr (especially p50α), which was originally proposed to be the cause of increased insulin sensitivity. However, mice lacking all p85 splice isoforms are also hypoglycemic and insulin-sensitive,[101] so that this explanation is no longer tenable. Unlike the p85-deficient mice, the pan-Pi3k1r knockout mice die shortly after birth with multiple abnormalities, consistent with the pleiotropic role of Pi3k.[101] These experiments provide evidence for a role of Pi3k in insulin action, but do not address the mechanism by which p85 regulates Pi3k activity. Interestingly, haploinsufficiency of *p85* protects both *Insr* and *Irs1* heterozyotes from insulin resistance and diabetes by improving the efficiency of insulin signaling.[102]

## Ship2

The SH2 domain-containing inositol 5-phosphatase Ship2 is able to dephosphorylate phosphoinositol. Decreased expression of this protein enhances insulin action. *Ship2* homozygous null mice develop fatal hypoglycemia as neonates.[103] The low insulin level in these mice indicates that this effect is secondary to increased insulin sensitivity, rather than increased insulin production. This hypothesis is supported by the finding of decreased levels of gluconeogenic enzymes in the liver. Mice heterozygous for the *Ship2* null allele live to adulthood and reveal increased glucose disposal during a glucose tolerance test as well as increased Glut4 translocation and glycogen storage in muscle.

## c-Jun Amino-Terminal Kinase (Jnk)

Serine phosphorylation of Irs and insulin receptors has been shown to dampen insulin signaling, potentially contributing to the onset of the metabolic syndrome.[104-107] Identification of the serine kinases that mediate this event has proved to be elusive. Evidence from knockout mice has begun to shed light on this controversial issue. For example, the c-Jun amino-terminal kinases (Jnk) had been proposed to impair insulin action in response to activation by cytokines and free fatty acids.[108] Mice lacking Jnk1 appear to be protected against genetic (*obese* mutation) and environmental (diet) insulin resistance.[109] Interestingly, Irs1 phosphorylation on Ser$^{307}$ is associated with decreased interaction with the insulin receptor,[110] and is reduced in *Jnk1*$^{-/-}$ mice.[109] Thus, Jnk1 appears to be one of the Irs serine kinases that are involved in down-regulation of insulin signaling.

## Salicylates, Ikkβ and the Metabolic Syndrome

Salicylates improve glucose disposal and increase insulin action. High-dose aspirin treatment decreases fasting plasma glucose and improves the lipid profile of type 2 diabetics by reducing hepatic glucose output and increasing insulin-stimulated glucose uptake.[111] The mechanism by which salicylates increase insulin sensitivity involves activation of I kappa B kinase beta (Ikkβ) as a critical step in lipid-induced insulin resistance. Overexpression of *Ikkβ* in cultured cells leads to altered insulin action, whereas inhibition of this gene improves insulin resistance. *Ikkβ* knockout mice do not develop insulin resistance in the setting of a lipid infusion, while haploinsufficiency at this locus prevents peripheral insulin resistance in the setting of high fat diet or leptin deficiency in *obese* mice.[112,113]

## The Forkhead Transcription Factor (Foxo1)

Foxo proteins are Akt substrates. Phosphorylation byAkt is associated with nuclear exclusion and suppression of their transcriptional activity.[99,114,115]

### Foxo1 in Liver

Foxo1 plays an important role in insulin control of hepatic glucose production by regulating transcription of *glucose-6-phosphatase* and *phospho-enolpyruvate carboxykinase*,[116] in concert with the Pparγ coactivator Pgc1.[117] Haploinsufficiency of the *Foxo1* gene restores insulin sensitivity and rescues the diabetic phenotype in insulin-resistant mice by reducing hepatic expression of glucogenetic genes and increasing adipocyte expression of insulin-sensitizing genes. Conversely, a gain-of-function Foxo1 mutant targeted to liver and pancreatic β-cells results in diabetes arising from a combination of increased hepatic glucose production and impaired β-cell compensation due to decreased Pdx1 expression.[118]

### Foxo1 in β-Cells

The localization of Foxo1 immunoreactivity to pancreatic β-cells has led to studies of its role in terminal differentiation and proliferation in a variety of cell systems. Mice lacking Irs2 develop β-cell failure. Haploinsufficiency of *Foxo1* reverses β-cell failure in *Irs2*–/– mice through partial restoration of β-cell proliferation and increased expression of the pancreatic transcription factor Pdx1. Based on the localization of Foxo1 to a subset of cells abutting pancreatic ducts, it has been suggested that insulin/IGFs regulate β-cell proliferation by relieving Foxo1 inhibition of Pdx1 expression in these cells.[53]

### Foxo1 in Adipose Cells

Foxo1 haploinsufficiency has been shown to decreased adipocyte size and increase expression of genes that promote lipid metabolism.[119] Recent evidence indicates that the effect of *Foxo1* haploinsufficiency in fat can be explained by a direct action of this transcription factor in adipocytes. In vitro studies of adipocyte differentiation indicate that Foxo1 expression is induced in the early stages of adipogenesis in 3T3-F442A cells. Expression of a phosphorylation-defective Foxo1 prevents adipocyte differentiation by altering expression patterns of genes involved in cell cycle control and adipogenesis. In contrast, a dominant negative Foxo1 restores adipocyte differentiation of murine embryonic fibroblasts from insulin receptor-deficient mice. The ability of Foxo1 to control adipogenesis may depend on its regulation of genes involved in cell cycle control, such as p21 and the retinoblastoma gene product.[119] Thus, the emerging model of Foxo1 function in fat cells is that it may control the coupling of extracellular (hormonal) cues that activate adipocyte differentiation to the cell cycle machinery, thus providing the integration of hormone-activated signaling pathways with the transcriptional cascade that promotes adipogenesis.[119] A similar paradigm appears to be at play in myoblast differentiation in response to IGFs.[120]

## Conclusions

Techniques of targeted mutagenesis have enabled investigators to introduce virtually any mutation in mice and study its metabolic effects. The studies reviewed here indicate both the complexity of pathways regulating insulin action, and the opportunities arising from the discovery of new potential targets for drug intervention.

## References

1. Saltiel AR, Kahn CR. Insulin signalling and the regulation of glucose and lipid metabolism. Nature 2001; 414(6865):799-806.
2. Bell GI, Polonsky KS. Diabetes mellitus and genetically programmed defects in beta-cell function. Nature 2001; 414(6865):788-791.
3. Louvi A, Accili D, Efstratiadis A. Growth-promoting interaction of IGF-II with the insulin receptor during mouse embryonic development. Dev Biol 1997; 189(1):33-48.
4. Accili D, Drago J, Lee EJ et al. Early neonatal death in mice homozygous for a null allele of the insulin receptor gene. Nat Genet Jan 1996; 12(1):106-109.
5. Joshi RL, Lamothe B, Cordonnier N et al. Targeted disruption of the insulin receptor gene in the mouse results in neonatal lethality. EMBO J 1996; 15(7):1542-1547.
6. Taylor SI. Lilly Lecture: Molecular mechanisms of insulin resistance. Lessons from patients with mutations in the insulin-receptor gene. Diabetes 1992; 41(11):1473-1490.
7. Kitamura T, Nakae J, Kahn C et al. Lipoatrophy, postnatal growth retardation and diabetes resulting from variable conditional inactivation of the insulin receptor gene in mice. Diabetes 2002; 51(Suppl 2):1337, (Abstract).
8. Frystyk J, Gronbaek H, Skjaerbaek C et al. Developmental changes in serum levels of free and total insulin-like growth factor I (IGF-I), IGF-binding protein-1 and -3, and the acid-labile subunit in rats. Endocrinology 1998; 139(10):4286-4292.
9. Frystyk J, Grofte T, Skjaerbaek C et al. The effect of oral glucose on serum free insulin-like growth factor-I and -II in health adults. J Clin Endocrinol Metab 1997; 82(9):3124-3127.
10. Clemmons DR. Role of insulin-like growth factor binding proteins in controlling IGF actions. Mol Cell Endocrinol 1998; 140(1-2):19-24.
11. Cherrington AD. Banting Lecture 1997. Control of glucose uptake and release by the liver in vivo. Diabetes 1999; 48(5):1198-1214.
12. Michael MD, Kulkarni RN, Postic C et al. Loss of insulin signaling in hepatocytes leads to severe insulin resistance and progressive hepatic dysfunction. Mol Cell 2000; 6(1):87-97.

13. Fisher SJ, Kahn CR. Insulin signaling is required for insulin's direct and indirect action on hepatic glucose production. J Clin Invest 2003; 111(4):463-468.
14. Rosen ED, Walkey CJ, Puigserver P et al. Transcriptional regulation of adipogenesis. Genes Dev 2000; 14(11):1293-1307.
15. Bluher M, Michael MD, Peroni OD et al. Adipose tissue selective insulin receptor knockout protects against obesity and obesity-related glucose intolerance. Dev Cell 2002; 3(1):25-38.
16. Lowell BB, S SV, Hamann A et al. Development of obesity in transgenic mice after genetic ablation of brown adipose tissue. Nature 1993; 366(6457):740-742.
17. Guerra C, Navarro P, Valverde AM et al. Brown adipose tissue-specific insulin receptor knockout shows diabetic phenotype without insulin resistance. J Clin Invest 2001; 108(8):1205-1213.
18. Cline GW, Petersen KF, Krssak M et al. Impaired glucose transport as a cause of decreased insulin-stimulated muscle glycogen synthesis in type 2 diabetes. N Engl J Med 1999; 341(4):240-246.
19. Kim JK, Michael MD, Previs SF et al. Redistribution of substrates to adipose tissue promotes obesity in mice with selective insulin resistance in muscle. J Clin Invest 2000; 105(12):1791-1797.
20. Bruning JC, Michael MD, Winnay JN et al. A muscle-specific insulin receptor knockout exhibits features of the metabolic syndrome of NIDDM without altering glucose tolerance. Mol Cell 1998; 2(5):559-569.
21. Lauro D, Kido Y, Castle AL et al. Impaired glucose tolerance in mice with a targeted impairment of insulin action in muscle and adipose tissue. Nat Genet 1998; 20(3):294-298.
22. Shefi-Friedman L, Wertheimer E, Shen S et al. Increased IGFR activity and glucose transport in cultured skeletal muscle from insulin receptor null mice. Am J Physiol Endocrinol Metab 2001; 281(1):E16-24.
23. Wojtaszewski JF, Higaki Y, Hirshman MF et al. Exercise modulates postreceptor insulin signaling and glucose transport in muscle-specific insulin receptor knockout mice. J Clin Invest 1999; 104(9):1257-1264.
24. Belke DD, Betuing S, Tuttle MJ et al. Insulin signaling coordinately regulates cardiac size, metabolism, and contractile protein isoform expression. J Clin Invest 2002; 109(5):629-639.
25. Shiojima I, Yefremashvili M, Luo Z et al. Akt signaling mediates postnatal heart growth in response to insulin and nutritional status. J Biol Chem 2002; 277(40):37670-37677.
26. Hu P, Zhang D, Swenson L et al. Minimally invasive aortic banding in mice: Effects of altered cardiomyocyte insulin signaling during pressure overload. Am J Physiol Heart Circ Physiol 2003; 285(3):H1261-1269.
27. Baron AD. Hemodynamic actions of insulin. Am J Physiol 1994; 267(2 Pt 1):E187-202.
28. King GL, Johnson SM. Receptor-mediated transport of insulin across endothelial cells. Science 1985; 227(4694):1583-1586.
29. Steil GM, Ader M, Moore DM et al. Transendothelial insulin transport is not saturable in vivo. No evidence for a receptor-mediated process. J Clin Invest 1996; 97(6):1497-1503.
30. Baron AD, Steinberg H, Brechtel G et al. Skeletal muscle blood flow independently modulates insulin-mediated glucose uptake. Am J Physiol 1994; 266(2 Pt 1):E248-253.
31. Bonadonna RC, Saccomani MP, Del Prato S et al. Role of tissue-specific blood flow and tissue recruitment in insulin-mediated glucose uptake of human skeletal muscle. Circulation 1998; 98(3):234-241.
32. Yang YJ, Hope ID, Ader M et al. Insulin transport across capillaries is rate limiting for insulin action in dogs. J Clin Invest 1989; 84(5):1620-1628.
33. Kondo T, Vicent D, Suzuma K et al. Knockout of insulin and IGF-1 receptors on vascular endothelial cells protects against retinal neovascularization. J Clin Invest 2003; 111(12):1835-1842.
34. Havrankova J, Roth J, Brownstein M. Insulin receptors are widely distributed in the central nervous system of the rat. Nature 1978; 272(5656):827-829.
35. Bruning JC, Gautam D, Burks DJ et al. Role of brain insulin receptor in control of body weight and reproduction. Science 2000; 289(5487):2122-2155.
36. Burks DJ, de Mora JF, Schubert M et al. IRS-2 pathways integrate female reproduction and energy homeostasis. Nature 2000; 407(6802):377-382.
37. Obici S, Feng Z, Morgan K et al. Central administration of oleic acid inhibits glucose production and food intake. Diabetes 2002; 51(2):271-275.
38. Obici S, Zhang BB, Karkanias G et al. Hypothalamic insulin signaling is required for inhibition of glucose production. Nat Med 2002; 8(12):1376-1382.
39. Polonsky KS, Sturis J, Bell GI. Noninsulin-dependent diabetes mellitus - a genetically programmed failure of the beta cell to compensate for insulin resistance. N Engl J Med 1996; 334(12):777-783.

40. Kulkarni RN, Bruning JC, Winnay JN et al. Tissue-specific knockout of the insulin receptor in pancreatic beta cells creates an insulin secretory defect similar to that in type 2 diabetes. Cell 1999; 96(3):329-339.
41. Zhao AZ, Zhao H, Teague J et al. Attenuation of insulin secretion by insulin-like growth factor 1 is mediated through activation of phosphodiesterase 3B. Proc Natl Acad Sci USA 1997; 94(7):3223-3228.
42. Leahy JL, Vandekerkhove KM. Insulin-like growth factor-I at physiological concentrations is a potent inhibitor of insulin secretion. Endocrinology 1990; 126(3):1593-1598.
43. Kitamura T, Kido Y, Nef S et al. Preserved pancreatic beta-cell development and function in mice lacking the insulin receptor-related receptor. Mol Cell Biol 2001; 21(16):5624-5630.
44. Kulkarni RN, Holzenberger M, Shih DQ et al. Beta-cell-specific deletion of the Igf1 receptor leads to hyperinsulinemia and glucose intolerance but does not alter beta-cell mass. Nat Genet 2002; 31(1):111-115.
45. Xuan S, Kitamura T, Nakae J et al. Defective insulin secretion in pancreatic beta cells lacking type 1 IGF receptor. J Clin Invest 2002; 110(7):1011-1019.
46. Kulkarni RN, Winnay JN, Daniels M et al. Altered function of insulin receptor substrate-1-deficient mouse islets and cultured beta-cell lines. J Clin Invest 1999; 104(12):R69-75.
47. Withers DJ, Sanchez-Gutierrez J, Towery H et al. Disruption of IRS-2 causes type 2 diabetes in mice. Nature 1998; 391:900-904.
48. Tuttle RL, Gill NS, Pugh W et al. Regulation of pancreatic beta-cell growth and survival by the serine/threonine protein kinase Akt1/PKBalpha. Nat Med 2001; 7(10):1133-1137.
49. Bernal-Mizrachi E, Wen W, Stahlhut S et al. Transgenic mice expressing a constitutively active Akt1/PKBα in pancreatic islet β-cells exhibit striking hypertrophy, hyperplasia and hyperinsulinemia. J Clin Invest 2001; 108:1631-1638.
50. Scheuner D, Song B, McEwen E et al. Translational control is required for the unfolded protein response and in vivo glucose homeostasis. Mol Cell 2001; 7(6):1165-1176.
51. Harding HP, Zeng H, Zhang Y et al. Diabetes mellitus and exocrine pancreatic dysfunction in perk-/- mice reveals a role for translational control in secretory cell survival. Mol Cell 2001; 7(6):1153-1163.
52. Accili D. A kinase in the life of the beta cell. J Clin Invest 2001; 108(11):1575-1576.
53. Kitamura T, Nakae J, Kitamura Y et al. The forkhead transcription factor Foxo1 links insulin signaling to Pdx1 regulation of pancreatic beta cell growth. J Clin Invest 2002; 110(12):1839-1847.
54. Zawalich WS, Zawalich KC. A link between insulin resistance and hyperinsulinemia: Inhibitors of phosphatidylinositol 3-kinase augment glucose-induced insulin secretion from islets of lean, but not obese, rats. Endocrinology 2000; 141(9):3287-3295.
55. Eto K, Yamashita T, Tsubamoto Y et al. Phosphatidylinositol 3-kinase suppresses glucose-stimulated insulin secretion by affecting post-cytosolic [Ca(2+)] elevation signals. Diabetes 2002; 51(1):87-97.
56. Leibiger IB, Leibiger B, Moede T et al. Exocytosis of insulin promotes insulin gene transcription via the insulin receptor/PI-3 kinase/p70 s6 kinase and CaM kinase pathways. Mol Cell 1998; 1(6):933-938.
57. Leibiger B, Leibiger IB, Moede T et al. Selective insulin signaling through A and B insulin receptors regulates transcription of insulin and glucokinase genes in pancreatic beta cells. Mol Cell 2001; 7(3):559-570.
58. Aspinwall CA, Qian WJ, Roper MG et al. Roles of insulin receptor substrate-1, phosphatidylinositol 3-kinase, and release of intracellular Ca2+ stores in insulin-stimulated insulin secretion in beta-cells. J Biol Chem 2000; 275(29):22331-22338.
59. White MF. The IRS-signalling system: A network of docking proteins that mediate insulin and interleukin signalling. Mol Cell Biochem 1998; 182(1-2):3-11.
60. Araki E, Lipes MA, Patti ME et al. Alternative pathway of insulin signalling in mice with targeted disruption of the IRS-1 gene. Nature 1994; 372(6502):186-190.
61. Tamemoto H, Kadowaki T, Tobe K et al. Insulin resistance and growth retardation in mice lacking insulin receptor substrate-1. Nature 1994; 372(6502):182-186.
62. Bruning JC, Winnay J, Bonner-Weir S et al. Development of a novel polygenic model of NIDDM in mice heterozygous for IR and IRS-1 null alleles. Cell 1997; 88(4):561-572.
63. Kido Y, Burks DJ, Withers D et al. Tissue-specific insulin resistance in mice with mutations in the insulin receptor, IRS-1, and IRS-2. J Clin Invest 2000; 105(2):199-205.
64. Kubota N, Tobe K, Terauchi Y et al. Disruption of insulin receptor substrate 2 causes type 2 diabetes because of liver insulin resistance and lack of compensatory beta-cell hyperplasia. Diabetes 2000; 49(11):1880-1889.

65. Previs SF, Withers DJ, Ren JM et al. Contrasting effects of IRS-1 versus IRS-2 gene disruption on carbohydrate and lipid metabolism in vivo. J Biol Chem 2000; 275(50):38990-38994.
66. Kushner JA, Ye J, Schubert M et al. Pdx1 restores beta cell function in Irs2 knockout mice. J Clin Invest 2002; 109(9):1193-1201.
67. Smith-Hall J, Pons S, Patti ME et al. The 60 kDa insulin receptor substrate functions like an IRS protein (pp60IRS3) in adipose cells. Biochemistry 1997; 36(27):8304-8310.
68. Xu P, Jacobs AR, Taylor SI. Interaction of insulin receptor substrate 3 with insulin receptor, insulin receptor-related receptor, insulin-like growth factor-1 receptor, and downstream signaling proteins. J Biol Chem 1999; 274(21):15262-15270.
69. Lavan BE, Lane WS, Lienhard GE. The 60-kDa phosphotyrosine protein in insulin-treated adipocytes is a new member of the insulin receptor substrate family. J Biol Chem 1997; 272(17):11439-11443.
70. Sciacchitano S, Taylor SI. Cloning, Tissue expression, and chromosomal localization of the mouse IRS-3 gene. Endocrinology 1997; 138(11):4931-4940.
71. Liu SCH, Wang Q, Lienhard GE et al. Insulin receptor substrate 3 is not essential for growth or glucose homeostasis. J Biol Chem 1999; 274:18093-18099.
72. Fantin VR, Wang Q, Lienhard GE et al. Mice lacking insulin receptor substrate 4 exhibit mild defects in growth, reproduction, and glucose homeostasis. Am J Physiol Endocrinol Metab 2000; 278(1):E127-E133.
73. Withers DJ, Burks DJ, Towery HH et al. Irs-2 coordinates Igf-1 receptor-mediated beta-cell development and peripheral insulin signalling. Nature Genet 1999; 23(1):32-40.
74. Miki H, Yamauchi T, Suzuki R et al. Essential role of insulin receptor substrate 1 (IRS-1) and IRS-2 in adipocyte differentiation. Mol Cell Biol 2001; 21(7):2521-2532.
75. Laustsen PG, Michael MD, Crute BE et al. Lipoatrophic diabetes in Irs1(-/-)/Irs3(-/-) double knockout mice. Genes Dev 2002; 16(24):3213-3222.
76. Kohn AD, Takeuchi F, Roth RA. Akt, a pleckstrin homology domain containing kinase, is activated primarily by phosphorylation. J Biol Chem 1996; 271(36):21920-21926.
77. Kohn AD, Summers SA, Birnbaum MJ et al. Expression of a constitutively active Akt Ser/Thr kinase in 3T3-L1 adipocytes stimulates glucose uptake and glucose transporter 4 translocation. J Biol Chem 1996; 271(49):31372-31378.
78. Cho H, Mu J, Kim JK et al. Insulin resistance and a diabetes mellitus-like syndrome in mice lacking the protein kinase Akt2 (PKB beta). Science 2001; 292(5522):1728-1731.
79. Garofalo RS, Orena SJ, Rafidi K et al. Severe diabetes, age-dependent loss of adipose tissue, and mild growth deficiency in mice lacking Akt2/PKB beta. J Clin Invest 2003; 112(2):197-208.
80. Katz EB, Stenbit AE, Hatton K et al. Cardiac and adipose tissue abnormalities but not diabetes in mice deficient in GLUT4. Nature 1995; 377(6545):151-155.
81. Stenbit AE, Tsao TS, Li J et al. GLUT4 heterozygous knockout mice develop muscle insulin resistance and diabetes. Nat Med 1997; 3(10):1096-1101.
82. Joost HG, Bell GI, Best JD et al. Nomenclature of the GLUT/SLC2A family of sugar/polyol transport facilitators. Am J Physiol Endocrinol Metab 2002; 282(4):E974-976.
83. Zisman A, Peroni OD, Abel ED et al. Targeted disruption of the glucose transporter 4 selectively in muscle causes insulin resistance and glucose intolerance. Nat Med 2000; 6(8):924-928.
84. Kim JK, Zisman A, Fillmore JJ et al. Glucose toxicity and the development of diabetes in mice with muscle- specific inactivation of GLUT4. J Clin Invest 2001; 108(1):153-160.
85. Kaur J, Singh P, Sowers JR. Diabetes and cardiovascular diseases. Am J Ther 2002; 9(6):510-515.
86. Abel ED, Kaulbach HC, Tian R et al. Cardiac hypertrophy with preserved contractile function after selective deletion of GLUT4 from the heart. J Clin Invest 1999; 104(12):1703-1714.
87. Tian R, Abel ED. Responses of GLUT4-deficient hearts to ischemia underscore the importance of glycolysis. Circulation 2001; 103(24):2961-2966.
88. Abel ED, Peroni O, Kim JK et al. Adipose-selective targeting of the GLUT4 gene impairs insulin action in muscle and liver. Nature 2001; 409(6821):729-733.
89. Yang C, Coker KJ, Kim JK et al. Syntaxin 4 heterozygous knockout mice develop muscle insulin resistance. J Clin Invest 2001; 107(10):1311-1318.
90. Shulman GI. Cellular mechanisms of insulin resistance in humans. Am J Cardiol 1999; 84(1A):3J-10J.
91. Brady MJ, Bourbonais FJ, Saltiel AR. The activation of glycogen synthase by insulin switches from kinase inhibition to phosphatase activation during adipogenesis in 3T3-L1 cells. J Biol Chem 1998; 273(23):14063-14066.
92. Printen JA, Brady MJ, Saltiel AR. PTG, a protein phosphatase 1-binding protein with a role in glycogen metabolism. Science 1997; 275(5305):1475-1478.

93. Crosson SM, Khan A, Printen J et al. PTG gene deletion causes impaired glycogen synthesis and developmental insulin resistance. J Clin Invest 2003; 111(9):1423-1432.
94. Najjar SM, Philippe N, Suzuki Y et al. Insulin-stimulated phosphorylation of recombinant pp120/HA4, an endogenous substrate of the insulin receptor tyrosine kinase. Biochemistry 1995; 34(29):9341-9349.
95. Najjar SM, Blakesley VA, Li Calzi S et al. Differential phosphorylation of pp120 by insulin and insulin-like growth factor-1 receptors: Role for the C-terminal domain of the beta- subunit. Biochemistry 1997; 36(22):6827-6834.
96. Poy MN, Yang Y, Rezaei K et al. CEACAM1 regulates insulin clearance in liver. Nat Genet 2002; 30(3):270-276.
97. Elchebly M, Payette P, Michaliszyn E et al. Increased insulin sensitivity and obesity resistance in mice lacking the protein tyrosine phosphatase-1B gene. Science 1999; 283(5407):1544-1548.
98. Klaman LD, Boss O, Peroni OD et al. Increased energy expenditure, decreased adiposity, and tissue-specific insulin sensitivity in protein-tyrosine phosphatase 1B-deficient mice. Mol Cell Biol 2000; 20(15):5479-5489.
99. Nakae J, Park BC, Accili D. Insulin stimulates phosphorylation of the forkhead transcription factor FKHR on serine 253 through a Wortmannin-sensitive pathway. J Biol Chem 1999; 274(23):15982-15985.
100. Terauchi Y, Tsuji Y, Satoh S et al. Increased insulin sensitivity and hypoglycaemia in mice lacking the p85 alpha subunit of phosphoinositide 3-kinase. Nat Genet 1999; 21(2):230-235.
101. Fruman DA, Mauvais-Jarvis F, Pollard DA et al. Hypoglycaemia, liver necrosis and perinatal death in mice lacking all isoforms of phosphoinositide 3-kinase p85alpha. Nat Genet 2000; 26(3):379-382.
102. Mauvais-Jarvis F, Ueki K, Fruman DA et al. Reduced expression of the murine p85alpha subunit of phosphoinositide 3-kinase improves insulin signaling and ameliorates diabetes. J Clin Invest 2002; 109(1):141-149.
103. Clement S, Krause U, Desmedt F et al. The lipid phosphatase SHIP2 controls insulin sensitivity. Nature 2001; 409:92-97.
104. Kanety H, Feinstein R, Papa MZ et al. Tumor necrosis factor alpha-induced phosphorylation of insulin receptor substrate-1 (IRS-1). Possible mechanism for suppression of insulin- stimulated tyrosine phosphorylation of IRS-1. J Biol Chem 1995; 270(40):23780-23784.
105. Paz K, Hemi R, LeRoith D et al. A molecular basis for insulin resistance. Elevated serine/threonine phosphorylation of IRS-1 and IRS-2 inhibits their binding to the juxtamembrane region of the insulin receptor and impairs their ability to undergo insulin-induced tyrosine phosphorylation. J Biol Chem 1997; 272(47):29911-29918.
106. De Fea K, Roth RA. Protein kinase C modulation of insulin receptor substrate-1 tyrosine phosphorylation requires serine 612. Biochemistry 1997; 36(42):12939-12947.
107. Rui L, Aguirre V, Kim JK et al. Insulin/IGF-1 and TNF-alpha stimulate phosphorylation of IRS-1 at inhibitory Ser307 via distinct pathways. J Clin Invest 2001; 107(2):181-189.
108. Aguirre V, Uchida T, Yenush L et al. The c-Jun NH(2)-terminal kinase promotes insulin resistance during association with insulin receptor substrate-1 and phosphorylation of Ser(307). J Biol Chem 2000; 275(12):9047-9054.
109. Hirosumi J, Tuncman G, Chang L et al. A central role for JNK in obesity and insulin resistance. Nature 2002; 420(6913):333-336.
110. Aguirre V, Werner ED, Giraud J et al. Phosphorylation of Ser307 in insulin receptor substrate-1 blocks interactions with the insulin receptor and inhibits insulin action. J Biol Chem 2002; 277(2):1531-1537.
111. Hundal RS, Petersen KF, Mayerson AB et al. Mechanism by which high-dose aspirin improves glucose metabolism in type 2 diabetes. J Clin Invest 2002; 109(10):1321-1326.
112. Kim JK, Kim YJ, Fillmore JJ et al. Prevention of fat-induced insulin resistance by salicylate. J Clin Invest 2001; 108(3):437-446.
113. Yuan M, Konstantopoulos N, Lee J et al. Reversal of obesity- and diet-induced insulin resistance with salicylates or targeted disruption of Ikkbeta. Science 2001; 293(5535):1673-1677.
114. Brunet A, Bonni A, Zigmond MJ et al. Akt promotes cell survival by phosphorylating and inhibiting a forkhead transcription factor. Cell 1999; 96:857-868.
115. Kops GJ, de Ruiter ND, De Vries-Smits AM et al. Direct control of the Forkhead transcription factor AFX by protein kinase B. Nature 1999; 398(6728):630-634.
116. Nakae J, Kitamura T, Silver DL et al. The forkhead transcription factor Foxo1 (Fkhr) confers insulin sensitivity onto glucose-6-phosphatase expression. J Clin Invest 2001; 108(9):1359-1367.

117. Puigserver P, Rhee J, Donovan J et al. Insulin-regulated hepatic gluconeogenesis through FOXO1-PGC-1alpha interaction. Nature 2003; 423(6939):550-555.
118. Nakae J, Biggs WH, Kitamura T et al. Regulation of insulin action and pancreatic beta-cell function by mutated alleles of the gene encoding forkhead transcription factor Foxo1. Nat Genet 2002; 32(2):245-253.
119. Nakae J, Kitamura T, Kitamura Y et al. The forkhead transcription factor foxo1 regulates adipocyte differentiation. Dev Cell 2003; 4(1):119-129.
120. Hribal ML, Nakae J, Kitamura T et al. Regulation of insulin-like growth factor-dependent myoblast differentiation by Foxo forkhead transcription factors. J Cell Biol 2003; 162(4):535-541.
121. Bluher M, Kahn BB, Kahn CR. Extended longevity in mice lacking the insulin receptor in adipose tissue. Science 2003; 299(5606):572-574.
122. Vicent D, Ilany J, Kondo T et al. The role of endothelial insulin signaling in the regulation of vascular tone and insulin resistance. J Clin Invest 2003; 111(9):1373-1380.

# CHAPTER 10

# Insulin Resistance

### C. Hamish Courtney and Jerrold M. Olefsky*

Insulin resistance can be said to exist "whenever normal concentrations of hormone produce a less than normal biological response".[1] In the 1930s, Himsworth first differentiated patients with diabetes mellitus into "insulin sensitive" and "insulin insensitive" based on the ability of subcutaneous insulin administration to dispose of an oral glucose load.[2] He further suggested that this differentiation corresponded to the clinical presentation of diabetes: that of either young ketosis-prone insulin sensitive or middle aged, nonketotic, insulin insensitive patients. The former is now classified as type 1 diabetes mellitus with the latter "insulin insensitive" classified as type 2 diabetes mellitus. Upon the development of the radioimmunoassay technique in 1960, Yalow and Berson demonstrated that patients with the adult-onset form of diabetes had, on average, higher circulating insulin levels than nondiabetic subjects.[3] It was thus concluded that "the tissues of the maturity onset diabetic do not respond to insulin as well as the tissues of the nondiabetic subjects respond to insulin."

The term "Syndrome X" or "Metabolic syndrome" has been coined to refer to subjects exhibiting features of insulin resistance[4] and this has been further defined by the National Cholesterol Education Program (Table 1) and modified by the WHO to add the requirement of hyperinsulinemia (upper quartile of the nondiabetic population) or elevated fasting plasma glucose (110 mg/dl, but <126 mg/dl).[5] Besides insulin resistance, associated manifestations of the syndrome include hypertension, dyslipidemia and obesity. Given its increasing prevalence and association with subsequent cardiovascular disease, insulin resistance represents a condition of considerable importance.

## Methods of Assessing Insulin Sensitivity

In view of the significance of insulin resistance it is important that insulin action be accurately assessed. Several in vivo techniques have emerged over recent years, some of which are discussed below.

### Glucose Clamp Technique

The hyperinsulinemic euglycemic clamp is regarded as the "gold standard" method of determining insulin sensitivity.[6] In this technique, while a fixed amount of insulin is infused, the blood glucose is "clamped" at a predetermined level by the titration of a variable rate glucose infusion. The underlying principle is that upon reaching steady state, by definition, glucose disposal is equivalent to glucose appearance. During hyperinsulinemia, glucose disposal ($R_d$) is primarily accounted for by glucose uptake into skeletal muscle, and glucose appearance is equal to the sum of the exogenous glucose infusion rate plus the rate of hepatic glucose output (HGO).

---

*Corresponding Author: Jerrold M. Olefsky—University of California San Diego, San Diego, California, U.S.A. Email: olefsky@popmail.ucsd.edu

*Mechanisms of Insulin Action*, edited by Alan R. Saltiel and Jeffrey E. Pessin.
©2007 Landes Bioscience and Springer Science+Business Media.

*Table 1. NCEP definition of metabolic syndrome*

| |
|---|
| **At least 3 of the following:** |
| Fasting plasma glucose >110 mg/dl |
| Abdominal obesity (waist girth >102 cm (men), >88 cm (women)) |
| Serum triglycerides >150 mg/dl |
| Serum HDL cholesterol <40 mg/dl (men), <50 mg/dl (women) |
| Blood pressure >130/85 mmHg (or medication) |

The addition of isotope dilution permits measurement of HGO ($R_a$), thus allowing determination of insulin action in the liver by the degree of suppression of HGO. Peripheral glucose disposal is also measured ($R_d$) and this largely represents insulin action in skeletal muscle, since 80-90% of overall in vivo glucose disposal is into skeletal muscle. Typically, labeled glucose is given as a constant infusion in addition to the insulin and variable rate glucose infusions. Regular samples are obtained for the determination of specific activity from which overall rates of glucose appearance ($R_a$) can be calculated. As the rate of glucose appearance is the sum of HGO and the exogenous glucose infusion rate, HGO can be calculated as the total $R_a$ - exogenous glucose infusion rate.[7] Additionally, by performing several studies at different insulin levels in the same subject, the dose-response curve for insulin-stimulated glucose disposal and suppression of hepatic glucose output can be constructed.[8]

Assessment of insulin action in adipose tissue can be inferred from the rate and extent of the reduction of circulating free fatty acids (FFA) upon commencement of hyperinsulinemia. The circulating FFA concentration represents the balance between FFA release and uptake, both of which are influenced by insulin. Adipocyte FFA release is suppressed by the action of insulin on hormone sensitive lipase and stimulation of lipoprotein lipase by insulin promotes cellular FFA uptake which could increase FFA removal. Reduction of the circulating FFA concentration by insulin is thus an integrated measure of the action of insulin on both processes.

The use of indirect calorimetry, in combination with the glucose clamp technique, allows estimation of carbohydrate and fat oxidation by measuring oxygen consumption and carbon dioxide production.[9] It is possible therefore to assess the proportions of glucose undergoing oxidative versus nonoxidative disposal, permitting conclusions to be drawn regarding the route of intracellular glucose metabolism.

## *Minimal Model - Frequently Sampled Intravenous Glucose Tolerance Test (FSIVGTT)*

The glucose clamp technique described above is experimentally complex requiring intravenous lines, the use of radioisotopes and supervision of the subject over several hours. Alternative means have been sought therefore to simplify assessment of insulin sensitivity. The minimal model method provides a measure of insulin sensitivity ($S_I$) using computer modeling to analyze glucose and insulin levels following injection of intravenous glucose.[10] Injection of glucose stimulates insulin release causing glucose uptake, with a consequent decline in plasma glucose levels. By modeling the dynamics of the relationship between insulin concentration and the change in glucose concentration, an insulin sensitivity index ($S_I$) can be calculated. In patients with type 2 diabetes who have impaired insulin secretion, the test has been adapted by the addition of an injection of exogenous insulin 20 minutes after the glucose bolus.

The insulin sensitivity index ($S_I$) determined from the minimal model method is equivalent to measures of insulin sensitivity derived from the glucose clamp technique[11] and thus it provides an experimental method of assessing insulin action that is relatively easy to perform. Furthermore, $S_I$ can be used in outpatient, epidemiological studies and also at multiple time points in the same patient.

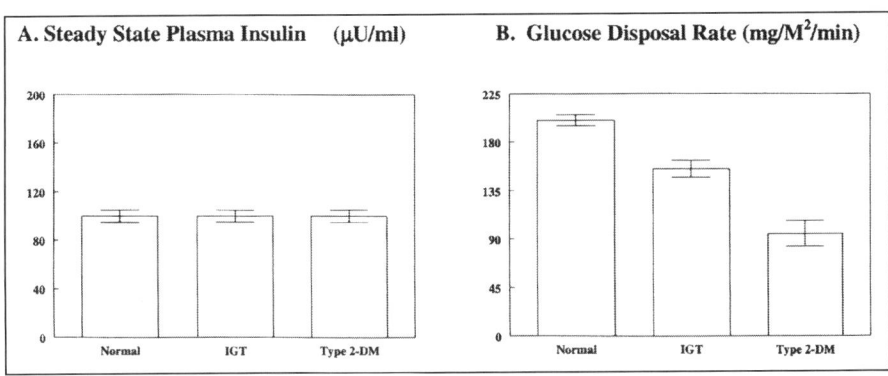

Figure 1. Mean steady-state glucose disposal rates (Panel B) and plasma insulin levels (Panel A) for control subjects, nonobese subjects with impaired glucose tolerance, and type 2 diabetic subjects (T2DM) during euglycemic glucose clamp studies performed at an insulin infusion rate of 40 mU/m$^2$/min. Results are plotted as means ± SEM. Figure created from data in Kolterman OG, Gray RS, Griffin J et al. J Clin Invest 1981; 68:957-969.[204]

Although the minimal model method provides an integrated assessment of overall insulin action in an individual, it is however only an estimate. In comparison, the glucose clamp technique is able to differentiate insulin action in muscle (glucose $R_d$), liver (suppression of HGO) and fat (suppression of FFA) and furthermore allows insulin dose responses to be characterized.

### Nuclear Magnetic Resonance (NMR) Studies

NMR spectroscopy provides a noninvasive means of repeatedly measuring intracellular metabolite concentrations in the same tissues, allowing substrate fluxes through insulin-mediated pathways to be monitored. Recent studies have utilized the technique to assess pathways such as glycogen synthesis and breakdown in both liver and skeletal muscle.[12,13] The utility of this technique to contribute to understanding the pathophysiology of insulin resistance will be outlined later.

As will be discussed in detail subsequently, altered fat metabolism is important in the pathophysiology of insulin resistance. Accumulation of triglyceride in insulin responsive tissues, such as skeletal muscle, has been found to strongly correlate with measures of insulin resistance.[14] NMR spectroscopy permits the noninvasive measurement of intramyocellular triglyceride content, with the added advantage that intramyocellular triglyceride and triglyceride in adipocytes between the muscle fibers can be distinguished.[15,16] This technique therefore allows convenient further investigation into the role of altered muscle triglyceride metabolism in insulin resistance.

## Insulin Resistance in Type 2 Diabetes Mellitus and Obesity

Type 2 diabetes mellitus is characterized in almost all cases by insulin resistance. This has been clearly demonstrated by the glucose clamp technique, as shown by the data presented in Figure 1, in which glucose clamps were performed in normal subjects, subjects with impaired glucose tolerance (IGT) and subjects with type 2 diabetes. Despite similar steady-state insulin levels, the glucose disposal rate was decreased by 24% in the subjects with IGT and by 58% in those with type 2 diabetes compared with normals.[17]

By plotting mean glucose disposal rates at multiple steady-state plasma insulin levels, insulin dose-response curves can be generated (Fig. 2). The rightward shift of the dose-response curve from normal controls, through subjects with IGT, to those with diabetes is clearly demonstrated indicating increasing insulin resistance. Furthermore, the presence of obesity

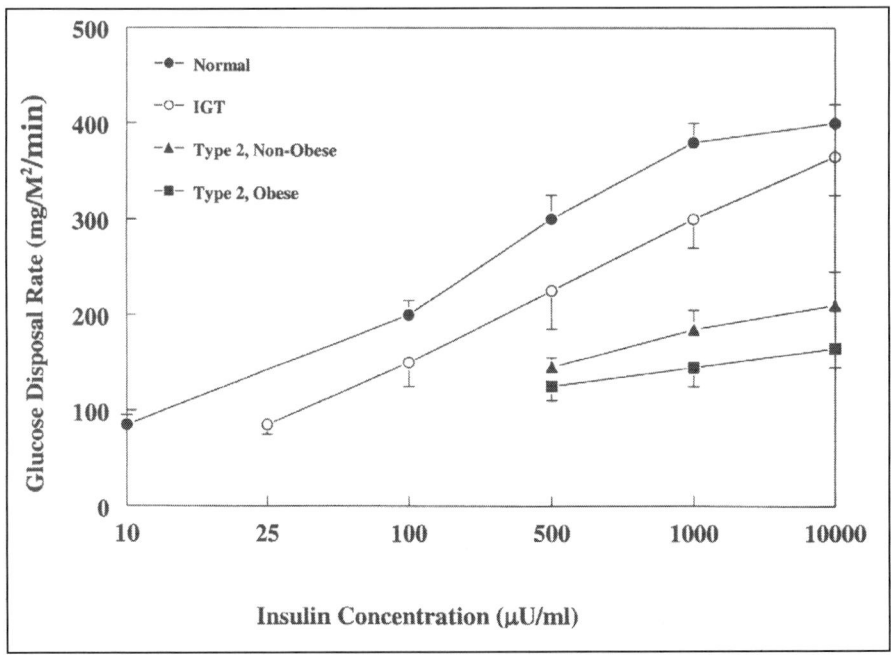

Figure 2. Mean insulin dose-response curves for glucose disposal in control subjects (●), subjects with impaired glucose tolerance (○), and nonobese (▲) and obese (■) type 2 diabetic subjects. Reprinted with permission from Kolterman OG, Gray RS, Griffin J et al. J Clin Invest 1981; 68:957-969.[204]

confers an additional degree of insulin resistance as exemplified by the further rightward shift of the dose-response curve in the obese versus lean diabetic subjects.

Obese subjects with normal glucose tolerance are also insulin resistant, as indicated by a rightward shift in the dose-response curve for insulin-stimulated glucose disposal during a glucose clamp.[8] Potential mechanisms to explain the deleterious effect of adipose tissue on insulin action will be discussed later.

In insulin resistant states, insulin action is impaired in the liver, skeletal muscle and adipose tissue. Each of these organ abnormalities will be dealt with separately.

## *Hepatic Glucose Metabolism*

In the fasting state, glucose is produced from the liver by both glycogenolysis and gluconeogenesis such that HGO accounts for approximately 90% of the glucose released into the circulation. Conversely, in the postprandial state, HGO is suppressed to help limit the rise in plasma glucose levels and furthermore the liver stores fuel by conversion of glucose carbons to glycogen. These effects are mediated in the fasting state by an increase in gluconeogenic substrate supply, a reduction in insulin concentration and an increase in other hormones such as glucagon with the converse changes occurring postprandially.

The ability of insulin to suppress HGO occurs by both direct and indirect means. Insulin directly reduces HGO by inhibition of gluconeogenic enzymes, such as phosphoenolpyruvate carboxykinase.[18] Indirectly, insulin may reduce HGO via its antilipolytic action, as a strong correlation exists between plasma FFA levels and HGO.[19] FFAs stimulate gluconeogenesis by increasing ATP and NADH production, generated from their oxidation in the liver.[20] The effect of insulin to inhibit HGO has been shown to closely relate to the ability of systemic insulin to suppress plasma FFA levels.[21]

Increased basal HGO is a characteristic feature of both obese and nonobese patients with type 2 diabetes, who by definition have fasting hyperglycemia. Conversely, subjects with IGT, who by definition have normal fasting glucose levels, have normal basal HGO values.[17,22,23] In the basal state, insulin-mediated glucose disposal accounts for only ~30% of overall glucose disposal, whereas noninsulin-mediated glucose disposal (primarily in the CNS) compromises ~70%. Therefore impairment in insulin-mediated glucose disposal due to insulin resistance will have little effect on overall basal glucose disposal or fasting glucose levels. As the fasting glucose level reflects the balance between HGO and glucose disposal, it follows that if reduced glucose disposal does not contribute to significant fasting hyperglycemia, then increased glucose entry into the circulation (increased HGO) is the factor most directly responsible for fasting hyperglycemia.

The elevation of HGO in type 2 diabetes is largely due to increased gluconeogenesis.[24,25] This is secondary to several factors, including a decrease in the ability of insulin to reduce gluconeogenic precursor flux[26] and suppress FFA levels.[19] Additional factors such as increased glucagon concentration also play a role.[27]

In addition to increased basal HGO, patients with type 2 diabetes also have a defect in postprandial hepatic glucose metabolism. In normal individuals, HGO is suppressed postprandially by insulin release and while high physiologic or supraphysiologic insulin levels will completely suppress HGO in type 2 diabetics, there is resistance to suppression of HGO at lower insulin concentrations, resulting in a higher total quantity of glucose entering the circulation.[28] This contributes to the postprandial hyperglycemia. Postprandial hyperglycemia is further compounded by impaired hepatic uptake of glucose, which, in itself, is a manifestation of hepatic insulin resistance.[29] Despite normal basal HGO, obese subjects demonstrate hepatic insulin resistance as evidenced by reduced suppression of HGO by insulin.[8]

## Skeletal Muscle Glucose Metabolism

As indicated previously, in the basal state, 30% of glucose uptake is insulin mediated, whereas in the post-prandial state, insulin-mediated glucose disposal increases to ~85%. Limb catheterization studies have shown that 80-90% of this increased insulin-mediated glucose disposal is into skeletal muscle.[30] Consequently, in insulin resistant states, an inability to respond to insulin stimulation with an adequate increase in glucose disposal largely contributes to post-prandial hyperglycemia.

While most quantitative assessments of in vivo insulin resistance report impaired insulin action based on steady-state measurements, kinetic defects in insulin action in obesity have also been demonstrated thus the rate of activation of insulin's effect to stimulate glucose disposal is decreased and the rate of deactivation of insulin's effect is increased.[31] Given that under physiologic post-prandial conditions insulin is secreted in a phasic rather than steady state manner in response to meal ingestion, it is likely that the kinetic defects in insulin action are of functional importance and that steady-state measurements of insulin action underestimate the functional defect in insulin sensitivity. This has been demonstrated by phasic administration of insulin during a glucose clamp, mimicking the time course and height of the mean insulin levels, as determined during a prior oral glucose tolerance test. Total insulin-stimulated glucose disposal during the "phasic" clamp was reduced by 64% in obese subjects compared to lean controls.[32] This is greater than the 20-50% decrease in steady-state insulin-mediated glucose disposal observed in glucose clamp studies in these same subjects[8,31] confirming the functional importance of kinetic abnormalities in insulin action.

Defects in muscle glycogen synthesis have been demonstrated in insulin resistant states, with a 50% defect in insulin-stimulated muscle glycogen synthesis in subjects with type 2 diabetes compared to normal subjects.[33] As glycogen synthesis is known to account for the majority of nonoxidative glucose metabolism, a defect in glucose incorporation into glycogen is an important manifestation of insulin resistance. The impairment in skeletal muscle glycogen synthesis has been attributed to defects in glucose transport,[34] hexokinase II[35] and glycogen synthase[36] (Fig. 3). To determine the rate-controlling step in the pathway, in vivo skeletal

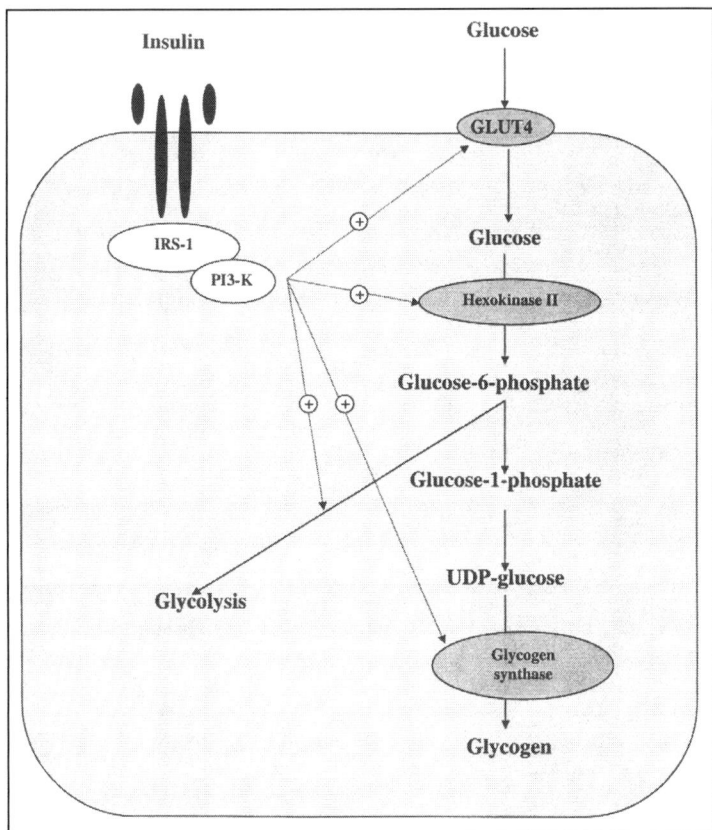

Figure 3. Pathway of glucose uptake, transport and glycogen synthesis.

muscle NMR has been utilized to measure intracellular free glucose and glucose-6-phosphate levels during insulin stimulated glucose clamp conditions. If a defect in glycogen synthase activity were the rate-controlling step this would lead to an increase in intracellular glucose-6-phosphate concentration. No increase in glucose-6-phosphate upon insulin stimulation was observed in diabetic subjects compared to normal controls, implying either a defect in glucose transport or hexokinase II activity.[34] Furthermore, in offspring of type 2 diabetic subjects, who while insulin resistant, are lean and normoglycemic, a similar defect in insulin-stimulated intramuscular glucose-6-phosphate was shown.[37] This implies that reduced glucose transport or hexokinase II activity is an early defect in development of type 2 diabetes and not secondary to factors such as glucotoxicity.

Glucose transport is recognized as a vital step in the action of insulin to cause skeletal muscle glucose uptake and is indeed rate-limiting for whole body glucose metabolism.[38] To distinguish between a defect in glucose transport and hexokinase II in muscle glycogen synthesis in insulin resistant states, NMR has been utilized to measure intracellular free glucose levels. A defect in hexokinase II would be anticipated, under hyperinsulinemic conditions, to result in increased intracellular free glucose relative to glucose-6-phosphate, whereas a defect in glucose transport would be expected, under similar conditions, to result in proportional changes in free glucose and glucose-6-phosphate. In diabetic subjects, the insulin-stimulated increase in free glucose was attenuated, indicating the primacy of a defect in glucose transport in the reduced glycogen synthesis and impaired insulin-stimulated glucose disposal of type 2 diabetes.[39]

In insulin resistant states in humans, many cellular defects in insulin action have been described, although as it is unclear whether these are primary or secondary, the principal defects in insulin resistance remain undetermined. This area has been reviewed recently in detail.[40]

## Adipose Tissue and Lipid Metabolism

Adipose tissue exists principally to store energy in the form of triglyceride, which in the post-absorptive state can then provide fuel for the body as FFA and glycerol following lipolysis. Lipolysis is markedly sensitive to suppression by insulin, with half-maximal suppression of FFA levels occurring in normal subjects at an insulin concentration of approximately 20 μU/ml.[41] The increase in FFA release, associated with an expanded fat mass results in increased circulating FFA levels, particularly in the post-prandial period, in subjects with obesity and type 2 diabetes.[41,42]

Randle and coworkers demonstrated many years ago that FFAs compete with glucose for oxidative metabolism in skeletal and cardiac muscle[43] and hypothesized that elevated FFA levels could therefore impair peripheral glucose use. It was originally proposed that enhanced cellular FFA uptake and oxidation would result in inhibition of pyruvate dehydrogenase by lipid-derived acetyl-CoA. This in turn would increase glucose-6-phosphate levels, resulting in impairment of phosphorylation of incoming glucose and hence glucose uptake. FFA oversupply can indeed cause impaired insulin-mediated glucose disposal, as shown in glucose clamp studies in which FFA levels were elevated by lipid/heparin infusion.[44,45] The decrease in carbohydrate oxidation however occurs rapidly (1-2 h), whereas, the reduction in glucose disposal takes longer to develop (4-5 h), suggesting that the latter effect is not acutely related to changes in FFA oxidation. In addition, the increased intracellular glucose-6-phosphate predicted by Randle has not been observed; indeed skeletal muscle glucose-6-phosphate levels decrease during glucose clamp studies when circulating FFA levels are elevated by lipid/heparin infusion.[44] Furthermore, intracellular free glucose levels were also lower in these studies, implying additional effects of the lipid infusion to impair insulin signaling to glucose transport. Thus, some aspect of intracellular lipid metabolism leads to insulin resistance.

Elevated FFAs do not appear to influence skeletal muscle insulin receptor autophosphorylation[46,47] but other defects in insulin signaling distal to the receptor have been demonstrated. In rats and man, lipid infusion led to a decrease in both insulin-stimulated tyrosine phosphorylation of IRS-1 and activation of PI3K in skeletal muscle.[48]

To speculate as to the mechanism by which FFA elevation may impair insulin signaling, it is necessary to review FFA metabolism within the cell. Upon intake into the cell, FFAs are converted to long-chain fatty acyl-CoAs (LCFA-CoA), which are transported into the mitochondria by carnitine palmitoyltransferases (CPT-1) prior to oxidation. Alternatively, if not transported into the mitochondria, LCFA-CoAs may be reesterified via diacylglycerol (DAG) to form triglycerides and phospholipids. Palmitoyl-CoA may also be converted into ceramide. Elevated circulating FFA levels lead to increased uptake of FFA into the cell, whereupon intracellular levels of LCFA-CoAs, intermediates such as DAG and ceramide, and triglyceride are increased (Fig. 4).[49,50]

Intramyocellular triglyceride content, measured either by muscle biopsy or NMR spectroscopy, is increased in obesity and type 2 diabetes[51] and is a strong predictor of insulin resistance in both animals and humans.[52,53] It is likely that increased intramyocellular triglyceride content may not in itself impair insulin signaling, but act as a marker of increased intracellular LCFA-CoAs and lipid intermediates. A strongly negative correlation has been demonstrated between whole body insulin sensitivity, as determined by the glucose clamp, and the content of LCFA-CoAs measured in muscle biopsy samples.[54]

There are several mechanisms by which fatty-acid intermediates can induce insulin resistance. Both LCFA-CoAs and DAG can activate protein kinase C (PKC), especially novel PKC isozymes such as PKC θ.[48] IRS-1 can be serine phosphorylated by PKC θ, impairing its ability to associate with the insulin receptor,[55] and interfering with PI3K activation and insulin signaling.

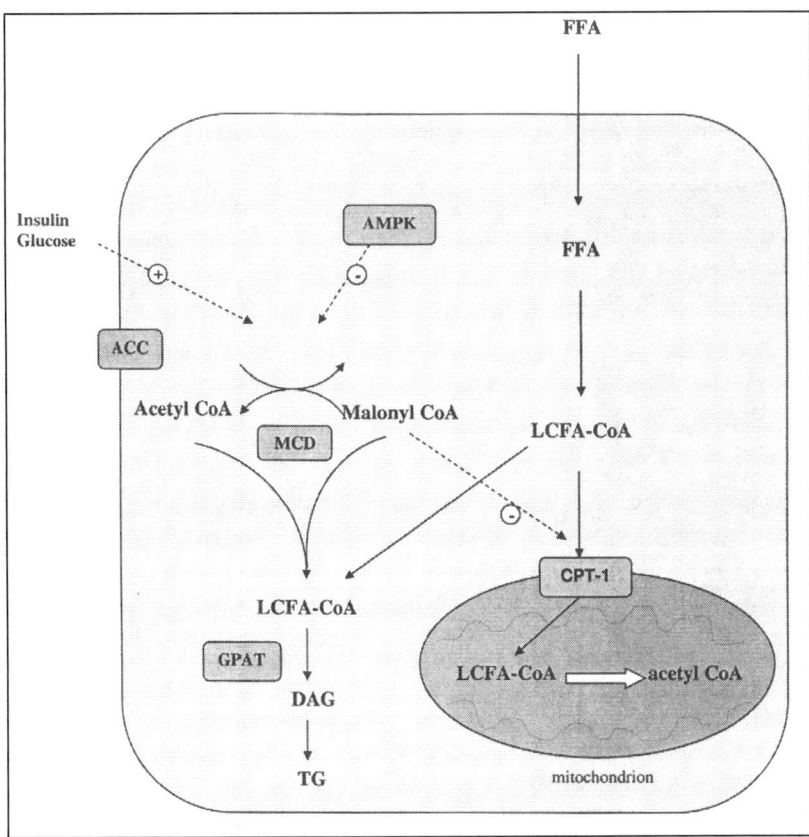

Figure 4. Pathways of fatty acid disposal in skeletal muscle cells. *ACC* acetyl CoA carboxylase, *AMPK* adenosine monophosphate activated protein kinase, *CPT-1* carnitine palmitoyltransferase-1, *DAG* diacylglycerol, *GPAT* glycerol-3-phosphate transferase, *MCD* malonyl CoA decarboxylase, *TG* triglyceride. Reprinted with permission from Endocrinology. DeGroot L, ed. Elsevier, 2005:1108. ©2005 Elsevier.

Further evidence implicating PKC θ in the development of fat-induced skeletal muscle insulin resistance comes from studies of PKC θ knockout mice which are protected from fat infusion induced insulin resistance.[56] Additionally, other intracellular fatty acid intermediates such as ceramide may impair insulin signaling. Thus direct inhibition of Akt with decreased insulin-stimulated glucose transport has been reported in 3T3-L1 adipocytes exposed to ceramide analogues.[57]

Lipid oversupply also activates the inflammatory cascade (Fig. 5). PKC θ is an upstream activator of IKKβ, a serine threonine kinase that activates the NFκB system. IKKβ phosphorylates the natural NFκB inhibitor IκB, which upon phosphorylation dissociates from NFκB, allowing NFκB to translocate to the nucleus where it functions as a transcription factor modulating target gene expression, including genes involved in the inflammatory response, such as inducible nitric oxide synthase (iNOS).[58-60] Evidence supportive of the involvement of this pathway comes from data showing that lipid-induced insulin resistance can be reversed by inhibition of IKKβ, either by salicylate administration or by IKKβ gene deletion.[61,62] Furthermore, mice with targeted disruption of iNOS are protected from high-fat feeding induced insulin resistance further implicating this pathway in the pathogenesis of lipid-induced insulin resistance.[63] The mechanism by which iNOS may affect insulin action remains unclear,

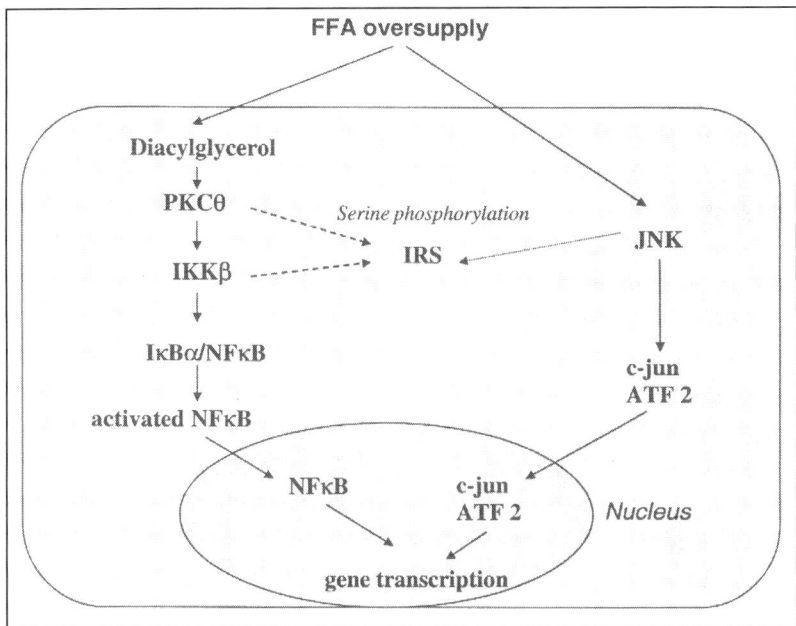

Figure 5. Activation of inflammatory cascade and JNK pathway by lipid oversupply.

although iNOS induction in obese wild-type mice resulted in impaired insulin stimulated PI3-kinase and Akt activation in muscle, defects which were prevented in obese iNOS knockout mice.[63]

Recent data has also implicated c-Jun amino-terminal kinase (JNK) as a mediator of FFA and inflammatory cytokine induced insulin resistance.[64] JNK is activated by both FFA and inflammatory cytokines[65] and influences gene transcription by transcription factors such as c-jun and ATF 2 (activating transcription factor 2). JNK activity is increased in various animal models of obesity (Fig. 5).[64] In addition, genetic knockout of JNK1 resulted in reduced adiposity, enhanced insulin action and improved insulin receptor signaling in obese mouse models.[64] It is postulated that JNK might impair insulin signaling via serine phosphorylation of IRS-1.[66]

## Impact of Regional Fat Distribution and Adipocyte Size

Intraperitoneal (visceral) adipose tissue may be particularly deleterious. Due to its anatomical location, visceral fat drains directly to the liver via the portal vein, therefore exposing the liver to high concentrations of FFA from this depot. Furthermore, visceral adipocytes appear to be more responsive to catecholamine-stimulated lipolysis and less responsive to suppression of lipolysis by insulin.[67,68] It has long been recognized that excess fat in the upper part of the body (central or abdominal) termed "android" obesity is associated with increased risk for type 2 diabetes, dyslipidemia and increased mortality compared to lower body (gluteo-femoral) or "gynoid" obesity.[69-71] While the relationship between visceral fat and cardiovascular risk is established, the association of insulin sensitivity with visceral versus subcutaneous truncal adipose tissue remains controversial. Visceral fat area, as determined by CT scan is correlated with decreased insulin action as measured by the glucose clamp.[72] On the other hand, using similar techniques, the total volume of subcutaneous truncal adipose tissue is a better predictor of insulin resistance than visceral fat.[14] It is possible that subcutaneous truncal adipose tissue contributes more FFA to the systemic circulation than visceral fat and, therefore, may have a more important influence on peripheral insulin action.

Subjects with deficiency of adipose tissue, as in lipodystrophy or lipoatrophy, are also insulin resistant, with excess triglyceride deposition in skeletal muscle and the liver.[73] In transgenic animal models with absence of white adipose tissue, insulin resistance is also associated with lipid infiltration of skeletal muscle and the liver,[74,75] a phenotype which can be reversed by surgical implantation of adipose tissue.[76] These findings suggest that adipose tissue plays a pivotal role in the buffering of fatty acid flux, with insufficient fat tissue leading to "ectopic triglyceride" storage in muscle and the liver resulting in deleterious metabolic effects.[77,78] The more common scenario however is of excess adipose tissue in obesity, and in this situation the antilipolytic effects of insulin are impaired. This could result in increased FFA flux into muscle and liver contributing to the increased intramyocellular and hepatic triglyceride content and insulin resistance observed in this condition.[79]

Another aspect of lipid metabolism that influences insulin action is that of adipocyte size. Larger adipocytes are more resistant to insulin stimulated glucose uptake and to insulin suppression of lipolysis[80,81] and larger subcutaneous abdominal adipocytes may predict the development of type 2 diabetes, independent of insulin resistance.[82] Smaller fat cells may be more efficient at fatty acid uptake and better able to buffer lipid flux. Indeed, it has been hypothesized that failure of adipogenic precursor cells to differentiate into adipocytes results in glucose intolerance,[83] which may be due to inefficient buffering of lipid flux by the remaining large adipocytes.

## *Cross-Talk between Adipocytes and Other Tissues*

Another way in which adipose cells "talk" to other tissues is through an endocrine mechanism. Adipocytokines are peptides secreted by adipose tissue which have diverse effects on food intake, energy expenditure, insulin sensitivity and systemic metabolism. These adipocyte-derived factors include leptin, tumor necrosis factor (TNF)- $\alpha$, adiponectin and resistin. In this manner adipocytes carry out an endocrine function, and indeed the adipose tissue is the largest endocrine organ in the body.

### Leptin

Leptin, the product of the *ob* gene, exerts wide-ranging effects, involving not only food intake and energy expenditure, but also aspects of neuro-endocrinology such as menstruation and fertility.[84-86] The effect of leptin on energy homeostasis, mediated predominantly via the ventrobasal hypothalamus, is to inhibit food intake and increase thermogenesis. Leptin deficient *ob/ob* mice are hyperphagic, massively obese, severely insulin resistant and diabetic, with reversal of this phenotype upon replacement of leptin.[87] The diabetic phenotype was corrected even in mice receiving low dose leptin, in whom significant weight loss did not occur, suggesting that the effects of leptin on glucose metabolism are not solely due to changes in body mass. Humans with rare mutations resulting in leptin deficiency are similarly hyperphagic and obese, and these symptoms are ameliorated by leptin administration.[88] It has also been suggested that leptin has an antilipogenic role, increasing FA oxidation and reducing nonoxidative FA metabolism.[89] In the leptin-resistant Zucker diabetic fatty rat (*fa/fa*), triglyceride accumulation has been demonstrated in pancreatic islets, with consequential lipotoxicity and development of glucose intolerance.[89]

Furthermore, an insulin-sensitizing effect of leptin administration has been reported both in studies in normal rodents[90,91] and in animals rendered insulin resistant by a high-fat diet[92,93] although it remains unclear whether this effect is mediated centrally or via a direct effect on insulin target tissues.[85,94]

In most cases of obesity-related insulin resistance in humans, leptin levels are elevated rather than reduced, suggesting the presence of leptin resistance.[95] This may be at a central level, as reduced leptin transport into the CNS has been reported in obesity,[96] although evidence of peripheral leptin resistance has also emerged. In isolated human skeletal muscle preparations from lean subjects, the addition of leptin promoted partitioning of FA from storage towards oxidation, a phenomenon not observed in muscle from obese subjects.[97] This reduced ability of leptin to stimulate fat oxidation in skeletal muscle in obese subjects may contribute to the development of increased intramyocellular triglyceride content found in this condition. It is unclear whether the mechanism by which leptin influences FA partitioning is solely by a direct effect on some aspect of intracellular FA flux or oxidative capacity or whether indirect effects may also be contributory. Furthermore, the mechanisms by which peripheral resistance to leptin might arise are also unclear, although SOCS-3 (suppressors of the cytokine signaling family) may play a role. SOCS-3 is a member of the SOCS family of cytokine-inducible intracellular proteins that feedback to inhibit cytokine receptors and cytoplasmic signaling adaptor molecules. Forced expression of SOCS-3 has been shown to block leptin receptor mediated signal transduction in mammalian cell lines.[98]

The presence of leptin resistance has been postulated to limit the utility of leptin as a therapeutic anti-obesity agent. Studies examining administration of leptin to obese humans have largely reported no effect on body weight or metabolic parameters although appetite was reduced.[99,100] One study did observe modest but significant weight loss and reduction of body fat, but without any changes in glycemic control or insulin action.[101] Concern has however been raised that an effect may have been missed in some studies due to inadequate dosing, an inappropriate dosage schedule or insufficient study power.[102]

Leptin administration to animal models and humans with lipodystrophy has also been examined. In lipodystrophy there is loss of adipose tissue with subsequent severe defects in insulin action in both liver and skeletal muscle and in addition, leptin levels are low. On leptin administration, insulin action, both in the liver and muscle is improved, in association with reduced hepatic and intramyocellular triglyceride content.[103,104]

In summary, leptin may influence insulin action directly or indirectly via its effects to modulate appetite and weight. The importance of its role both in the pathophysiology of insulin resistance and as a therapeutic agent remain less clearly defined.

## TNF Alpha

In animal models of obesity and insulin resistance and in human subjects with obesity and impaired glucose tolerance, both TNF-α mRNA and protein in adipose tissue and circulating TNF-α levels are increased.[105,106] Animals with an aP2 null mutation, who fail to synthesize TNF-α on high fat feeding, develop a similar degree of obesity as fat-fed controls, although without the accompanying hyperinsulinemia and hyperglycemia, suggesting a role for TNF-α in carbohydrate metabolism.[107] Furthermore, TNF-α infusion in rats has been reported to reduce peripheral insulin-mediated glucose disposal and insulin suppression of HGO,[108] a situation completely prevented by pretreatment with the PPAR γ agonist, troglitazone.[109] Furthermore, in vivo neutralization of TNF-α improves insulin action in Zucker *fa/fa* rats,[110] implicating TNF-α as a contributor to insulin resistance.

Several possible mechanisms could explain a negative effect of TNF-α on insulin action. An effect of TNF-α on insulin signaling has been postulated as TNF-α treatment of 3T3-L1 adipocytes increases serine phosphorylation of IRS-1 thus impairing its ability to associate with the insulin receptor and interfering with PI3K activation and insulin signaling.[55] This effect however follows chronic (6h to 5 days) TNF-α treatment, whereas rapid (< 1h) effects of TNF-α on adipocyte gene expression have been reported.[111,112] This involves suppression of genes known to be involved in insulin action such as adiponectin, PPAR γ and GLUT4 with an increase in inflammatory pathway genes such as NF-κB.

Evidence implicating TNF-α as a factor in the pathogenesis of insulin resistance from knockout mice deficient in TNF-α or one or both of its receptors however has been less consistent and has shown at best only partial protection from obesity-induced insulin resistance.[113,114] Further doubt as to the extent of the role played by TNF-α in humans comes from the failure of TNF-α neutralizing antibody to alter insulin sensitivity in obese[115] or type 2 diabetic subjects.[116]

## Resistin

Resistin is a recently discovered adipocytokine that is elevated in animal models of obesity.[117,118] Administration of resistin to normal mice leads to modest impairment of glucose tolerance[117] and severe hepatic but not peripheral insulin resistance in rats[119] suggesting that resistin antagonizes the action of insulin. Furthermore, neutralization of resistin by an anti-resistin antibody enhances insulin-mediated glucose uptake in adipocytes and treatment of *ob/ob* mice with the insulin-sensitizing PPAR γ agonist, rosiglitazone causes reduction in resistin levels.[117] However, other studies in mice have reported different results, arguing against a role for resistin as a modulator of insulin action.[120] In humans the role of resistin in determining insulin action has been questioned, as no relationship has been found between adipocyte resistin expression and body weight, insulin sensitivity or other metabolic parameters.[121,122] However, substantial sequence differences exist between mouse and human resistin.

## Adiponectin

Adiponectin, also known as AdipoQ or Acrp30 (in rodents) has received considerable interest in view of its influence on insulin action. It is a protein expressed exclusively in adipose tissue and circulates at relatively high concentrations. It is composed of an N-terminal end, a short hypervariable region, a collagen-like sequence and a C-terminal globular region.[123] The globular head group may be proteolytically cleaved from the full-length molecule and as described below also may have biological effects. Furthermore, adiponectin combines to form higher order complexes such as trimers and hexamers, all of which circulate in human serum.[124] The regulation and relative bioactivity of the various forms of adiponectin remain unclear.

Circulating levels of adiponectin are reduced in insulin resistant *ob/ob* mice and in humans with insulin resistant states such as obesity and type 2 diabetes mellitus. Concentrations of adiponectin positively correlate with measures of insulin sensitivity as determined by the glucose clamp, independent of the degree of adiposity.[125] In a longitudinal study of obese Rhesus monkeys genetically predisposed to develop insulin resistance and type 2 diabetes, circulating adiponectin levels decreased in parallel with the progression of insulin resistance.[126] In mice high fat feeding leads to the development of insulin resistance and is associated with a reduction in circulating adiponectin.[127]

These findings indicate decreased adiponectin levels in states of insulin resistance suggesting a possible role for adiponectin in maintaining insulin sensitivity. Further support for this concept comes from studies examining the effects of administration of adiponectin. Administration of the full-length molecule to *db/db*, KKA$^y$ (KK mice overexpressing agouti) and high-fat fed wild type mice, in whom pretreatment serum adiponectin levels were low, resulted in amelioration of the insulin resistant phenotype in each case.[127] This was associated with a reduction in both triglyceride content in skeletal muscle and liver and in circulating FFA and triglycerides. Additionally, adiponectin resulted in increased expression and activity of Acyl-CoA oxidase in skeletal muscle with consequently enhanced FA oxidation. Adiponectin administration has also been shown to activate 5'-AMP-activated protein kinase (AMPK) in both skeletal muscle and hepatocytes[128,129] with subsequent phosphorylation of acetyl CoA carboxylase (ACC) resulting in reduced malonyl CoA in skeletal muscle. Malonyl-CoA inhibits CPT 1, which as indicated previously is necessary for transfer of LCFA-CoA molecules into the mitochondria prior to oxidation. Reduced malonyl-CoA with resultant increased mitochondrial lipid oxidation may possibly thus explain the lowered intramyocellular triglyceride content produced by adiponectin.

Adiponectin has been shown to inhibit TNF-α induced IκB-α phosphorylation and subsequent NF-κB activation in human aortic endothelial cells.[130] While this has not yet been examined in insulin responsive tissues, it is possible that a similar effect may ameliorate the deleterious effect of activation of the inflammatory cascade on insulin action.

In mice models with diabetes (*ob/ob* and NOD) a single intraperitoneal injection of full-length adiponectin, but not the globular head alone, caused an acute reduction of blood glucose level with no corresponding elevation in insulin levels.[131] In a subsequent study, the same group using the glucose clamp technique attributed the glucose lowering effect to a reduction in HGO with no effect demonstrated on peripheral glucose disposal.[132] Hepatic expression of the gluconeogenic enzymes phosphoenolpyruvate carboxykinase and glucose-6-phosphatase were significantly reduced by adiponectin administration.

The active moiety of adiponectin producing these effects is unclear. Both the globular head and full-length molecule induced muscle effects in one study[129] whereas only the globular head was effective in another.[128] The full-length molecule only, and not the globular head, produced hepatic effects.[129] The reasons for these differences are unclear and furthermore it has become apparent that the circulating full-length protein is composed of both low molecular weight (LMW) trimer-dimer complexes and high molecular weight (HMW) complexes[133] although the relative functional importance of these complexes are as yet unknown.

Further attempts to elucidate the role of adiponectin have utilized genetic manipulations of mice. Differing results have, however, been obtained. Mice homozygous for adiponectin KO have been reported as insulin resistant in some studies.[134] Others have found a normal phenotype on regular chow, with insulin resistance only on a high fat/high sucrose diet[135] while Ma et al reported normal insulin sensitivity even on high fat feeding.[136] The reason for these differences is unclear. Support for a role for adiponectin in modulation of insulin action comes from transgenic mice overexpressing globular region adiponectin. On high fat feeding, transgenic mice have increased glucose tolerance and insulin sensitivity compared with wild type controls.[137] Additionally, globular head adiponectin transgenic mice crossed with *ob/ob* mice demonstrate partial amelioration of diabetes, despite having similar body weight as *ob/ob* controls.

A significant increase in plasma adiponectin levels occurs upon treatment with the insulin-sensitizing thiazolidinediones (TZD) and indeed, in vitro, adiponectin has been shown to be a direct TZD-response gene.[138,139] Analysis of the oligomeric forms of adiponectin has shown that the TZD-induced increase is largely accounted for by an increase in the HMW form, reflected by an increase in the $S_A$ index, which is defined as the ratio HMW:(HMW+LMW).[140] The $S_A$ index is reduced in insulin resistant type 2 diabetic subjects compared to insulin sensitive controls and the TZD-induced increase in $S_A$ strongly correlates with improvements in hepatic insulin sensitivity.

Differential alteration in the oligomeric forms of adiponectin may explain the recognized phenomenon of a uniform TZD-induced increase in total adiponectin levels in all subjects, including normal controls, in whom no other metabolic effect of TZDs are observed. Total adiponectin levels are also observed to increase in response to TZDs in diabetic patients termed "TZD nonresponders" who have had no insulin-sensitizing or hypoglycemic response to the drug.[139]

It is unlikely that increased adiponectin is the sole mediator of TZD-induced insulin sensitization, although adiponectin would seem to be a convenient biomarker for TZD administration.

Adiponectin appears to be a significant modulator of insulin action and thus may be promising as a therapeutic tool in the future. Further clarification of its role is awaited.

## *Malonyl-CoA*

A role for malonyl-CoA as a modulator of insulin action and hence possible contributor to insulin resistance has been described. Malonyl-CoA is found in insulin-responsive tissues such as liver and skeletal muscle and acts as an inhibitor of CPT-1, the enzyme regulating FA oxidation by controlling LCFA-CoA entry into the mitochondria.

Malonyl-CoA increases with excess carbohydrate supply or muscle inactivity, resulting in inhibition of CPT-1 and hence FA oxidation. Inhibition of oxidation of LCFA-CoAs causes their accumulation in the cytoplasm with the subsequent deleterious effects on insulin action that have been detailed earlier in this chapter. Conversely, starvation and exercise cause a reduction in skeletal muscle malonyl-CoA, with a corresponding increase in FA oxidation.[141,142] It is possible therefore that malonyl-CoA acts as a fuel-sensor but also by virtue of its effects on FA oxidation, may modulate insulin action. In rodent models with insulin resistance, skeletal muscle malonyl-CoA is elevated with a corresponding increase in LCFA-CoA concentration.[143]

Amelioration of insulin resistance with the PPAR γ agonist, pioglitazone, is associated with a reduction in skeletal muscle malonyl-CoA but without any effect on muscle triglyceride levels.[144] However, increased malonyl-CoA is not always observed in insulin resistant conditions, as high-fat feeding in rodents, while inducing insulin resistance, has not been associated with a corresponding increase in malonyl-CoA.[145] Additionally, data in humans has not shown any difference in skeletal muscle malonyl-CoA levels in type 2 diabetic patients versus controls.[146]

Malonyl-CoA appears to be regulated, at least in part, by ACC which catalyzes its formation from acetyl-CoA and by malonyl CoA decarboxylase (MCD) which catalyzes its degradation (Fig. 4). Insulin and glucose increase the activity of ACC by decreasing its phosphorylation[147] and LCFA-CoAs themselves may allosterically inhibit ACC activity.[148] Furthermore, citrate also activates ACC and may directly increase malonyl-CoA concentration given that it is a precursor of acetyl-CoA and hence malonyl-CoA.[149] The relative importance of each of these factors in vivo however remains controversial. ACC is also phosphorylated by AMPK, leading to decreased ACC activity and this will be discussed in more detail later in this chapter.

Supporting the effect of insulin and glucose to increase malonyl-CoA, infusion of insulin and glucose in rats stimulates skeletal muscle malonyl-CoA production with a corresponding decrease in FA oxidation.[141] In humans during euglycemic-hyperinsulinemia, skeletal muscle FA oxidation is reduced secondary to inhibition of LCFA-CoA entry into the mitochondria.[150] This is consistent with an inhibitory effect of malonyl-CoA on CPT-1 and indeed during a high-dose insulin clamp, an increase in malonyl-CoA concentration in skeletal muscle, associated with a reduction in FA oxidation has been reported.[151] Low dose insulin infusion however also decreased FA oxidation without any corresponding increase in malonyl-CoA, suggesting that increased malonyl-CoA is at best only partially responsible for the insulin-induced inhibition of FA oxidation. In another study however in which a similar insulin level was achieved, but with hyperglycemia rather than euglycemia and maintenance of circulating FFA levels by an exogenous infusion, reduced functional CPT-1 activity and muscle FA oxidation was reported in association with increased skeletal muscle malonyl-CoA.[152] ACC may therefore not only be directly activated by insulin/glucose, but also may be regulated by citrate. Stimulation of carbohydrate metabolism by glucose and insulin administration will increase citrate, which as indicated above activates ACC and also increases substrate supply for the generation of malonyl-CoA.

Citrate also participates in the glucose-fatty acid cycle as in situations of FA excess, phosphofructokinase and thus glycolysis are inhibited.[43] Citrate therefore potentially links both the malonyl-CoA fuel-sensing mechanism and the glucose-fatty acid cycle, acting as a general signal to the muscle cell, indicating conditions of excess fuel.[149] A surplus of glucose will increase intracellular citrate, which via increased ACC and malonyl-CoA will inhibit FA oxidation and via inhibition of phosphofructokinase further use of glucose in glycolysis. An excess of FFA supply will also increase citrate and hence inhibit glycolysis. Malonyl-CoA also increases to inhibit FA oxidation, but this is limited by the concomitant increase in LCFA-CoA which will inhibit ACC and also compete with malonyl-CoA for binding on CPT-1.[153,154] The relevance of this system to the development of insulin resistance remains unclear. As indicated earlier, during human clamp studies in which insulin resistance was induced by fat infusion, a disparity exists in the time course between the reduction in carbohydrate oxidation and the decrease in insulin-mediated glucose disposal, suggesting additional factors are involved. Furthermore, increasing circulating FFA has not been associated with an increase in skeletal muscle citrate levels.[155] No difference has been observed between post-absorptive skeletal muscle citrate levels in type 2 diabetic patients and normal controls,[146] although during hyperinsulinemia the increase in citrate in the diabetic subjects was much less than in controls. Despite this, both groups had a similar increase in malonyl-CoA, suggesting the possibility of an abnormality in the regulation of malonyl-CoA in diabetes.

## AMP-Activated Protein Kinase (AMPK)

AMPK is a regulatory protein kinase that phosphorylates all isozymes of ACC resulting in enzyme inactivation thus increasing FA oxidation.[156] AMPK itself is activated by an increase in the intracellular AMP:ATP ratio, and has therefore been proposed as an important link between cellular energy charge and intracellular fuel metabolism.

Muscle contraction increases intramuscular AMP:ATP ratio and activates AMPK.[157] This inactivates ACC, thus reducing malonyl-CoA and thereby increasing FA oxidation (Fig. 4). Activation of muscle AMPK with a concomitant decrease in ACC activity and malonyl-CoA content has been reported in rats exercised on a treadmill, with the level of AMPK activation proportional to the intensity of exercise.[158] The effect of these changes on fat oxidation has been examined in perfused rat hindlimbs exposed to AICAR (5-aminoimidazole-4-carboxamide-riboside), an activator of AMPK, which resulted in reduction of malonyl-CoA and increased palmitate oxidation.[159] Additionally, glucose uptake was also enhanced by AICAR administration and by contraction rather than the reduction that would have been predicted by the glucose-fatty acid cycle given the increase in intracellular FA oxidation. Furthermore, the glucose uptake was not mediated by PI-3 kinase activation[160] and therefore may be the consequence of phosphorylation of target proteins not yet identified. AMPK is important in the contraction-induced utilization of both carbohydrate and FA as fuel. Demonstration of these effects however has been less clear in humans. During exercise, phosphorylation and inactivation of ACC in skeletal muscle has been reported, but a corresponding decrease in malonyl-CoA has not always been found.[161,162]

In hepatocytes, AMPK has similar effects on ACC, also resulting in increased FA oxidation.[163] Furthermore, AMPK also appears to inhibit glycerol-3-phosphate acyltransferase (GPAT), the enzyme which catalyzes the initial step in glycerolipid biosynthesis.[164]

The overall effect of AMPK activation is stimulation of FA oxidation and glucose uptake in skeletal muscle and stimulation of FA oxidation with inhibition of cholesterol and triglyceride synthesis in the liver. It has been hypothesized that a defect in AMPK signaling could account for many of the abnormalities observed in insulin resistant states.[165] Activation of AMPK has been reported to be necessary for the glucose lowering effects of metformin[166] and incubation of myoblasts with the TZD rosiglitazone increased AMP:ATP ratio with consequent activation of AMPK.[167] Little data yet exists however on the function of AMPK in either insulin resistant animal models or humans.

## Thiazolidinediones

TZDs are a class of drugs which act as agonists for the peroxisome proliferator-activated receptor γ (PPAR γ) nuclear receptor.[168] Administration of TZDs to a wide variety of insulin resistant animal models results in insulin sensitization with lowered plasma glucose levels and concomitant reduction of hyperinsulinemia, indicating improvement in insulin action. This phenomenon occurs regardless of the mechanism of the underlying insulin resistance.[109,169-172] Similarly, administration of TZDs to insulin resistant humans with type 2 diabetes causes insulin sensitization with reduction in both fasting and postprandial plasma glucose levels and circulating insulin levels.[173] Amelioration of insulin resistance has also been demonstrated in nondiabetic insulin resistant conditions such as obesity, impaired glucose tolerance and women with polycystic ovarian syndrome.[174-176] The improvement in insulin action is associated with increased insulin-stimulated glucose disposal, which is usually in the order of 20-40%, thus representing only a partial recovery of insulin resistance, unlike animal studies in which a complete reversal is often seen. An effect of TZDs to improve insulin-mediated suppression of hepatic glucose output has been reported in some[173] but not all studies.[177] The beneficial effects on glucose homeostasis are often associated with improvements in other aspects of Syndrome X, such as decreased blood pressure, modest increases in HDL cholesterol and reductions in PAI-1 levels.[174,178,179]

In view of these effects on insulin action, intense interest has arisen in understanding the targets and mechanism of TZD action. TZDs behave as agonists for the PPAR γ receptor, a member of the nuclear receptor superfamily of transcription factors. Ligand binding causes dissociation of a corepressor complex from the receptor and recruitment of a coactivator complex resulting in activation of target genes. The activation of PPAR γ by ligands correlates well with the insulin sensitizing actions of TZDs implicating activation of PPAR-response genes as the means of action of TZDs. The actual genes responsible for the insulin sensitizing effects however remain unclear. One difficulty is distinguishing between primary effects causing insulin sensitization, effects that are secondary to improvements in glucose homeostasis and changes unrelated to glucose lowering action.

Descriptions of the phenotypes resulting from genetic gain or loss of PPAR γ function have added complexity. Mice, homozygous for PPAR γ knockout are nonviable, but heterozygous PPAR γ knockout mice have been studied.[180,181] Contrary to expectation, these mice, with 50% reduction in whole body PPAR γ receptors, had marked enhancement of insulin sensitivity during glucose clamp studies compared to wild type controls.[181] Similarly, Kubota and colleagues reported protection from high fat feeding induced insulin resistance in mice with heterozygous for PPAR γ knockout.[182]

Human subjects with PPARγ mutations have been described. Subjects with PPAR γ mutations within the ligand-binding domain, which act in a dominant negative manner, develop lipodystrophy and severe insulin resistance that is apparent even in early childhood.[183,184] While such mutations are rare, the Pro12Ala polymorphism is however more commonly found in the population. It also confers loss-of-function, although it has less potent effects on PPARγ receptor action. In the original report of this mutation, subjects carrying the Ala allele rather than being insulin resistant, displayed decreased insulin levels, improved insulin sensitivity and amelioration of other features of Syndrome X.[185] However, these subjects also had a lower body mass index which may have confounded the results somewhat. Subsequent studies of the Pro12Ala polymorphism have reported conflicting results in terms of glucose metabolism,[186,187] although a meta-analysis of published studies suggested a modest protection from diabetes in those carrying the Ala allele.[188]

Conversely, the Pro115Glu PPARγ mutation results in a modest increase in receptor activity. It appears to promote obesity and insulin resistance in human subjects, although only a small number of patients have to date been described.[189]

Gurnell and colleagues have attempted to reconcile the differing phenotypic expressions of PPARγ genotypes by suggesting a sinusoidal relationship between insulin sensitivity and PPARγ activity.[190] In this model, marked impairment of PPAR activity, resulting from loss-of-function mutations in the PPARγ receptor ligand binding domain, is associated with severe insulin resistance. Conversely, a modest reduction in PPARγ activity increases insulin sensitivity, as seen in heterozygous PPARγ knockout mice and possibly human subjects with a Pro12Ala polymorphism. A modest increase in receptor activity, such as in the Pro115Glu variant may be associated with a decrease in insulin sensitivity, whereas enhanced activation of PPARγ receptors by TZDs improves insulin sensitivity.

The principle site of TZD action has also been difficult to elucidate. As indicated previously, the majority of insulin stimulated glucose disposal is into skeletal muscle and thus improvement of insulin action in skeletal muscle is necessary for the insulin sensitizing effects of TZDs. PPAR γ is expressed however at much higher levels in adipose tissue than in skeletal muscle,[191-192] suggesting that the insulin-sensitizing effect of TZDs on muscle is either via a direct effect of activation of muscle PPAR γ receptors or via the indirect action of PPAR γ activation in adipose tissue, or indeed a combination of both. There are several means by which TZDs may act on adipose tissue and cross-talk to skeletal muscle resulting in improved insulin action. It has been proposed that reduced FFA flux may be an important mechanism in this regard. TZDs may reduce FFA flux by firstly altering body fat distribution causing an accumulation of subcutaneous fat with a concomitant reduction in visceral fat.[193] As indicated previously, visceral fat is less responsive to suppression of lipolysis by insulin and more responsive to catecholamine-stimulated lipolysis. TZDs also induce adipocyte differentiation[194] resulting in smaller adipocytes in rat models,[195] which are likely to be more insulin sensitive, thus decreasing FFA flux.[196] Additionally, TZDs alter expression of other adipocytokines including that of leptin,[197] TNF-α,[198] resistin[117] and adiponectin[138] all of which may have a secondary effect on skeletal muscle insulin action.

Evidence supportive of a direct role for muscle PPAR γ receptors in skeletal muscle insulin action has come from studies of mice with muscle specific PPAR γ gene deletion. These animals have postprandial hyperglycemia, hyperinsulinemia and are strikingly insulin resistant with no amelioration of this phenotype by TZD treatment.[199] This indicates the primacy of muscle PPAR γ receptors in both skeletal muscle insulin action and in the insulin sensitizing effect of TZDs.

Additional effects of TZDs in skeletal muscle include a reduction in intramyocellular triglyceride and LCFA-CoA concentration reported in insulin resistant high fat fed rats,[200] although an effect on intramyocellular triglycerides was not observed in humans with type 2 diabetes.[201] As indicated above, the TZD pioglitazone reduces skeletal muscle malonyl-CoA and LCFA-CoA in insulin resistant rodent tissue[144] and myoblasts cultured with rosiglitazone show increased AMPK activation.[167]

An effect of TZDs on aspects of the inflammatory cascade has also been described. Treatment of type 2 diabetic subjects with troglitazone increased IκB concentration and decreased NFκB binding activity in mononuclear cell nuclear extracts.[202] Furthermore, in vitro liganded PPAR γ inhibits lipopolysaccharide-induced iNOS gene transcription.[203]

Although much remains to be learned about PPAR γ receptors and TZD action, TZDs have contributed significantly to elucidating the pathophysiology of insulin resistance. Further discoveries in the mechanism of TZD action will undoubtedly advance understanding of the causes and treatment of insulin resistance.

## Conclusion

In conclusion, it is well established that insulin resistance is fundamental to the pathogenesis of type 2 diabetes mellitus and thus is an important cause of morbidity in the Western and developed world. Insulin resistance however is not simply a defect in glucose disposal but underlies a much more widespread dysregulation of metabolism that significantly contributes to the development of cardiovascular disease.

An understanding of the pathogenetic mechanisms behind insulin resistance is of great interest as it potentially allows design of appropriate therapeutic interventions. Substantial progress has been achieved in recent years in our understanding of the intracellular signaling pathways mediating insulin's varied biologic effects. A further area of progress is the understanding of "cross-talk" that exists between metabolically active tissues as exemplified by the adipocytokines, which can influence glucose homeostasis by acting on nonadipose tissues. Also of profound importance has been studies of the role played by PPAR γ receptors in insulin signaling and the effect of their agonists, the TZDs. These agents represent a major advance in this field, as they are the first direct pharmacologic means available to treat insulin resistance. While much remains to be learned about their exact mode of action, a more complete understanding of this could permit development of more efficacious treatments for insulin resistance.

## References

1. Kahn CR. Insulin resistance, insulin insensitivity and insulin unresponsiveness: A necessary distinction. Metabolism 1978; 27:1893-1902.
2. Himsworth HP. Diabetes mellitus: Its differentiation into insulin-sensitive and insulin-insensitive types. Lancet 1936; i:127-130.
3. Yalow RS, Berson SA. Plasma insulin concentrations in nondiabetic and early diabetic subjects: Determination by a new sensitive immunoassay technique. Diabetes 1960; 9:254-260.
4. Reaven GM. Role of insulin resistance in human disease. Diabetes 1988; 37:1595-1607.

5. Laaksonen DE, Lakka HM, Niskamen LK et al. Metabolic syndrome and development of diabetes mellitus: Application and validation of recently suggested definitions of the metabolic syndrome in a prospective cohort study. Am J Epidemiol 2002; 156:1070-1077.
6. DeFronzo RA, Tobin JD, Andres R. Glucose clamp technique: A method for quantifying insulin secretion and resistance. Am J Physiol 1979; 237:E214-223.
7. Finegood DT, Bergman RN, Vranic M. Estimation of endogenous glucose production during hyperinsulinemic euglycemic glucose clamps: Comparison of unlabelled and labeled exogenous glucose infusates. Diabetes 1987; 36:914-924.
8. Kolterman OG, Insel LJ, Saekow M et al. Mechanisms of insulin resistance in human obesity. Evidence for receptor and post-receptor defects. J Clin Invest 1980; 65:1272-1284.
9. Thiébaud D, Jacot E, DeFronzo RA et al. The effect of graded doses of insulin on total glucose uptake, glucose oxidation and glucose storage in man. Diabetes 1982; 31:957-963.
10. Bergman RN, Ider YZ, Bowden CR et al. Quantitative estimation of insulin sensitivity. Am J Physiol 1979; 236:E667-677.
11. Bergman RN, Prager R, Volund A et al. Equivalence of the insulin sensitivity index in man derived by the minimal model method and the euglycemia glucose clamp. J Clin Invest 1987; 79:790-800.
12. Roden M, Petersen KF, Shulman GI. Nuclear magnetic resonance studies of hepatic glucose metabolism in humans. Recent Prog Horm Res 2001; 56:219-237.
13. Petersen KF, Shulman GI. Pathogenesis of skeletal muscle insulin resistance in type 2 diabetes mellitus. Am J Cardiol 2002; 90(Suppl):11G-18G.
14. Goodpaster BH, Thaete FL, Simoneau J-A et al. Subcutaneous abdominal fat and thigh muscle composition predict insulin sensitivity independently of visceral fat. Diabetes 1997; 46:1579-1585.
15. Schick F, Eismann B, Jung WI et al. Comparison of localized proton NMR signals of skeletal muscle and fat tissue in vivo: Two lipid compartments in muscle tissue. Magn Reson Med 1993; 29:158-167.
16. Krssak M, Falk Petersen K, Dresner A et al. Intramyocellular lipid concentrations are correlated with insulin sensitivity in humans: A $^1$H NMR spectroscopy study. Diabetologia 1999; 42:113-116.
17. Kolterman OG, Gray RS, Griffin J et al. Receptor and postreceptor defects contribute to the insulin resistance in noninsulin dependent diabetes mellitus. J Clin Invest 1981; 68:957-969.
18. O'Brien RM, Granner DK. PEPCK gene as model of inhibitory effects of insulin on gene transcription. Diabetes Care 1990; 13:327-339.
19. Rebrin K, Steil GM, Mittelman S et al. Causal linkage between insulin regulation of lipolysis and liver glucose output. J Clin Invest 1996; 98:741-749.
20. Williamson JR, Browning ET, Scholz R. Control mechanisms of gluconeogenesis and ketogenesis. I Effects of oleate on gluconeogenesis in perfused rat liver. J Biol Chem 1969; 224:4607-4616.
21. Prager R, Wallace P, Olefsky JM. Direct and indirect effects of insulin to inhibit hepatic glucose output in obese subjects. Diabetes 1987; 36:607-611.
22. Ferrannini E, Groop LC. Hepatic glucose production in insulin resistant states. Diabetes Metab Rev 1989; 5:711-726.
23. Dinneen S, Gerich J, Rizza R. Carbohydrate metabolism in noninsulin-dependent diabetes mellitus. N Engl J Med 1992; 327:707-713.
24. Magnusson I, Rothman DL, Katz LD et al. Increased rate of gluconeogenesis in type II diabetes mellitus: A $^{13}$C nuclear magnetic resonance study. J Clin Invest 1992; 90:1323-1327.
25. Tayek JA, Katz J. Glucose production, recycling and gluconeogenesis in normals and diabetics: A mass isotopomer [U-$^{13}$C] glucose study. Am J Physiol 1996; 270:E709-717.
26. McGuiness OP, Ejiofor J, Audoly LP et al. Regulation of glucose production by NEFA and gluconeogenic precursors during chronic glucagon infusion. Am J Physiol 1998; 275:E432-439.
27. Baron AD, Schaeffer L, Shragg P et al. Role of hyperglucagonemia in maintenance of increased rates of hepatic glucose output in type II diabetics. Diabetes 1987; 36:274-283.
28. Basu A, Basu R, Shah P et al. Effects of type 2 diabetes on the ability of insulin and glucose to regulate splanchnic and muscle glucose metabolism: Evidence for a defect in hepatic glucokinase activity. Diabetes 2000; 49:272-283.
29. Ludvik B, Nolan JJ, Roberts A et al. Evidence for decreased splanchnic glucose uptake after oral glucose administration in noninsulin dependent diabetes mellitus. J Clin Invest 1997; 100:2354-2361.
30. DeFronzo RA, Gunnarsson R, Bjorkman O et al. Effects of insulin on peripheral and splanchnic glucose metabolism in noninsulin dependent (type II) diabetes mellitus. J Clin Invest 1985; 76:149-155.
31. Prager R, Wallace P, Olefsky JM. In vivo kinetics of insulin action on peripheral glucose disposal and hepatic glucose output in normal and obese subjects. J Clin Invest 1986; 78:472-481.
32. Prager R, Wallace P, Olefsky JM. Hyperinsulinemia does not compensate for peripheral insulin resistance in obesity. Diabetes 1987; 36:327-334.

33. Shulman GI, Rothman DL, Jue T et al. Quantitation of muscle glycogen synthesis in normal subjects and subjects with noninsulin dependent diabetes by $^{13}$C nuclear magnetic resonance spectroscopy. N Engl J Med 1990; 322:223-228.
34. Rothman DL, Shulman RG, Shulman GI. $^{31}$P nuclear magnetic resonance measurements of muscle glucose-6-phosphate: Evidence for reduced insulin -dependent muscle glucose transport or phosphorylation activity in noninsulin-dependent diabetes mellitus. J Clin Invest 1992; 89:1069-1075.
35. Kruszynska YT, Mulford MI, Baloga J et al. Regulation of skeletal muscle hexokinase II by insulin in nondiabetic and NIDDM subjects. Diabetes 1998; 47:1107-1113.
36. Bogardus C, Lillioja S, Stone K et al. Correlation between muscle glycogen synthase activity and in vivo insulin action in man. J Clin Invest 1984; 73:1185-1190.
37. Rothman DL, Magnusson I, Cline G et al. Decreased muscle glucose transport/phosphorylation is an early defect in the pathogenesis of noninsulin-dependent diabetes mellitus. Proc Natl Acad Sci USA 1995; 92:983-987.
38. Kahn BB. Glucose transport: Pivotal step in insulin action. Diabetes 1996; 45:1644-1654.
39. Cline G, Petersen KF, Krssak M et al. Impaired glucose transport as a cause of decreased insulin stimulated muscle glycogen synthesis in type 2 diabetes. N Engl J Med 1999; 341:240-246.
40. Zierath JR, Wallberg-Henriksson H. From receptor to effector: Insulin signal transduction in skeletal muscle from type 2 diabetic patients. Ann NY Acad Sci 2002; 967:120-134.
41. Swislocki ALM, Chen Y-DI, Golay A et al. Insulin suppression of plasma free-fatty acid concentration in normal individuals and patients with Type 2 (noninsulin-dependent) diabetes mellitus. Diabetologia 1987; 30:622-626.
42. Puhakainen I, Koivisto VA, Yki-Jarvinen H. Lipolysis and gluconeogenesis from glycerol are increased in patients with noninsulin-dependent diabetes mellitus. J Clin Endocrinol Metab 1992; 75:789-794.
43. Randle PJ, Hales CN, Garland PB et al. The glucose fatty-acid cycle. Its role in insulin sensitivity and the metabolic disturbances of diabetes mellitus. Lancet 1963; i:785-789
44. Roden M, Price TB, Perseghin G et al. Mechanism of free fatty acid induced insulin resistance in humans. J Clin Invest 1996; 97:2859-2865.
45. Boden G. Free fatty acids, insulin resistance and type 2 diabetes mellitus. Proc Assoc Am Physicians 1999; 111:241-248.
46. Gumbiner B, Mucha JF, Lindstrom JE et al. Differential effects of acute hypertriglyceridemia on insulin action and insulin receptor autophosphorylation. Am J Physiol 1996; 270:E424-429.
47. Kruszynska YT, Worrall DS, Ofrecio J et al. Fatty acid-induced insulin resistance: Decreased muscle PI3-kinase activation but unchanged Akt phosphorylation. J Clin Endocrinol Metab 2002; 87:226-234.
48. Griffin ME, Marcucci MJ, Cline GW et al. Free fatty acid-induced insulin resistance is associated with activation of protein kinase C theta and alterations in the insulin signaling cascade. Diabetes 1999; 48:1270-1274.
49. Oakes ND, Cooney GJ, Camilleri S et al. Mechanisms of liver and muscle insulin resistance induced by chronic high-fat feeding. Diabetes 1997; 46:1768-1774.
50. Schmitz-Peiffer C, Browne CL, Oakes ND et al. Alterations in the expression and cellular localization of protein kinase C isozymes epsilon and theta are associated with insulin resistance in skeletal muscle of the high-fat-fed rat. Diabetes 1997; 46:169-178.
51. Anderwald C, Bernroider E, Krssak M et al. Effect of insulin treatment in type 2 diabetic patients on intracellular lipid content in liver and skeletal muscle. Diabetes 2002; 51:3025-3032.
52. Kraegen EW, Clark PW, Jenkins AB et al. Development of muscle insulin resistance after liver insulin resistance in high-fat-fed rats. Diabetes 1991; 40:1397-1403.
53. Pan DA, Lillioja S, Kriketos AD et al. Skeletal muscle triglyceride levels are inversely related to insulin action. Diabetes 1997; 46:983-988.
54. Ellis BA, Poynten A, Lowy AJ et al. Long-chain acyl-CoA esters as indicators of lipid metabolism and insulin sensitivity in rat and human muscle. Am J Physiol 2000; 279:E554-560.
55. Paz K, Hemi R, LeRoith D et al. A molecular basis for insulin resistance. Elevated serine/threonine phosphorylation of IRS-1 and IRS-2 inhibits their binding to the juxtamembrane region of the insulin receptor and impairs their ability to undergo insulin-induced tyrosine phosphorylation. J Biol Chem 1997; 272:29911-29918.
56. Kim JK, Fillmore JJ, Sunshine MJ et al. Transgenic mice with inactivation of PKC θ are protected from lipid-induced defects in insulin action and signaling in skeletal muscle. Diabetes 2001; 50(Suppl 2):A58.
57. Summers SA, Garza LA, Zhou H et al. Regulation of insulin-stimulated glucose transported GLUT4 translocation and Akt kinase activity by ceramide. Mol Cell Biol 1998; 18:5457-5464.
58. Barnes PJ, Karin M. Nuclear factor-κB: A pivotal transcription factor in chronic inflammatory diseases. N Engl J Med 1997; 336:1066-1071.

59. DiDonato JA, Hayakawa M, Rothwarf DM et al. A cytokine-responsive IκB kinase that activates the transcription factor NFκB. Nature 1997; 388:548-554.
60. Zandi E, Rothward DM, Delhase M et al. The IκB kinase complex (IKK) contains two kinase subunits, IKKα and IKKβ necessary for IκB phosphorylation and NF-κB activation. Cell 1997; 91:243-252.
61. Kim JK, Kim YJ, Fillmore JJ et al. Prevention of fat-induced insulin resistance by salicylate. J Clin Invest 2001; 108:437-446.
62. Yuan M, Konstantopoulos N, Lee J et al. Reversal of obesity- and diet-induced insulin resistance with salicylates or targeted disruption of IKKβ. Science 2001; 293:1673-1677.
63. Perreault M, Marette A. Targeted disruption of inducible nitric oxide synthase protects against obesity-linked insulin resistance in muscle. Nat Med 2001; 7:1138-1143.
64. Hirosumi J, Tuncman G, Chang L et al. A central role for JNK in obesity and insulin resistance. Nature 2002; 420:333-336.
65. Uysal KT, Wiesbrock SM, Marino MW et al. Protection from obesity-induced insulin resistance in mice lacking TNF- α, function. Nature 1997; 389:610-614.
66. Aguirre V, Uchida T, Yenush L et al. The c-Jun $NH_2$-terminal kinase promotes insulin resistance during association with insulin receptor substrate-1 and phosphorylation of $Ser^{307}$. J Biol Chem 2000; 275:9047-9054.
67. Rebuffe-Scrive M, Andersson B, Olbe L et al. Metabolism of adipose tissue in intraabdominal depots of nonobese men and women. Metabolism 1989; 38:453-458.
68. Bolinder J, Krager L, Ostman J et al. Differences at the receptor and post-receptor levels between human omental and subcutaneous adipose tissue in the action of insulin on lipolysis. Diabetes 1983; 32:117-123.
69. Vague J. La differénciation sexuelle, facteur determinant des formes de l'obésité. Presse méd 1947; 55:339-340.
70. Ladipus L, Bengtsson C, Larsson B et al. Distribution of adipose tissue and risk of cardiovascular disease and death: 12 year follow-up of participants in the study of women in Gothenburg, Sweden. Br Med J 1984; 289:1257-1261.
71. Ohlson LO, Larsson B, Svärdsudd K et al. The influence of body fat distribution on the incidence of diabetes mellitus - 13.5 years of follow-up of the participants in the study of men born in 1913. Diabetes 1985; 34:1055-1058.
72. Park KS, Rhee BD, Lee K-U et al. Intra-abdominal fat is associated with decreased insulin sensitivity in healthy young men. Metabolism 1991; 40:600-603.
73. Robbins DC, Horton ES, Tulp O et al. Familial partial lipodystrophy: Complications of obesity in the nonobese? Metabolism 1982; 31:445-452.
74. Reitman ML, Mason MM, Moitra J et al. Transgenic mice lacking white fat: Models for understanding human lipoatrophic diabetes. Ann NY Acad Sci 1999; 892:289-296.
75. Kim JK, Gavrilova O, Chen Y et al. Mechanism of insulin resistance in A-ZIP/F-1 fatless mice. J Biol Chem 2000; 275:8456-8460.
76. Gavrilova O, Marcus-Samuels B, Graham D et al. Surgical implantation of adipose tissue reverses diabetes in lipoatrophic mice. J Clin Invest 2000; 105:271-278.
77. Frayn KN. Adipose tissue as a buffer for daily lipid flux. Diabetologia 2002; 45:1201-1210.
78. Ravussin E, Smith SR. Increased fat intake, impaired fat oxidation, and failure of fat cell proliferation result in ectopic fat storage, insulin resistance and type 2 diabetes mellitus. Ann NY Acad Sci 2002; 967:363-378.
79. Groop LC, Saloranta C, Shank M et al. The role of free fatty acid metabolism in the pathogenesis of insulin resistance in obesity and noninsulin-dependent diabetes mellitus. J Clin Endocrinol Metab 1991; 72:96-107.
80. Czech MP. Cellular basis of insulin insensitivity in large rat adipocytes. J Clin Invest 1976; 57:1523-1532.
81. Olefsky JM. Insensitivity of large rat adipocytes to the antipolytic effects of insulin. J Lipid Res 1977; 18:459-464.
82. Weyer C, Foley JE, Bogardus C et al. Enlarged subcutaneous abdominal adipocyte size, but not obesity itself, predicts Type II diabetes independent of insulin resistance. Diabetologia 2000; 43:1498-1506.
83. Danforth Jr E. Failure of adipocyte differentiation causes Type II diabetes mellitus? Nat Genet 2000; 26:13
84. Barash IA, Cheung CC, Weigle DS et al. Leptin is a metabolic signal to the reproductive system. Endocrinology 1996; 137:3144-3147.
85. Elmquist JK, Maratos-Flier E, Saper CB et al. Unraveling the central nervous system pathways underlying responses to leptin. Nat Neurosci 1998; 1:445-450.

86. Flier JS. What's in a name? In search of leptin's physiologic role. J Clin Endocrinol Metab 1998; 83:1407-1413.
87. Pelleymounter MA, Cullen MJ, Baker MB et al. Effects of the obese gene product on body weight regulation in ob/ob mice. Science 1995; 269:540-543.
88. Farooqi IS, Jebb SA, Langmack G et al. Effects of recombinant leptin therapy in a child with congenital leptin deficiency. N Eng J Med 1999; 341:879-884.
89. Unger RH, Zhou YT, Orci L. Regulation of fatty acid homeostasis in cells: Novel role of leptin. Proc Natl Acad Sci USA 1999; 96:2327-2332.
90. Sivitz WI, Walsh SA, Morgan DA et al. Effects of leptin on insulin sensitivity in normal rats. Endocrinology 1997; 138:3395-3401.
91. Wang JL, Chinookoswong N, Scull S et al. Differential effects of leptin in regulation of tissue glucose utilization in vivo. Endocrinology 1999; 140:2117-2124.
92. Buettner R, Newgard CB, Rhodes CJ et al. Correction of diet-induced hyperglycemia, hyperinsulinemia and skeletal muscle insulin resistance by moderate hyperleptinemia. Am J Physiol 2000; 278:E563-569.
93. Yaspelis IIIrd BB, Davis JR, Saberi M et al. Leptin administration improves skeletal muscle insulin responsiveness in diet-induced insulin resistant rats. Am J Physiol 2001; 280:E130-142.
94. Kim Y-B, Uotani S, Pierroz DD et al. In vivo administration of leptin activates signal transduction directly in insulin sensitive tissues: Overlapping but distinct pathways from insulin. Endocrinology 2000; 141:2328-2339.
95. Maffei M, Halaas J, Ravussin E et al. Leptin levels in human and rodent: Measurement of plasma leptin and ob RNA in obese and weight-reduced subjects. Nat Med 1995; 1:1155-1161.
96. Caro JF, Kolaczynski JW, Nyce MR et al. Decreased cerebrospinal-fluid/serum leptin ratio in obesity: A possible mechanism for leptin resistance. Lancet 1996; 348:159-161.
97. Steinberg GR, Parolin ML, Heigenhauser GJF et al. Leptin increases FA oxidation in lean but not obese human skeletal muscle: Evidence of peripheral leptin resistance. Am J Physiol 2002; 283:E187-192.
98. Bjorbaek C, El-Haschimi K, Frantz JD et al. The role of SOCS-3 in leptin signaling and insulin resistance. J Biol Chem 1999; 274:30059-30065.
99. Hukshorn CJ, Saris WH, Westerterp-Plantenga MS et al. Weekly subcutaneous pegylated recombinant native human leptin (PEG-OB) administration in obese men. J Clin Endocrinol Metab 2000; 85:4003-4009.
100. Westerterp-Plantenga MA, Saris WH, Hukshorn CJ et al. Effects of weekly administration of pegylated recombinant human OB protein on appetite profile and energy metabolism in obese men. Am J Clin Nutr 1999; 74:426-434.
101. Heymsfield SB, Greenberg AS, Fujioka K et al. Recombinant leptin for weight loss in obese and lean adults: A randomized, controlled, dose-escalation trial. JAMA 1999; 282:1568-1575.
102. Mantzoros CS, Flier JS. Leptin as a therapeutic agent - Trials and tribulations. J Clin Endocrinol Metab 2000; 85:4000-4002.
103. Shimomura I, Hammer RE, Ikemoto S et al. Leptin reverses insulin resistance and diabetes mellitus in mice with congenital lipodystrophy. Nature 1999; 401:73-76.
104. Petersen KF, Oral EA, Dufour S et al. Leptin reverses insulin resistance and hepatic steatosis in patients with severe lipodystrophy. J Clin Invest 2002; 109:1345-1350.
105. Hotamisligil G, Spiegelman B. Tumor necrosis factor alpha: A key component of obesity-diabetes link. Diabetes 1994; 43:1271-1278.
106. Zinman B, Hanley AJ, Harris SB et al. Circulating tumor necrosis factor- $\alpha$, in a native Canadian population with high rates of type 2 diabetes mellitus. J Clin Endocrinol Metab 1999; 84:272-278.
107. Hotamisligil G, Johnson RS, Distel RJ et al. Uncoupling of obesity from insulin resistance through a targeted mutation in aP2, the adipocyte fatty acid binding protein. Science 1996; 274:1377-1379.
108. Lang CH, Dobrescu C, Bagby GJ. Tumor necrosis factor impairs insulin action on peripheral glucose disposal and hepatic glucose output. Endocrinology 1992; 130:43-52.
109. Miles PD, Romeo OM, Higo K et al. TNF-alpha-induced insulin resistance in vivo and its prevention by troglitazone. Diabetes 1997; 46:1678-1683.
110. Hotamisligil GS, Shargill NS, Spiegelman BM. Adipose tissue expression of tumor necrosis factor alpha: Direct role in obesity-linked insulin resistance. Science 1993; 259:87-91.
111. Ruan H, Hacohen N, Golub TR et al. Tumor necrosis factor-alpha suppresses adipocyte-specific genes and activates expression of preadipocyte genes in 3T3-L1 adipocytes: Nuclear factor-kappaB activation by TNF-alpha is obligatory. Diabetes 2002; 51:1319-1336.
112. Ruan H, Miles PD, Ladd CM et al. Profiling gene transcription in vivo reveals adipose tissue as an immediate target of tumor necrosis factor-alpha: Implications for insulin resistance. Diabetes 2002; 51:3176-3188.

113. Ventre J, Doebber T, Wu M et al. Targeted disruption of the tumor necrosis factor-α gene: Metabolic consequences in obese and nonobese mice. Diabetes 1997; 46:1526-1531.
114. Schreyer SA, Chua Jr SC, LeBoeuf RC. Obesity and diabetes in TNF-α receptor-deficient mice. J Clin Invest 1998; 102:402-411.
115. Paquot N, Castillo MJ, LeFebvre PJ et al. No increased insulin sensitivity after a single intravenous administration of a recombinant human tumor necrosis factor receptor: Fc fusion protein in obese insulin-resistant patients. J Clin Endocrinol Metab 2000; 85:1316-1319.
116. Ofei F, Hurel S, Newkirk J et al. Effects of an engineered human anti-TNF-α antibody (CDP571) on insulin sensitivity and glycemia control in patients with NIDDM. Diabetes 1996; 45:881-885.
117. Steppan CM, Bailey ST, Bhat S et al. The hormone resistin links obesity to diabetes. Nature 2001; 409:307-312.
118. Kim K-H, Lee K, Moon YS et al. A cysteine-rice adipose tissue-specific secretory factor inhibits adipocyte differentiation. J Biol Chem 2001; 276:11252-11256.
119. Rajala MW, Obici S, Scherer PE et al. Adipose-derived resistin and gut-derived resistin-like molecule-beta selectively impair insulin action on glucose production. J Clin Invest 2003; 111:225-230.
120. Way JM, Gorgun CZ, Tong Q et al. Adipose tissue resistin expression is severely suppressed in obesity and stimulated by peroxisome proliferator-activated receptor gamma agonists. J Biol Chem 2001; 276:25651-25653.
121. Savage DB, Sewter CP, Klenk ES et al. Resistin/Fizz 3 expression in relation to obesity and peroxisome proliferator-activated receptor-gamma action in humans. Diabetes 2001; 50:2199-2202.
122. Janke J, Engeli S, Gorzelniak K et al. Resistin gene expression in human adipocytes is not related to insulin resistance. Obesity Research 2002; 10:1-5.
123. Shapiro L, Scherer PE. The crystal structure of a complement-1q family protein suggests an evolutionary link to tumor necrosis factor. Curr Biol 1998; 8:335-338.
124. Scherer PE, Williams S, Fogliano M et al. A novel serum protein similar to C1q, produced exclusively in adipocytes. J Biol Chem 1995; 270:26746-26749.
125. Weyer C, Funahashi T, Tanaka S et al. Hypoadiponectinemia in obesity and type 2 diabetes: Close association with insulin resistance and hyperinsulinemia. J Clin Endocrinol Metab 2001; 86:1930-1935.
126. Hotta K, Funahashi T, Bodkin NL et al. Circulating concentrations of the adipocyte protein adiponectin are decreased in parallel with reduced insulin sensitivity during the progression to type 2 diabetes in rhesus monkeys. Diabetes 2001; 50:1126-1133.
127. Yamauchi T, Kamon J, Waki H et al. The fat-derived hormone adiponectin reverses insulin resistance associated with both lipoatrophy and obesity. Nat Med 2001; 7:941-946.
128. Tomas E, Tsao TS, Saha AK et al. Enhanced muscle fat oxidation and glucose transport by ACRP30 globular domain: Acetyl-CoA carboxylase inhibition and AMP-activated protein kinase activation. Proc Natl Acad Sci 2002; 99:16309-16313.
129. Yamauchi T, Kamon J, Minokoshi Y et al. Adiponectin stimulates glucose utilization and fatty-acid oxidation by activating AMP-activated protein kinase. Nat Med 2002; 8:1288-1295.
130. Ouchi N, Kihara S, Arita Y et al. Adiponectin, an adipocyte-derived plasma protein, inhibits endothelial NF-?B signaling though a cAMP-dependent pathway. Circulation 2000; 102:1296-1301.
131. Berg AH, Combs TP, Du X et al. The adipocyte-secreted protein Acrp30 enhances hepatic insulin action. Nat Med 2001; 7:947-953.
132. Combs TP, Berg AH, Obici S et al. Endogenous glucose production is inhibited by the adipose-derived protein Acrp30. J Clin Invest 2001; 108:1875-1881.
133. Pajvani UB, Du X, Combs TP et al. Structurefunction studies of the adipocyte-secreted hormone Acrp30/adiponectin: Implications for metabolic regulation and bioactivity. J Biol Chem 2002; 278:9073-9085.
134. Kubota N, Terauchi Y, Yamauchi T et al. Disruption of adiponectin causes insulin resistance and neointimal formation. J Biol Chem 2002; 277:25863-25866.
135. Maeda N, Shimomura I, Kishida K et al. Diet-induced insulin resistance in mice lacking adiponectin/ACRP30. Nat Med 2002; 8:731-737.
136. Ma K, Cabrero A, Saha PK et al. Increased beta-oxidation but no insulin resistance or glucose intolerance in mice lacking adiponectin. J Biol Chem 2002; 277:34658-34661.
137. Yamauchi T, Kamon J, Waki H et al. Globular adiponectin protected ob/ob mice from diabetes and ApoE-deficient mice from atherosclerosis. J Biol Chem 2003; 278:2461-2468.
138. Maeda N, Takahashi M, Funahashi T et al. PPARgamma ligands increase expression and plasma concentrations of adiponectin, an adipose-derived protein. Diabetes 2001; 50:2094-2099.
139. Yu JG, Jarvorschi S, Hevener AL et al. The effect of thiazolidinediones on plasma adiponectin levels in normal, obese, and type 2 diabetic subjects. Diabetes 2002; 51:2968-2974.

140. Pajvani UB, Hawkins M, Doebber T et al. Complex distribution, not absolute amounts, are critical for adiponectin-mediated improvement in insulin sensitivity. J Biol Chem 2004; 279:12152-12162.
141. Saha AK, Kurowski TG, Ruderman NB. A malonyl-CoA fuel-sensing mechanism in muscle: Effects of insulin, glucose and denervation. Am J Physiol 1995; 269:E283-E289.
142. Winder WW, Arogyasami J, Elayan IM et al. Time course of exercise-induced decline in malonyl-CoA in different muscle types. Am J Physiol 1990; 259:E266-271.
143. Ruderman NB, Saha AK, Vavvas D et al. Lipid abnormalities in muscle of insulin resistant rodents - The malonyl-CoA hypothesis. Ann NY Acad Sci 1997; 827:221-230.
144. Saha AK, Kurowski TG, Colca JR et al. Lipid abnormalities in tissues of the KKA$^y$ mouse: Effects of pioglitazone on malonyl-CoA and diacylglycerol. Am J Physiol 1994; 267:E95-101.
145. Oakes ND, Bell KS, Furler SM et al. Diet-induced muscle insulin resistance in rats is ameliorated by acute dietary lipid withdrawal or a single bout of exercise: Parallel relationship between insulin stimulation of glucose uptake and suppression of long chain fatty acyl-CoA. Diabetes 1997; 46:2022-2028.
146. Båvenholm PN, Kuhl J, Pigon J et al. Insulin resistance in type 2 diabetes: Association with truncal obesity, impaired fitness and atypical malonyl coenzyme A regulation. J Clin Endocrinol Metab 2003; 88:82-87.
147. Witters LA, Watts TD, Daniels DL et al. Insulin stimulates the dephosphorylation and activation of acetyl-CoA carboxylase. Proc Natl Acad Sci USA 1988; 85:5473-5477.
148. Hardie DG. Regulation of fatty acid synthesis via phosphorylation of acetyl CoA carboxylase. Prog Lipid Res 1989; 28:117-146.
149. Saha AK, Vavvas TG, Kurowski TG et al. Malonyl-CoA regulation in skeletal muscle: Its link to cell citrate and the glucose fatty acid cycle. Am J Physiol 1997; 272:E641-648.
150. Sidossis LS, Stuart CA, Shulman GI et al. Glucose plus insulin regulates fat oxidation by controlling the rate of fatty acid entry in to the mitochondria. J Clin Invest 1996; 98:2244-2250.
151. Båvenholm PN, Pigon JP, Saha AK et al. Fatty acid oxidation and the regulation of malonyl-CoA in human muscle. Diabetes 2000; 49:1078-1083.
152. Rasmussen BB, Holmback UC, Volpi E et al. Malonyl coenzyme A and the regulation of functional carnitine palmitoyltransferases-1 activity and fat oxidation in human skeletal muscle. J Clin Invest 2002; 110:1687-1693.
153. McGarry JD, Leatherman GF, Foster DW. Carnitine palmitoyltransferase I: The site of inhibition of hepatic fatty acid oxidation by malonyl-CoA. J Biol Chem 1978; 253:4128-4136.
154. Ruderman NB, Saha AK, Vavvas D et al. Fuel sensing and insulin resistance. Am J Physiol 1999; 276:E1-18.
155. Boden G, Jadali F, White J et al. Effects of fat on insulin-stimulated carbohydrate metabolism in normal men. J Clin Invest 1991; 88:960-966.
156. Winder WW, Hardie DG. Inactivation of acetyl-CoA carboxylase and activation of AMP-activated protein kinase in muscle during exercise. Am J Physiol 1996; 270:E299-E304.
157. Hutber CA, Hardie DG, Winder WW. Electrical stimulation inactivates muscle acetyl-CoA carboxylase and increases AMP-activated protein kinase. Am J Physiol 1997; 272:E262-266.
158. Rasmussen BB, Winder WW. Effect of exercise intensity on skeletal muscle malonyl-CoA and acetyl-CoA carboxylase. J Appl Physiol 1997; 83:1104-1109.
159. Merrill GF, Kurth EJ, Hardie DG et al. AICAriboside increases AMP-activated protein kinase, fatty acid oxidation and glucose uptake in rat muscle. Am J Physiol 1997; 273:E1107-1112.
160. Hayashi T, Hirshman MF, Kurth EJ et al. Evidence for 5'-AMP-activated protein kinase mediation of the effect of muscle contraction on glucose transport. Diabetes 1998; 47:1369-1373.
161. Dean D, Daugaard JR, Young ME et al. Exercise diminishes the activity of acetyl-CoA carboxylase in human muscle. Diabetes 2000; 49:1295-1300.
162. Odland LM, Heigenhauser GJ, Lopaschuk GD et al. Human skeletal muscle malonyl-CoA at rest and during prolonged submaximal exercise. Am J Physiol 1996; 270:E541-E544.
163. Velasco G, Geelen MJH, Guzman M. Control of hepatic fatty acid oxidation by 5'-AMP-activated protein kinase involves a malonyl-CoA-dependent and a malonyl-CoA-independent mechanism. Arch Biochem Biophys 1997; 337:169-175.
164. Muoio DM, Seefeld K, Witters LA et al. AMP-activated kinase reciprocally regulates triacyglycerol synthesis and fatty acid oxidation in liver and muscle: Evidence that sn-glycerol-3-phosphate acyltransferase is a novel target. Biochem J 1999; 338:783-791.
165. Winder WW, Hardie DG. AMP-activated protein kinase, a metabolic master switch: Possible roles in type 2 diabetes. Am J Physiol 1999; 277:E1-E10.
166. Zhou G, Myers R, Li Y et al. Role of AMP-activated protein kinase in mechanisms of metformin action. J Clin Invest 2001; 108:1167-1174.

167. Fryer LGD, Parbu-Patel A, Carling D. The anti-diabetic drugs rosiglitazone and metformin stimulate AMP-activated protein kinase through distinct signaling pathways. J Biol Chem 2002; 277:25226-25232.
168. Ibrahimi A, Teboul L, Gaillard D et al. Evidence for a common mechanism of action for fatty acids and thiazolidinedione antidiabetic agents on gene expression in preadipose cells. Mol Pharmacol 1994; 46:1070-1076.
169. Fujiwara T, Yoshioka S, Yoshioka T et al. Characterization of new oral antidiabetic agent CS-045: Studies in KK ad ob/ob mice and Zucker fatty rats. Diabetes 1988; 37:1549-1558.
170. Lee MK, Miles PD, Khoursheed M et al. Metabolic effects of troglitazone on fructose-induced insulin resistance in the rat. Diabetes 1994; 43:1435-1439.
171. Miles PD, Higo K, Romeo OM et al. Troglitazone prevents hyperglycemia-induced but not glucosamine-induced insulin resistance. Diabetes 1998; 47:395-400.
172. Kraegen EW, James DE, Jenkins AB et al. A potent in vivo effect of ciglitazone on muscle insulin resistance induced by high fat feeding of rats. Metabolism 1989; 38:1089-1093.
173. Suter S, Nolan J, Wallace P et al. Metabolic effects of a new oral hypoglycemic agent, CS-045, in noninsulin dependent diabetic subjects. Diabetes Care 1992; 15:193-203.
174. Nolan JJ, Ludvik B, Beerdsen P et al. Improvement in glucose tolerance and insulin resistance in obese subjects treated with troglitazone. N Engl J Med 1994; 331:1188-1193.
175. Berkowitz K, Peters R, Kjos SL et al. Effect of troglitazone on insulin sensitivity and pancreatic beta cell function in women at high risk for NIDDM. Diabetes 1996; 45:1572-1579.
176. Ehrmann DA, Schneider DJ, Sobel BE et al. Troglitazone improves defects in insulin action, insulin secretion, ovarian steroidogenesis, and fibrinolysis in women with polycystic ovarian syndrome. J Clin Endocrinol Metab 1997; 82:2108-2116.
177. Sironi AM, Vichi S, Gastaldelli A et al. Effects of troglitazone on insulin action and cardiovascular risk factors in patients with noninsulin-dependent-diabetes. Clin Pharmacol Ther 1997; 62:194-202.
178. Ginsberg HN. Insulin resistance and cardiovascular disease. J Clin Invest 2000; 106:453-458.
179. Kumar S, Boulton AJ, Beck-Nielsen H et al. Troglitazone, an insulin action enhancer, improves metabolic control in NIDDM patients. Troglitazone Study Group. Diabetologia 1996; 39:701-709.
180. Barak Y, Nelson MC, Ong ES et al. PPAR gamma is required for placental, cardiac, and adipose tissue development. Mol Cell 1999; 4:585-595.
181. Miles PDG, Barak Y, He W et al. Improved insulin sensitivity in mice heterozygous for PPAR gamma deficiency. J Clin Invest 2000; 105:287-292.
182. Kubota N, Terauchi Y, Miki H et al. PPARγ mediates high-fat diet-induced adipocyte hypertrophy and insulin resistance. Mol Cell 1999; 4:597-609.
183. Barroso I, Gurnell M, Crowley VEF et al. Dominant negative mutations in human PPARγ are associated with severe insulin resistance, diabetes and hypertension. Nature 1999; 402:880-883.
184. Savage DB, Tan GD, Acerini CL et al. Human metabolic syndrome resulting from dominant-negative mutations in the nuclear receptor PPARγ. Diabetes 2003; 52:910-917.
185. Deeb SS, Fajas L, Nemoto M et al. A Pro12Ala substitution in PPARγ2 associated with decreased receptor activity, lower body mass index and improved insulin sensitivity. Nat Genet 1998; 20:284-287.
186. Mancini FP, Vaccaro O, Sabatino L et al. Pro12Ala substitution on the peroxisome proliferator-activated receptor-gamma2 is not associated with type 2 diabetes. Diabetes 1999; 48:1466-1468.
187. Hara K, Okada T, Tobe K et al. The Pro12Ala polymorphism in PPAR gamma2 may confer resistance to type 2 diabetes. Biochem Biophys Res Commun 2000; 271:212-216.
188. Altshuler D, Hirschhorn JN, Klannemark M et al. The common PPARgamma Pro12Ala polymorphism is associated with decreased risk of type 2 diabetes. Nat Genet 2000; 26:76-80.
189. Ristow M, Muller-Wieland D, Pfeiffer A et al. Obesity associated with a mutation in a genetic regulator of adipocyte differentiation. N Engl J Med 1998; 339:953-959.
190. Gurnell M, Savage DB, Chatterjee KK et al. The metabolic syndrome: Peroxisome proliferator-activated receptory and its therapeutic modulation. J Clin Endocrinol Metab 2003; 88:2412-2421.
191. Kliewer SA, Forman BM, Blumberg B et al. Differential expression and activation of a family of murine peroxisome proliferator-activated receptors. Proc Natl Acad Sci USA 1994; 91:7355-7359.
192. Braissant O, Foufelle F, Scott C et al. Differential expression of peroxisome proliferator-activated receptors (PPARs): Tissue distribution of PPAR-alpha, beta and gamma in the adult rat. Endocrinology 1996; 147:354-366.
193. Kelly IE, Han TS, Walsh K et al. Effect of a thiazolidinedione compound on body fat and fat distribution of patients with type 2 diabetes. Diabetes Care 1999; 22:288-293.
194. Tontonoz P, Hu E, Devine J et al. Stimulation of adipogenesis in fibroblasts by PPARγ2, a lipid-activated transcription factor. Cell 1994; 79:1147-1156.
195. Okuno A, Tamemoto H, Tobe K et al. Troglitazone increases the number of small adipocytes without the change of adipose tissue mass in obese Zucker rats. J Clin Invest 1998; 101:1354-1361.

196. Racette SB, Davis AO, McGill JB et al. Thiazolidinediones enhance insulin-mediated suppression of fatty acid flux in type 2 diabetes mellitus. Metabolism 2002; 51:169-174.
197. Kallen CB, Lazar MA. Antidiabetic thiazolidinediones inhibit leptin (ob) gene expression in 3T3-L1 adipocytes. Proc Natl Acad Sci USA 1996; 93:5793-5796.
198. Hofmann C, Lorenz K, Braithwaite SS et al. Altered gene expression for tumor necrosis factor-$\alpha$, and its receptors during drug and dietary modulation of insulin resistance. Endocrinology 1994; 134:264-270.
199. Hevener AL, He W, Barak Y et al. Muscle specific PPAR$\gamma$ gene deletion causes insulin resistance. Nat Med 2003; 9:1491-1497.
200. Ye JM, Doyle PJ, Iglesias MA et al. Peroxisome proliferator-activated receptor (PPAR)-$\alpha$ activation lowers muscle lipids and improves insulin sensitivity in high fat-fed rats. Comparison with PPAR-$\gamma$ activation. Diabetes 2001; 50:411-417.
201. Mayerson AB, Hundal RS, Dufour S et al. The effects of rosiglitazone on insulin sensitivity, lipolysis, and hepatic and skeletal muscle triglyceride content in patients with type 2 diabetes. Diabetes 2002; 51:797-802.
202. Aljada A, Garg R, Ghanim H et al. Nuclear factor-$\kappa$B suppressive and inhibitor-$\kappa$B stimulatory effects of troglitazone in obese patients with type 2 diabetes: Evidence of anti-inflammatory action? J Clin Endocrinol Metab 2001; 86:3250-3256.
203. Li M, Pascual G, Glass CK. Peroxisome proliferator-activated receptor $\gamma$- Dependent repression of the inducible nitric oxide synthase gene. Mol Cell Biol 2000; 20:4699-4707.
204. Kolterman OG, Gray RS, Griffin J et al. Receptor and post-receptor defects contribute to the insulin resistance in non-insulin dependent diabetes mellitus. J Clin Invest 1981; 68:957-969.

# Index

## A

Acetyl CoA  98, 104, 193, 196
Actin  39, 40, 41, 61, 62, 117
Adaptor protein containing a PH and SH2 domain  38
Adipocytes  36-44, 52-58, 62, 78, 82, 97, 112, 117, 170, 174-177, 179, 182, 187, 192-196, 201
Adiponectin  111, 194-197, 201
AFX  115, 123, 183
Alanine scanning mutagenesis  9, 12, 15, 17, 32
AMP-activated protein kinase (AMPK)  121, 193, 196, 198, 199, 201
Animal models  62, 94, 159, 160, 193-196, 199
APS  34, 38
ARFs  35
ARNO  35
AS160  36, 46, 56

## B

β-cell-specific insulin receptor knockouts (bIRKO)  138, 143, 172
Beta cells  33, 52, 53, 58, 59, 60-62, 134, 181
Bladder carcinoma  141
Blood-brain barrier  154, 157
Botulinum toxin  59

## C

c-jun N-terminal protein kinase  113
*C. difficile*  40, 42
Cancer  26, 27, 141
Carnitine palmitoyltransferase I (CPT-I)  98, 104
Caveolin  38, 40, 115
Cbl adaptor proteins  38
Cbl associated protein  38
Cdc42  39, 40, 42, 62
Cdc42/Rac interactive binding (CRIB) domain  39
CEACAM1  183
Ceacam1  177
Cell-cell adhesion  141
cis/trans model  110

COP9 signalsome  77
Csk  34, 112
Cytochalasin  40

## D

DEAD box  77
Diabetes  24, 53, 60, 62, 70, 71, 76, 77, 90-95, 100, 101, 104, 105, 107-109, 124, 128-133, 135-141, 143, 144, 152, 159, 160, 163, 168-173, 175-183, 185-187, 189-191, 193, 194, 196-201, 203-209
DNA microarray  120, 124

## E

eEF1A  80, 81
eEF1B  80
eEF2  80, 81
eIF2  75-77, 80, 81, 172
eIF2B  76, 77
eIF4A  75, 77-79
eIF4E  75, 77-79
eIF4E-Binding Proteins  78
eIF4G  74, 75, 77-79
eIF5  80
Elongation factors  71
EMT  141, 143
Epidermal growth factor (EGF)  6, 8, 22, 23, 27, 28, 32, 141
Epithelial-to-mesenchymal transition  141
Exo70  39, 40, 41, 46, 58
Exocyst complex  40-42, 58
Exocytosis  34-36, 41-44, 52-55, 58-62, 181

## F

Fatty acid oxidation  98, 99, 104, 161, 162
Fibroblasts  35, 38, 57, 74, 112, 179
Fillopodia  40
FKHR  115, 123, 124, 183
FKHRL1  115
Forkhead transcription factor  100, 120, 176, 178, 181, 183, 184
FOXO  26, 115, 120, 121, 123, 124, 178, 184
FOXO1  100, 117, 145, 168, 176, 178, 179, 181, 183, 184

Free fatty acids (FFAs) 91, 94, 98, 99, 105, 177, 178, 186, 188, 191
Fructose-1,6-bisphosphatase 98-100
Fyn 34, 38, 112

## G

γ-catenins 141
G6Pase complex 98-101
Genetics 53, 168
Glucagon 90, 91, 94, 95, 97-101, 103, 111, 135, 143, 144, 188, 189
Glucocorticoid response unit (GRU) 120
Glucokinase (GK) 91, 92, 94, 95, 103, 111, 112, 118, 144, 181
Glucokinase regulatory protein (GKRP) 91, 92
Gluconeogenesis 90, 91, 97-100, 105, 115, 120, 122, 152, 155-161, 163, 169, 170, 176, 184, 188, 189
Glucose transport 33, 37, 45, 57, 71, 90, 91, 133, 171, 176, 177, 180, 189-192
Glucose transporter 33, 52, 53, 55, 90, 91, 104, 170, 182
Glucose-6-phosphatase 98-100, 155, 163, 178, 183, 197
GLUT2 58, 111, 140
GLUT4 33-41, 42-46, 52-58, 61, 111, 115, 168, 170, 171, 175, 176, 182, 195
GLUT4 vesicle translocation 52, 54-56, 58
Glyceraldehyde-3-phosphate dehydrogenase (GAPDH) 118
Glycogen synthesis 77, 90, 101-104, 112, 171, 176, 177, 180, 183, 187, 189-191
Golgi-localized, g-ear-containing, Arf-binding (GG 44
Grb2 34, 113

## H

$H_2O_2$ 114, 124
Helicases 77
Heme-controlled repressor (HCR) 76
Hepatocyte growth factor/scatter factor (HGF/SF) 141, 144
Hydrogen peroxide 114
Hyperphagia 104, 154, 160
Hypothalamus 105, 152, 154, 155, 157-161, 163, 194

## I

IGF 1-13, 15-19, 21-24, 33, 112, 115, 117, 119-124, 133-141, 143-145, 169-173, 179-183
IGF-1 receptor 136-138, 182
IGF-II 6, 18, 23-25, 121, 135-137, 141, 179
Ikkb 168, 178, 183, 192
Initiation factors 71, 74, 76, 78-80
Insulin 1, 2, 4, 6-24, 33-46, 52-62, 71-83, 89, 90, 92-95, 97-100, 102-105, 110-115, 117-124, 133, 135-141, 143, 152-161, 163, 168-198, 201
Insulin clearance 168, 169, 177, 183
Insulin granule trafficking 61
Insulin receptor 3, 4, 6-19, 21, 23, 33-35, 37, 38, 39, 42, 46, 54-56, 77, 81, 112, 114, 115, 120, 122-124, 133, 134, 136-138, 141, 143, 144, 152, 154, 155, 157, 159, 161, 168-171, 173, 174, 178-184, 191, 193, 195
Insulin resistance 7, 11, 21, 56, 57, 62, 104, 111, 113, 124, 136, 139, 141, 143, 152, 153, 157-159, 168-189, 191-201
Insulin response sequences or elements (IRSs/IREs) 111, 124
Insulin-like growth factor binding protein-1 121
Insulin-like growth factors 1, 2
IRS 7, 34, 37-39, 42, 55, 111, 112, 117-124, 133, 136-139, 143, 145, 154, 157, 170, 173, 174, 177, 178, 180-183, 191, 193, 195
Irs knockouts 174
Irs1 34, 117, 168, 172-176, 178, 182
Irs2 117, 168, 172-174, 178, 179, 182
Irs3 117, 168, 174, 175, 182
Irs4 168, 174, 175

Islet β-cells 92, 94, 133, 139, 181

## J

Jasplakinolide 40
JNK 113, 178, 183, 192, 193
Jun, fos, elk, fra 115

# Index

## K

Kidney 6, 77, 91, 97, 100, 112, 122
Knockout 4, 6, 55-58, 60, 61, 133, 135-137, 139, 140, 159, 168-172, 174-178, 180-182, 192, 193, 195, 200

## L

Lamellipodia 40
Latrunculin 40
Lectin 141, 143
Leucine 3, 8, 14, 37, 83, 117
Lipid metabolism 71, 95, 159, 161, 162, 179, 182, 191, 194
Lipid rafts 38, 39, 40, 41
Liver 6, 26, 28, 90-105, 110, 112, 117, 122, 133, 135-137, 139, 144, 152, 155-163, 169, 171, 174, 175, 177-179, 181-183, 186-188, 193-197, 199

## M

Malonyl-CoA 162, 196-199, 201
Maturity onset diabetes of the young 92, 124
Membrane trafficking 35, 36, 112
Mesenchyme-to-epithelial transition (MET) 141
Met-tRNAi 75-77, 80
Metabolic syndrome 105, 171, 174, 177, 178, 180, 185, 187
Metabolism 1, 3, 4, 58, 71, 90, 92, 95, 98, 99, 101-105, 110, 112, 114, 133, 137, 153, 157, 159-163, 170-172, 174, 175, 179, 180, 182, 183, 186-191, 194, 195, 198-201
MKK1 113
Mouse embryonic fibroblasts (MEF) 57
mRNA 40, 60, 71, 74-80, 83, 92, 100, 110, 117, 118, 155-157, 160, 172, 195
mTOR 71, 75, 76, 79, 81-83, 112, 115, 121, 123, 124
Munc18 52-54, 60, 62
Munc18 proteins 53, 54, 62
Myristoylation 35

## N

N-cadherin 141
N-WASP 39, 40
National Cholesterol Education Program 185
Nck 34, 77
Negative cooperativity 1, 9-12, 14, 15, 17, 19-21
Neurotransmitters 43

## O

Organogenesis 141

## P

p38 mitogen activated protein kinases (MAPKs) 45, 113
Pancreas 6, 56, 61, 135, 139, 141, 143, 144, 172
PBP10 45, 46
Phosphatase and tensin homolog 35
Phosphatases 114, 177
Phosphatidylinositol-3'-kinase (PI3K) 33, 75
Phosphenolpyruvate (PEP) 97, 99
Phosphoenolpyruvate Carboxykinase (PEPCK) 94, 97-112, 115, 117, 118, 120-124, 155-161, 188, 197
Phosphofructokinase 91, 94, 97, 198
Phosphoinositide-binding peptide 45
Phosphoinositide-dependent protein kinase (PDK1) 35
Phospholipase D (PLD) 35
PI 3-kinase 34, 35, 37-39, 42, 75, 79, 81, 82, 112, 136, 139, 170, 172, 177, 176
PKB 35-37, 40, 43, 46, 56, 58, 79, 82, 83, 112, 115, 117, 121, 139, 182
PKCl/z 37, 42, 43
Placental lactogen 139, 140
Polio virus 79
Potassium channels 60, 105, 157
PP-1 101-103
PPARγ 38, 178, 200
Proline-rich 34, 37, 38, 42
Protein kinase B 35, 56, 112, 183
Protein kinase C 42, 56, 114, 119, 183, 191
Protein phosphatase 1 101, 102, 177, 182
Protein targeting to glycogen (PTG) 102,103, 177, 182, 183
Protein tyrosine phosphatase (Ptp) 34, 111, 114, 139, 168, 176, 177, 183
Proto-Oncogenes 115
Pten 35, 139
Ptg1 168, 176, 177
Pyruvate carboxylase 97-99
Pyruvate dehydrogenase 98, 99, 191
Pyruvate kinase 91, 97

## R

Rab proteins 36, 46, 61
Ras 35, 36, 38, 40, 80, 83, 113, 121
Receptor tyrosine kinases 3, 5, 34, 140, 172
Recombinase 135
Release factors 71
Replication 76, 140, 141, 143
Resistin 111, 194, 196, 201
Rho 38, 39, 40, 42, 62
Ribosomes 71, 74, 77

## S

S6K1 74, 75, 79, 81-83
*Saccharomyces cerevisiae* phosphorylation 76
Salicylates 178, 183
Sec1 53, 54, 56, 57, 60
Sec3 40
Ser/Thr protein kinase 71, 75, 79
Serum response element (SRE) 117, 119
SH2 domains 34
Ship2 168, 176, 178, 183
siRNA 36, 55, 60, 61, 133, 138, 143
Skeletal muscle 6, 33, 52-54, 56, 57, 61, 62, 77, 82, 102, 104, 133, 161, 170, 174-177, 180, 185-201
SNAP-25 52-55, 58-60
SNARE 36, 41, 42, 52-62
SoHo 38
Sp1 117-119
SREBP-1 117, 118
SREBP-1c 117, 118, 121
Streptozotocin-induced diabetes 135
Swinholide 40
Syndrome X 185, 199, 200
Synip 36, 37, 42, 46, 57, 58
Syntaxin 36, 37, 52-62, 176, 182

## T

t-SNAREs 41, 52, 61
T1 translocase 98, 100, 101
Targeted mutagenesis 168, 179
TC10 38-42, 46, 58
TCA cycle 94, 98, 103
Tensin 35
Ternary complex factor 119
Thiazolidinediones 38, 197, 199
Thyroid Transcription Factor-2 119
Tomosyn 42, 57, 58
Trans-autophosphorylation 34
Trans-Golgi network (TGN) 44-46, 58
Transcription 77, 90, 92, 97, 99, 100, 110-112, 114, 115, 117-124, 135, 139, 143, 144, 176, 178, 179, 181, 183, 184, 192, 193, 200, 201
Transferrin receptor (TfR) 43, 56
Translation 1, 6, 40, 56, 71-81, 83, 110, 172, 181
Translocase 98, 100, 101
TUG 43
Type 2 diabetes 94, 95, 100, 101, 105, 124, 133, 140, 143, 144, 152, 163, 169, 170, 172, 177, 180, 181, 183, 185-187, 189, 190, 191, 193, 194, 196, 199, 201
Tyrosine aminotransferase (TAT) 115, 121-124
Tyrosine kinases 3, 5, 34, 38, 140, 172

## U

Ubiquitous factor (UFA) 119

## V

v-SNAREs 41
VAMP2 36, 41, 52, 54-60, 62
Vesicle exocytosis 34, 36, 41, 52, 59, 62
Vimentin 141

## W

Wiscott-Aldrich syndrome 39

## Z

Zinc ring finger 37
Zipper domains 37
Zucker diabetic fatty (ZDF) rats 100, 104